264

Topics in Current Chemistry

Topics in Current Chemistry

Recently Published and Forthcoming Volumes

Supramolecular Chirality
Volume Editors: Crego-Calama, M., Reinhoudt, D. N.
Vol. 265, 2006

Radicals in Synthesis II
Complex Molecules
Volume Editor: Gansäuer, A.
Vol. 264, 2006

Radicals in Synthesis I
Methods and Mechanisms
Volume Editor: Gansäuer, A.
Vol. 263, 2006

Molecular Machines
Volume Editor: Kelly, T. R.
Vol. 262, 2006

Immobilisation of DNA on Chips II
Volume Editor: Wittmann, C.
Vol. 261, 2005

Immobilisation of DNA on Chips I
Volume Editor: Wittmann, C.
Vol. 260, 2005

Prebiotic Chemistry
From Simple Amphiphiles to Protocell Models
Volume Editor: Walde, P.
Vol. 259, 2005

Supramolecular Dye Chemistry
Volume Editor: Würthner, F.
Vol. 258, 2005

Molecular Wires
From Design to Properties
Volume Editor: De Cola, L.
Vol. 257, 2005

Low Molecular Mass Gelators
Design, Self-Assembly, Function
Volume Editor: Fages, F.
Vol. 256, 2005

Anion Sensing
Volume Editor: Stibor, I.
Vol. 255, 2005

Organic Solid State Reactions
Volume Editor: Toda, F.
Vol. 254, 2005

DNA Binders and Related Subjects
Volume Editors:Waring, M. J., Chaires, J. B.
Vol. 253, 2005

Contrast Agents III
Volume Editor: Krause,W.
Vol. 252, 2005

Chalcogenocarboxylic Acid Derivatives
Volume Editor: Kato, S.
Vol. 251, 2005

New Aspects in Phosphorus Chemistry V
Volume Editor: Majoral, J.-P.
Vol. 250, 2005

Templates in Chemistry II
Volume Editors: Schalley, C. A.,Vögtle, F., Dötz, K. H.
Vol. 249, 2005

Templates in Chemistry I
Volume Editors: Schalley, C. A.,Vögtle, F., Dötz, K. H.
Vol. 248, 2004

Collagen
Volume Editors: Brinckmann, J., Notbohm, H., Müller, P. K.
Vol. 247, 2005

Radicals in Synthesis II

Complex Molecules

Volume Editor: Andreas Gansäuer

With contributions by

M. Albert · J. M. Cuerva · L. Fensterbank · S. G. Hansen
J. Justicia · E. Lacôte · M. Malacria · A. M. McGhee
J. L. Oller-López · J. E. Oltra · D. J. Procter · B. Quiclet-Sire
T. Skrydstrup · J. C. Walton · S. Z. Zard

Springer

The series *Topics in Current Chemistry* presents critical reviews of the present and future trends in modern chemical research. The scope of coverage includes all areas of chemical science including the interfaces with related disciplines such as biology, medicine and materials science. The goal of each thematic volume is to give the nonspecialist reader, whether at the university or in industry, a comprehensive overview of an area where new insights are emerging that are of interest to a larger scientific audience.
As a rule, contributions are specially commissioned. The editors and publishers will, however, always be pleased to receive suggestions and supplementary information. Papers are accepted for *Topics in Current Chemistry* in English.
In references *Topics in Current Chemistry* is abbreviated Top Curr Chem and is cited as a journal.
Visit the TCC content at springerlink.com

ISSN 0340-1022
ISBN-10 3-540-31325-7 Springer Berlin Heidelberg New York
ISBN-13 978-3-540-31325-0 Springer Berlin Heidelberg New York
DOI 10.1007/11615064

Springer is a part of Springer Science+Business Media

springer.com

Cover design: *Design & Production* GmbH, Heidelberg
Typesetting and Production: LE-TEX Jelonek, Schmidt & Vöckler GbR, Leipzig

Printed on acid-free paper 02/3100 YL – 5 4 3 2 1 0

Volume Editor

Prof. Dr. Andreas Gansäuer

Kekulé-Institut für Organische
Chemie und Biochemie
Gerhard-Domagk-Straße 1
53121 Bonn, Germany
andreas.gansaeuer@uni-bonn.de

Editorial Board

Topics in Current Chemistry Also Available Electronically

For all customers who have a standing order to Topics in Current Chemistry, we offer the electronic version via SpringerLink free of charge. Please contact your librarian who can receive a password or free access to the full articles by registering at:

springerlink.com

If you do not have a subscription, you can still view the tables of contents of the volumes and the abstract of each article by going to the SpringerLink Homepage, clicking on "Browse by Online Libraries", then "Chemical Sciences", and finally choose Topics in Current Chemistry.

You will find information about the

- Editorial Board
- Aims and Scope
- Instructions for Authors
- Sample Contribution

at springer.com using the search function.

Preface

"I didn't think that radical chemistry could be so mild and selective," is the nicer version of comments one often hears after seminars. What is the underlying reason for the misconception? Probably that radical transformations often seem counterintuitive to those brought up with classical retrosynthetic schemes. As a result, the use of radicals is considered by many synthetic chemists as a last resort only to be used when other more traditional methods have failed. Additionally, radical reactions are usually regarded as being unselective and involving toxic reagents.

This is, of course, false; such a conservative approach neglects the mild, selective, and original solutions available through using radical chemistry for demanding synthetic problems. Moreover, a solid physical organic understanding of the mechanism behind most radical reactions has now been established. This basis serves us well in predicting many results as well as in developing novel reactions. In short, radical chemistry has developed with amazing speed from a laboratory curiosity into an integral, predictable, and highly productive part of organic chemistry. This account is meant to further spread this point of view.

The first volume (*Methods and Mechanisms*) concentrates on the mechanistic aspects of radical chemistry and the development of novel methods, while the second volume (*Complex Molecules*) focuses on the use of radicals in synthetic applications. While such traditional separation (novel methods are increasingly aimed at preparing complex molecules and the synthesis of complex molecules requires careful planning) may seem a little outdated at the beginning of the 21st century, it is nevertheless employed for the sake of convenience.

The chapters, written by leading experts, provide state-of-the-art reviews of exciting and pertinent topics of current research in radical chemistry. These include a discussion of computed data concerning radical stabilities and their evaluation, the surprising chemistry of radical cations, modern concepts and reagents for enantioselective radical chemistry, the mechanistic aspects of epoxide opening via electron transfer, the evolution of ecologically benign and efficient tin-free radical reactions, the attractive novel reagents and radical traps for unusual cyclizations, the exciting possibilities of xanthate derived radical processes, the emerging field of radical chemistry on solid supports, the recent development of highly versatile radical tandem reactions, the mild and selective derivatization of amino acids and sugars through the use of radicals, and the increasing use of Cp_2TiCl-catalyzed and -mediated radical reactions in natural product synthesis.

Of course not all of the exciting recent developments in radical chemistry can be covered in depth in just two books. It is therefore planned to expand this series in the near future. I offer my apologies to the authors left out this time and ask them to contribute next time!

Hopefully this book will meet the challenge of convincing a large number of scientists of the benefits of radical chemistry and spark novel developments in the fields of new radical methodology and the application of radical reactions in the synthesis of complex molecules.

Bonn, February 2006 Andreas Gansäuer

Contents

Contents of Volume 263

Radicals in Synthesis I

Volume Editor: Andreas Gansäuer
ISBN: 3-540-31329-X

Top Curr Chem (2006) 264: 1–62
DOI 10.1007/128_026

Published online: 27 January 2006

Tandem Radical Reactions

Matthias Albert · Louis Fensterbank · Emmanuel Lacôte · Max Malacria (✉)

Laboratoire de chimie organique (UMR CNRS 7611),
Institut de chimie moléculaire (FR 2769), Université Pierre et Marie Curie,
4 place Jussieu, C.229, 75005 Paris, France
louis.fensterbank@upmc.fr, lacote@ccr.jussieu.fr, max.malacria@upmc.fr

Abstract Radical tandem reactions—and in a wider context radical dominos or cascades—have attracted a lot of attention because of their intrinsic elegance and the construction of a broad and sometimes unique array of molecular architectures they allow in a single step. This review focuses on the latest progress in the design and development of new tandem reactions. The first part is devoted to intramolecular processes; the second part covers tandem and domino processes involving both intra- and intermolecular steps. The third part introduces intermolecular-only reactions. Finally, the last part focuses on tandem reactions involving both radical and non-radical elementary steps.

Keywords Cascades · Domino processes · Polycyclizations · Radical reactions · Tandem reactions

Abbreviations

AIBN	Azo-*bis*-isobutyronitrile
BMDMS	bromomethyldimethylsilyl
Boc	*tert*-butoxycarbonyl
Cbz	carbobenzyloxy
DCE	1,2-Dichloroethane
DEPO	Diethylphosphine oxide
DLP	Dilauroyl peroxide
DMAP	4-Dimethylamino pyridine
DME	Ethylene glycol dimethyl ether
DMF	Dimethyl formamide
HMPA	Hexamethylphosphoramide
Ment	Menthyl
PTOC	pyridine thiocarbonyl
SET	Single electron transfer
SHi	intramolecular homolytic substitution
SM	Starting material
TBDPS	*tert*-butyldiphenyl silyl
TBS	*tert*-butyldimethyl silyl
TEMPO	2,2,6,6-tetramethyl-1-piperidinyloxy radical
TFE	trifluoroethanol
THF	tetrahydrofuran
TMP	2,2,6,6-tetramethyl piperidide
TMS	trimethylsilyl
TTMSS	*tris*-(trimethylsilyl)-silyl
TTMSSH	*tris*-(trimethylsilyl)-silane
V-40	1,1′-Azobis(cyclohexane-1-carbonitrile)
V-501	4,4′-azo-bis-(4-cyanopentanoic acid)
coll	collidine
dppe	diphenylphosphino ethane
dppp	diphenylphosphino propane

1 Introduction

Radical tandem reactions and in a wider context radical dominos or cascades have attracted a lot of attention because of their intrinsic elegance and the construction of a broad and sometimes unique array of molecular architectures they allow. Contrary to a long-standing idea, efficiency and selectivity requirements can also be met. A good illustration is the one-pot assembly of linear triquinane **2** from acyclic precursor **1** (Scheme 1) [1]. In this ten-elementary radical step process, five C – C bonds are created as well as three quaternary centers and four stereogenic centers, almost all completely controlled.

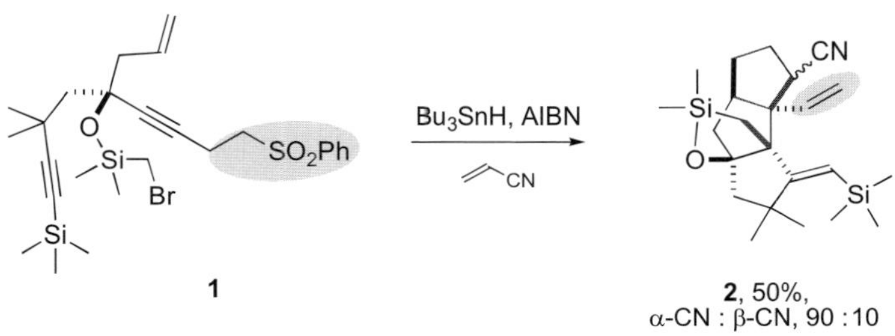

Scheme 1 Access to the linear triquinane framework

Mechanistically, this sequence is an interplay of intra- and intermolecular events featuring four cyclizations, one hydrogen transfer, one β-elimination and an intermolecular addition. In this chapter, we will deal with the recent developments of these different aspects in the literature since 2000, as the two volumes edited by Philippe Renaud and Mukund Sibi have covered the literature prior to 2000 [2].

2 Intramolecular Processes

2.1 Introduction

Owing to their high versatility, selectivity and compatibility with densely functionalized substrates, radical cyclizations are now frequently introduced in retrosynthetic strategies and have already led to the total synthesis of various and relevant natural products [3]. Two recent contributions fully illustrate this. Thus, Muratake and Natsume used a 5-*exo-trig* cyclization of a vinyl radical to provide a congested methylenebicyclo[2.2.2]octane as a key step in the total synthesis of (±)-nominine [4], and Reddy et al. exploited a sim-

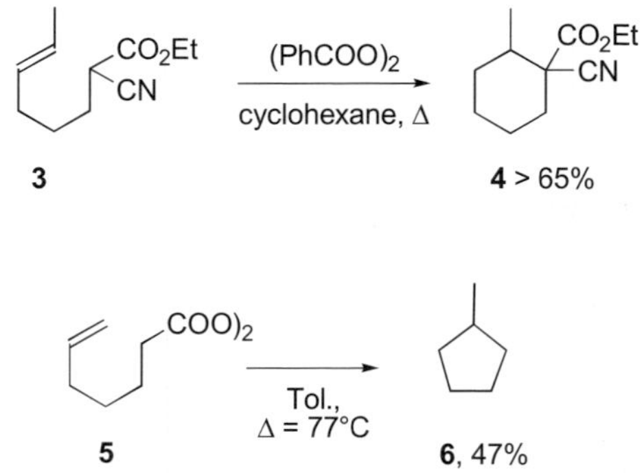

Scheme 2 First synthetic use of radical cyclizations

ple 6-*endo-trig* radical cyclization of a temporary silicon tethered methylene radical in the total synthesis of salinosporamide A [5].

From an historical point of view [6, 7], it is interesting to note that the first examples of radical cyclizations (5-*exo-trig*) have been described by polymer chemists [8, 9], and then by physical chemists in the gas phase [10] or in photochemical studies [11]. The first studies in solution phase which paved the way for further synthetic interests were reported by Julia et al. in 1960 [12] and by Lamb et al. in 1963 (Scheme 2) [13]. In Julia's seminal work, it was shown that unsaturated cyanoesters like **3** could undergo a 6-*endo-trig* cyclization (adduct **4**) when heated in the presence of peroxides in boiling cyclohexane. In parallel to this, Lamb observed that the thermal decomposition of 6-heptenoyl peroxide **5** gave methylcyclopentane **6** as the major adduct, something that was considered quite puzzling at that time, since it corresponded to the anti-Kharash adduct. The irreversibility of the reaction in these conditions was demonstrated and it was concluded that the 5-*exo* mode of cyclization originated from kinetic control. This also suggested that the cyclization involving the cyanoester was reversible, to afford the most stable cyclohexyl radical. All this was confirmed by physical organic studies [7] and the 5-hexenyl radical cyclization became the cornerstone of radical chemistry. Its kinetics [14, 15], modelization (Beckwith–Houk model) [16, 17] and stereochemical attributes [18] have been fully addressed.

Because a radical cyclization gives birth to a new radical species that can in turn be engaged in a new radical cyclization and so on, tandem and even cascade processes can be envisaged. It should be remembered that pioneering examples of this principle can be found in Julia's seminal work, which includes notable entries on the formation of polycyclic adducts, as well as probably one of the first examples of asymmetric radical cyclization involving a (–)-menthyl ester, yielding a product with a diastereoselectivity of 8 : 2 [6]. There are several ways to envisage a cascade process. While it can rely entirely on cyclization processes, it may also implement other radical processes such as hydrogen transfers, fragmentations or homolytic substitutions. This principle has been largely applied, and a review has dealt with it from an algorithmic point of view [19].

2.2 The Polycyclization Approach

Enchaining radical cyclizations steps to provide polycyclic compounds is the simplest approach. Because the 5-*exo* cyclization is the most straightforward mode of cyclization, polyquinanes have been the object of many studies [20]. Round-trip strategies as defined by Haney and Curran [21], from a vinyl radical such as **8** have been developed by Takasu et al. [22] and revisited by Tripp et al. (Scheme 3) [23]. Mixtures of diastereomers **9** are obtained from linear precursor **7**. The same group has examined the intramolecular addition of a vinyl radical onto a dienoate to produce [4 + 1]- or [4 + 2]-annulated compounds [24]. Tandem 5-*exo-trig* cyclizations have also been shown to yield tricyclo[6.2.1.0^{1,6}]undecan-4-one and related polycyclic compounds, using a nickel-catalyzed electroreduction [25].

Haney and Curran also studied the radical cyclization of vinyl iodide **10** in the presence of the fluorous tin hydride **14** (Scheme 4) [21]. After fluorous

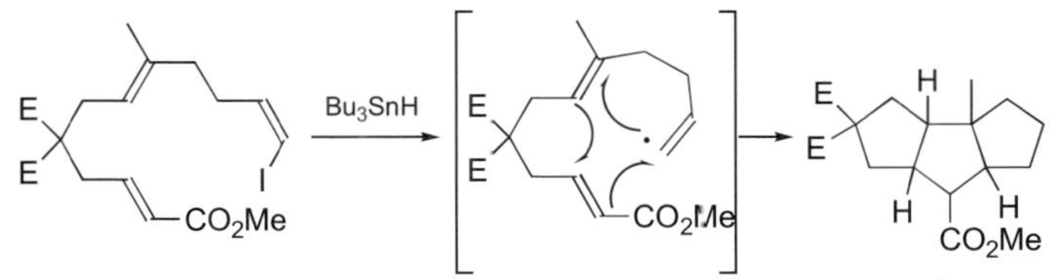

Scheme 3 Round-trip strategy for the construction of triquinanes

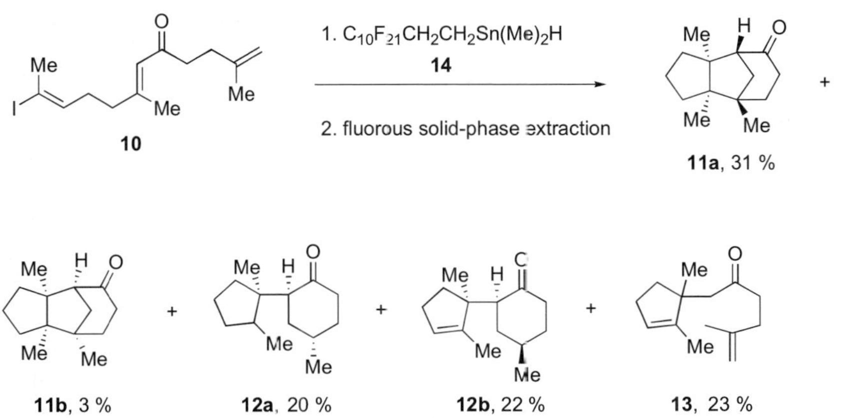

Scheme 4 Toward the synthesis of isogymnomitrene and gymnomitrene

solid-phase extraction of the crude mixture, products **11–13** were isolated as pure compounds, and ketone **11b** as a mixture with **11a**. Gratifyingly, after olefination of ketones **11a** and **11b**, isogymnomitrene and gymnomitrene were obtained.

Harrington-Frost and Pattenden have reported a new synthesis of pentalene, a natural angular triquinane, using a novel tandem cyclization involving a ketene radical intermediate [26].

In the first total synthesis of *epi*-illudol [27], a natural protoilludane produced by fungal basidiomycetes showing some antibacterial activity, the choice of a tandem radical transannular strategy from highly strained cycloundecadienyne substrate **15** was guided by biomimetic considerations (Scheme 5). A bromomethyldimethyl silyl (BMDMS) ether was chosen as the trigger for the cascade process and to create the bis-allylic diol moiety of the natural sesquiterpene after Tamao oxidation. This strategy is highly adaptable. We could assemble linear triquinane **17** from the same 11-membered ring base. Thus, on just moving the BMDMS link from one propargylic position to the other (**16**), the radical cascade proceeds as predicted [28, 29].

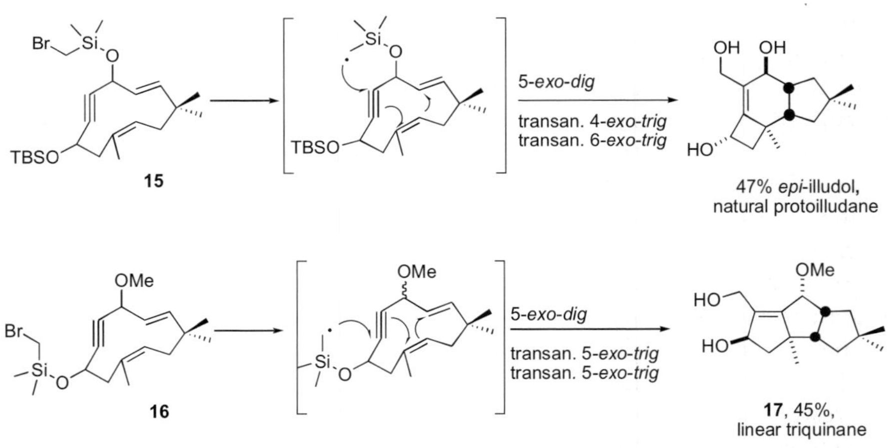

Scheme 5 Transannular*transan*cascades from BMSDMS ethers

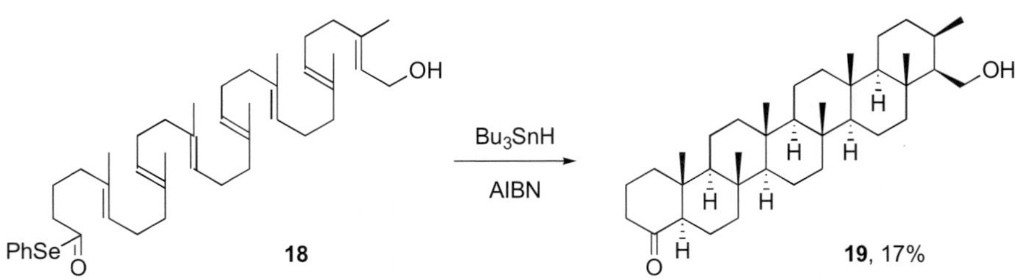

Scheme 6 Preparation of polycyclohexanes

Other polycyclic derivatives have been obtained, for instance polycylohexanes via multiple 6-*endo-trig* cyclizations [30] as in the formation of **19** from **18** (Scheme 6) [31].

Harrowven prepared a series of [5]helicenes by an approach relying on the use of an aryl ring as the acceptor. A fair yield of **21** was obtained from diiodo precursor **20** (Scheme 7) [32].

Joshi et al. have reported a diastereospecific synthesis of pentacyclic β-lactams by 6-*exo-trig*/7-*endo-dig* tandem radical cyclization [33]. On the occasion of the total synthesis of new sargine-type alkaloids, Takayama et al. have disclosed a formation of pentacyclic derivatives from a tryptophan derivative [34].

A variety of radicals can serve as triggers for cascade processes. To this end, bromomethylsilyl ethers proved to be versatile precursors that set the stage for an initial α-silyl radical [35]. Using this methodology, Lee et al. were able to prepare the unnatural (–) enantiomer of lasonolide A and revised its structure [36]. Lasonolide A is a macrolactone possessing two embedded tetrahydropyran rings and showing antitumoral activity against the highly lethal lung carcinoma. Both six-membered rings are introduced via radical cyclizations, the C19–C23 ring being installed via a tandem cyclization that introduces all the stereogenic centers with their correct configuration.

O-Stannyl ketyl radicals were generated from an amide carbonyl group and when incorporated in cinnamic enamides, indolizidinone rings can be assembled [37]. Du and Curran have shown that α-thioaminoalkyl radicals such as **24** resulting from the addition of a tris(trimethylsilyl)radical onto *N*-aryl thiocarbamates, thioamides or thioureas **22**, can be exploited in the context of the synthesis of carbocyclic and heterocyclic fused quinolines **23** (Scheme 8) [38].

Because of their relevance in health sciences, nitrogen-based heterocycles represent important targets. Several groups have addressed their synthesis by proposing new strategies. Ynamides are versatile precursors which can give birth to isoindolinones, as in the formation of **28** from **27** (Scheme 9) [39]. Interesting entries have also been reported by Shen and Hsung using allenamides [40]. Recently, a short synthesis of lennoxamine based on the cyclization of an aryl radical onto an enamide followed by homolytic aromatic substitution has been reported [41].

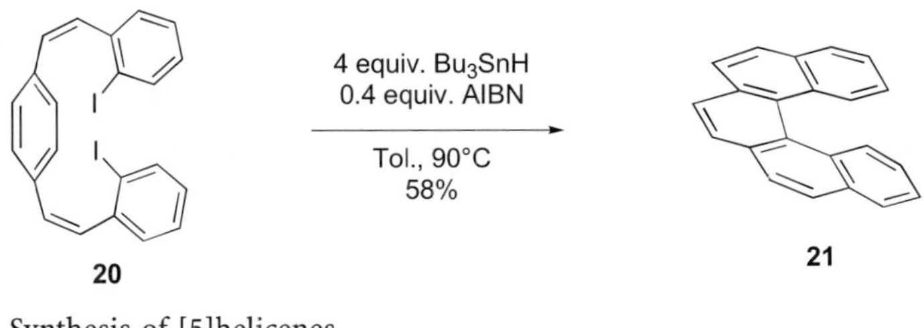

20 **21**

Scheme 7 Synthesis of [5]helicenes

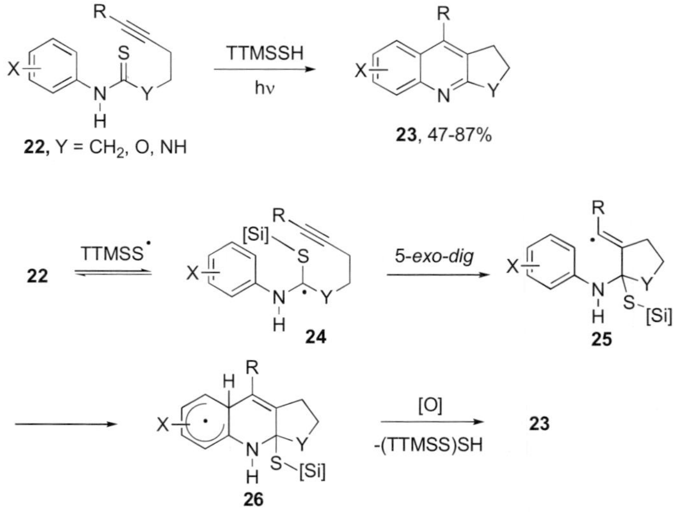

Scheme 8 Access to quinolines

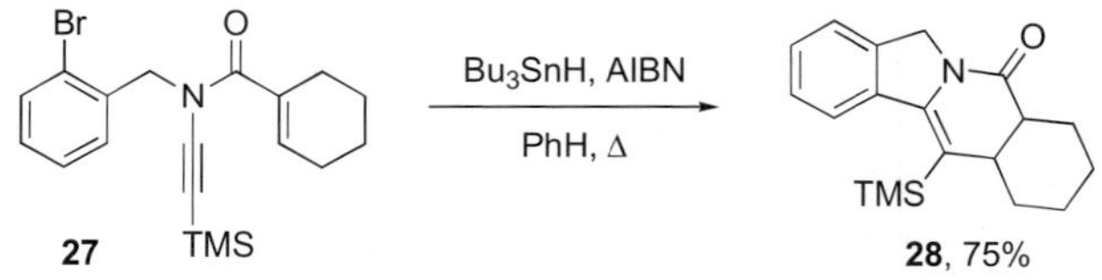

Scheme 9 Ynamides as radical acceptors

Azides are highly valuable radical acceptors that form nitrogen-centered radicals after addition onto them, as in the case of the transformation of **31** into **32**. This reaction opens new possibilities for making pyrrolidines. Murphy, for instance, disclosed the synthesis of (±)-horsfiline and (±)-coerulescine by tandem cyclization of iodoaryl alkenyl azides such as **29** [42]. By the same strategy and using precursor **33**, formal syntheses of (±)-vindoline [43] and (±)-aspidospermidine [44] have been rendered possible (Scheme 10).

N-centered radicals such as iminyl radicals can also be generated from a radical cyclization step onto nitrile groups. The fate of the iminyl radical depends on the nature of the α-substituent. When this latter is an alkyl group, further cyclization can take place [45]. A combination of both processes has been achieved: addition of a stannyl radical onto an azide moiety generated a *N*-stannylaminyl radical that cyclized onto a nitrile and underwent further cyclization, thus opening access to pyrrolopyrroles and pyrrolopyridines derivatives [46].

Tokuda and Senboku have studied the radical cyclization of aminyl radicals resulting from the reaction of tributyltin hydride with *N*-chloro pre-

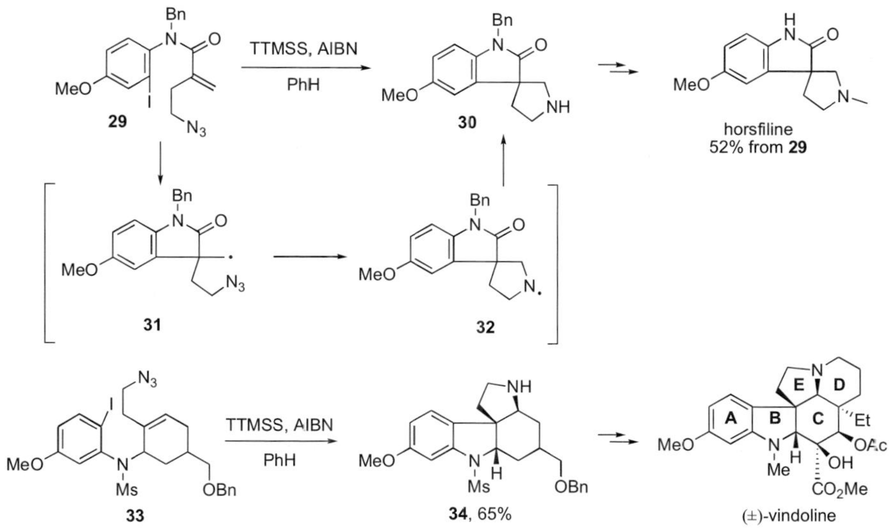

Scheme 10 Use of azides as radical acceptors

cursors. While 1,4-aryl migrations have been observed with benzylamine precursors [47], tandem cyclization of *N*-propargyl precursors allow for the formation of 2-methylenepyrrolizidines, as illustrated by the transformation of **35** into **36** (Scheme 11) [48].

Zard has developed the use of *N*-amidyl radicals. The precursors of the radical intermediates are *O*-benzoyl hydroxyamines such as **37**. Addition of a tributylstannyl radical to the carbonyl group of the benzoate moiety is followed by the cleavage of the weak N – O bond. A subsequent 5-*exo*/6-*endo* tandem cyclization takes place to yield the skeleton of the natural product deoxyserratine (Scheme 12) [49]. Later, the same group disclosed a tin-free source of amidyl radicals that relies on the use of *N*-(*O*-ethyl thiocarbonylsulfanyl)amides and lauryl peroxide as initiator. Examples of polycyclization were also given [50]. On the occasion of a model study toward the synthesis of kirkine, the use of thiosemicarbazide precursors gave access to the tetracyclic structure of the natural product [51].

Stereoselective polycyclization is a highly stimulating aim. Among several contributors in that field, Yang et al. focused on Lewis-acid- pro-

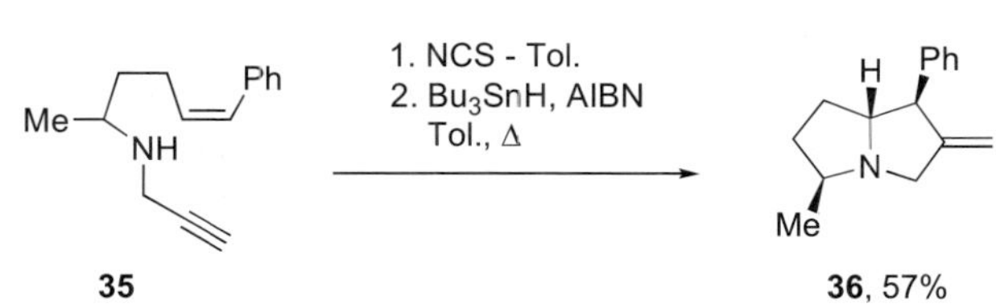

Scheme 11 Tandem reactions with aminyl radicals

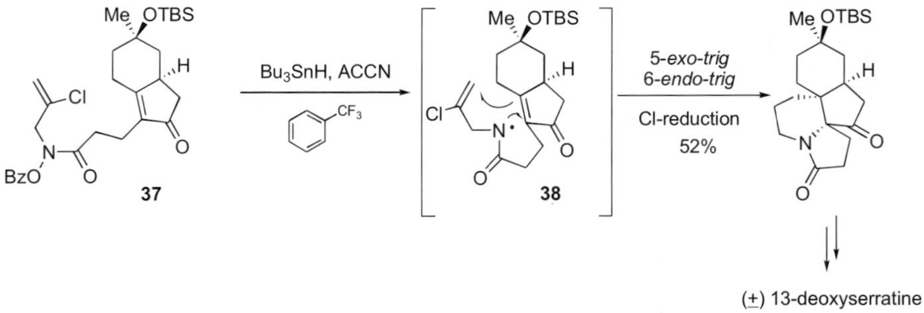

Scheme 12 Total synthesis of (±) deoxyserratine

moted phenylseleno transfers [52]—as in the **39** → **40** transformation— and bromine transfers [53]. In the latter case, enantioselective transfer can take place in the presence of the chiral pybox ligand **44**. It yields bromide bicyclic adduct **43** with 66% ee. With a $Mg(ClO_4)_2$-bisoxazoline chiral complex (Scheme 13), ee's of up to 84% have been obtained.

Based on Snider's manganese chemistry [54], Yang et al. have reported a concise and enantioselective synthesis of terpenoid (+)-triptocallol using a diastereoselective polycyclization as the key step (Scheme 14) [55]. In practice, either enantiomer can be prepared starting from a single epimer of the starting material by running the reaction in the presence or absence of a lanthanide additive ($Yb(OTf)_3$). As usual in cascade chemistry, a high increase in complexity is achieved within a single step, with few secondary steps being required to finish the construction of the target molecule. A similar

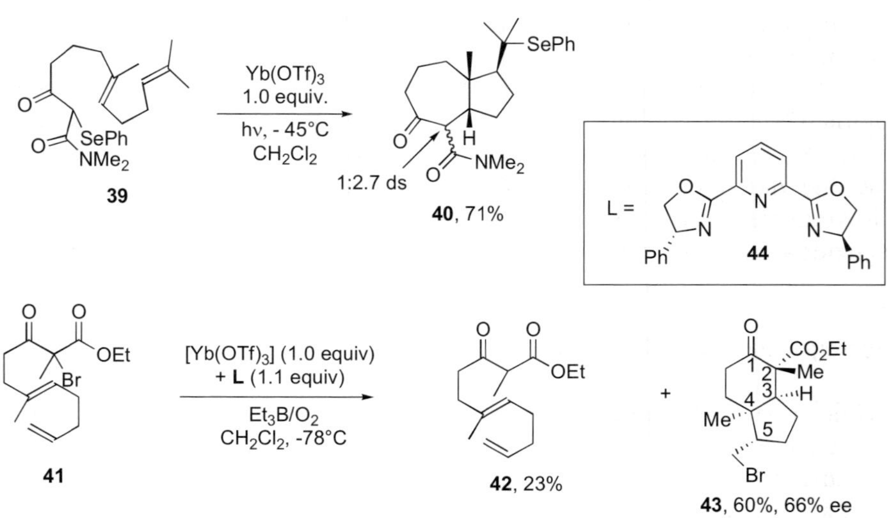

Scheme 13 Lewis-acid promoted atom transfer cascades

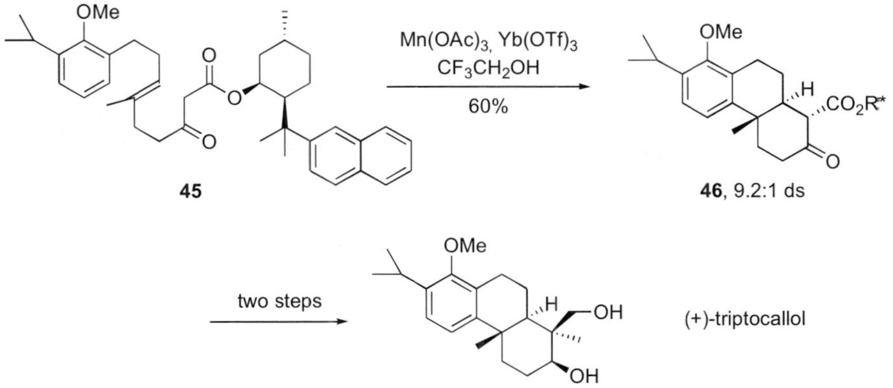

Scheme 14 Total synthesis of (+)-triptocallol

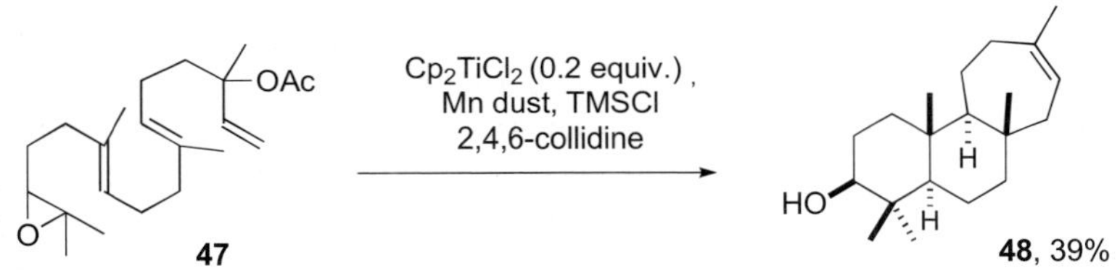

Scheme 15 Titanocene-catalyzed domino reactions

approach has been used for the first enantioselective syntheses of (+)- and (−)-wilforonide [56].

Metallic mediators obviously open up interesting opportunities for conducting radical tandem reactions. In this context, the homolytic opening of epoxides mediated by bis(cyclopentadienyl)titanium (III) chloride (Cp_2TiCl), introduced by Nugent and RajanBabu and rendered catalytic and asymmetric by Gansäuer and Bluhm, has proven to be a versatile tool [57]. Justicia et al. have very recently disclosed new catalytic tandem reactions that feature a final 7-*endo-trig* cyclization [58] followed by subsequent titanium-promoted β-elimination of acetate [59, 60]. The tricyclic compound **48** obtained from epoxide **47** is a precursor of the natural product valaparane (Scheme 15).

2.3 Hydrogen Transfer and Cyclization

While often considered undesirable pathways in the past, radical hydrogen transfers have also been shown to constitute intriguing possibilities for organic synthesis. Curran has for instance defined the concept of protection and radical translocation (PRT) groups. In this context, 2-*O*-(2-bromoaryl)dimethylsilyl ethers, aryl amides and 2-bromo-4-methoxyphenyl ethers have served, thanks to their dual role, in numerous synthetic applications [61]. Interestingly, all reports on hydrogen transfers seem to concur, suggesting that radical translocation originates from a balance of the C – H bond dissociation energy, the proximity of the reacting centers and the transition state geometry. Theoretical studies by Huang and Dannenberg show that activation enthalpies for 1,5- and 1,6-H transfers are the lowest, due to C – H – C angles only slightly distorted from linearity in the transition state [62]. Particularly favorable C – H – C angles are also involved in 1,7- and 1,8-H transfers, but conformational strain makes these transfers much less available. In the case of 1,3- and 1,4-H transfers, C – H – C angles are too distorted from linearity and these transfers are quite rare. Moreover, Dorigo and Houk have estimated that the entropic advantage of forming a six-membered ring over a seven-membered ring is sufficient to overcome the more favorable enthalpy of the 1,6-H transfer, thus justifying the generally observed preponderance of the 1,5-H transfer over the 1,6-H transfer [63].

We observed that vinyl radicals show exquisite reactivity for efficient and chemoselective H-abstraction [64]. The resulting translocated radical can then undergo various types of inter- or intramolecular transformations. Interestingly, we have also shown that the hydrogen transfer step can serve as a driving force for the unfavorable 4-*exo-dig* mode of cyclization [65].

Clive et al. have devised a 5-*exo-dig*/1,5-H transfer from a hydrosilane/5-*endo-trig* sequence that affords new entries to silafuran derivatives [66]. Beaufils et al. have developed a tandem which exploits vinyl radicals resulting from the intermolecular addition of a thiyl radical onto alkyne (such as **49**, Scheme 16) [67]. After radical translocation and 5-*exo-trig* cyclization, good yields of functionalized cyclopentanes **50** are obtained. It is worthy of note that this method gives higher yields than the related tin hydride method from vinyl bromides [68]. It also opens up the possibility of generating nonstabilized radicals. The same group has applied this process for the total synthesis of (–)-erythrodiene [69] and other diastereoselective processes [70].

The thiol methodology is highly satisfactory for a large range of substrate. However, the need for high dilutions and slow addition of thiophenol, together with some scope limitation on substrates undergoing slow hydrogen transfer led Renaud to search for a more effective mediator. In this context, phosphorus-based reagents proved to be very effective. Thus, dimethylphos-

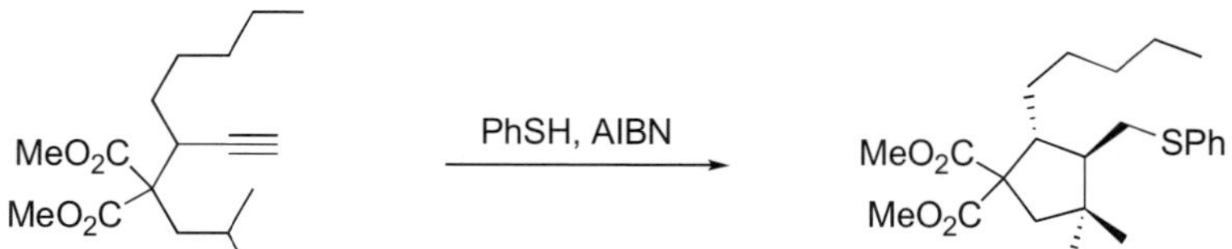

49 **50**, 90%

Scheme 16 Radical translocation-based cascades initiated by thiophenol

phite radical addition to the terminal alkyne of **51** followed by 1,5-H abstraction, 5-*exo-trig* cyclization and reduction by dimethylphosphite yielded the fused bicyclic system **52** in high yield (Scheme 17) [71].

Phosphorus-based mediators have now gained wide acceptance as tin hydride surrogates [72, 73]. They even allow running the reactions in water, as illustrated by the conversion of **53** into **54** which features a hydrogen abstraction by an aryl radical (Scheme 18) [74].

A bromobenzyl moiety is a versatile moiety for triggering radical translocations. When it is attached to a nitrogen atom, the resulting α-amino radical can be an interesting intermediate for alkaloid synthesis as demonstrated by Jones in an elegant approach to the ABCE-rings of the Aspidosperma and Strychnos families [75].

Our continuing interest in the radical cyclization processes involving carbonyl derivatives has brought new insights to the light [76]. While non-

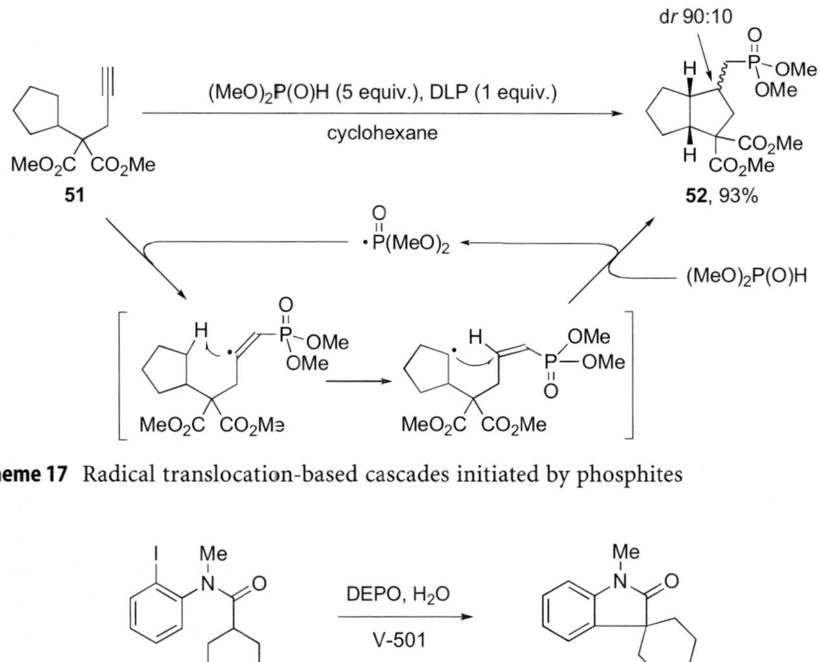

Scheme 17 Radical translocation-based cascades initiated by phosphites

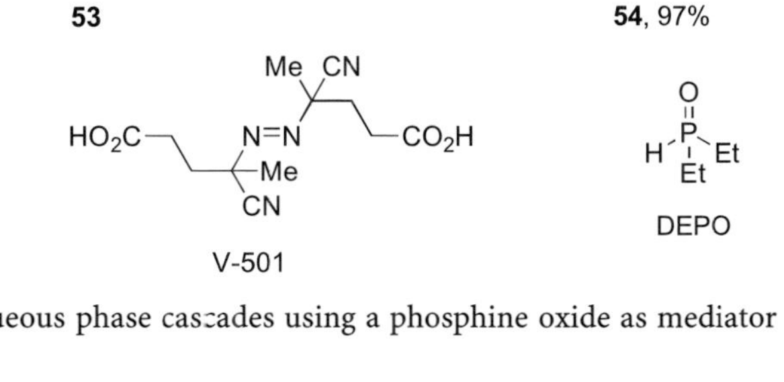

Scheme 18 Aqueous phase cascades using a phosphine oxide as mediator

sterically hindered vinyl radicals give high yield of cyclohexanols with no trace of 1,5-H transfer [77], bulky vinyl radicals like **57** will favor the hydrogen abstraction. This was demonstrated with an aldehyde precursor and was further applied with the enone substrate **55** to give a cascade process that assembles the 5-8-5 skeleton **56** (Scheme 19) [78]. This sequence relies notably on an 8-*endo-trig* cyclization of radical **58**.

As evidenced by these findings, 1,5-H transfers are dominating the field of intramolecular H-abstraction. Nevertheless, other types of H-transfers could also be used. Using Enders' chemistry, we could synthesize precursor **59**, which gave enantiopure triol **61** upon reaction with tributyltin hydride and acrylonitrile, and subsequent fluoride treatment (Scheme 20) [79]. The mech-

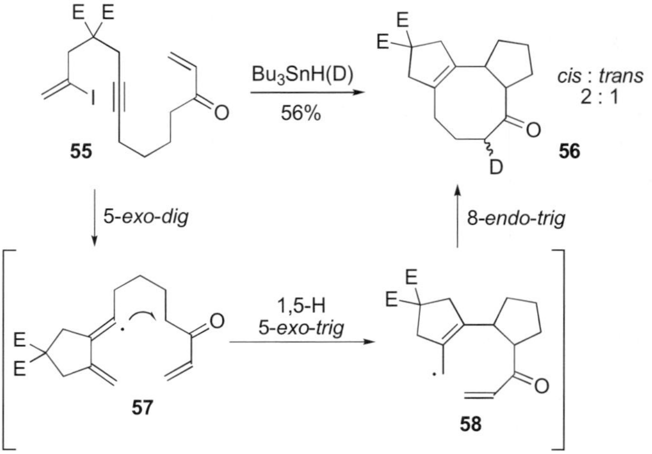

Scheme 19 Access to 5-8-5 tricyclic structures via an 8-*endo-trig* cyclization

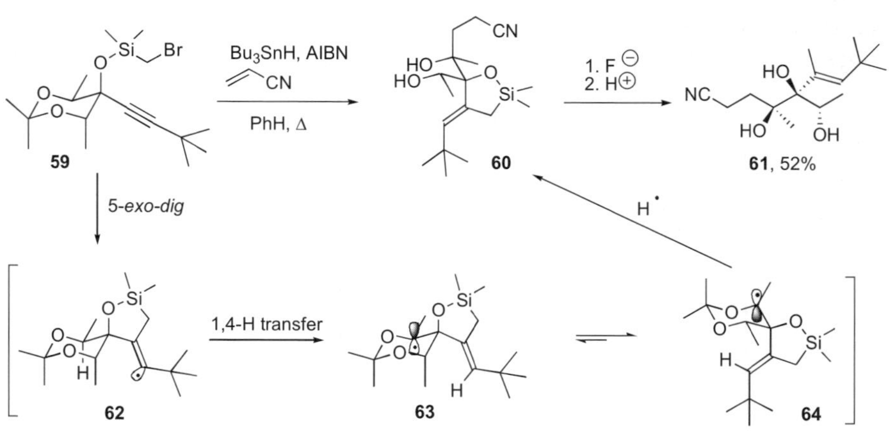

Scheme 20 1,4-H transfer-based domino processes

anism of this sequence involved a 5-*exo-dig* cyclization, a very rare 1,4-H transfer and a final trapping of the most reactive conformer **64** with acrylonitrile. Thus, the protected vinyl radical **62** underwent a completely chemo- and diastereoselective 1,4-H transfer to the α-oxygenated position of the ring, at the expense of a statistically more favorable 1,5-H transfer involving a methyl group. This generated an anomeric radical (**63**). As demonstrated by Dupuis et al. on mannosyl derivatives [80], such intermediates benefit from a twofold stabilization: a planar arrangement of the singly occupied *p*-orbital (SOMO) with the *p*-type lone pairs of the ring oxygen atom and delocalization of the SOMO into the σ^*-LUMO of the adjacent axial C – O bond.

2.4 Fragmentation and Cyclization

Controlled fragmentations can bring a high degree of elegance in natural product synthesis. They generally originate from highly sophisticated retrosynthetic plans. Most of the approaches exploit a cyclopropyl methyl radical rearrangement and several variants (**a–c**) have been encountered (Scheme 21). A variety of homoallylic radical rearrangements have also been developed.

A well designed simple cyclopropyl ring opening is a key step of a new total synthesis of (±)-estrone by Pattenden (Scheme 22). An initial 12-*endo-trig* macrocyclization was followed by the ring opening. Two additional cyclizations stereoselectively assembled the estrone skeleton **66**, albeit in low yield [81].

A much less well-known aspect of the α-cyclopropyl radical chemistry [82] is what happens when the initial radical is vinylic. We have shown that this opening is a valuable route to allenes [83]. We could also calibrate the rate of this opening (Scheme 23). We installed an unsaturation suited for a further 5-*exo-trig* cyclization from the initial vinyl radical (precursor **67**). A clean reaction was observed. It led to cyclopentenol **69** as a single diastereomer, suggesting that the 5-*exo-trig* cyclization of **68** was faster than the ring-

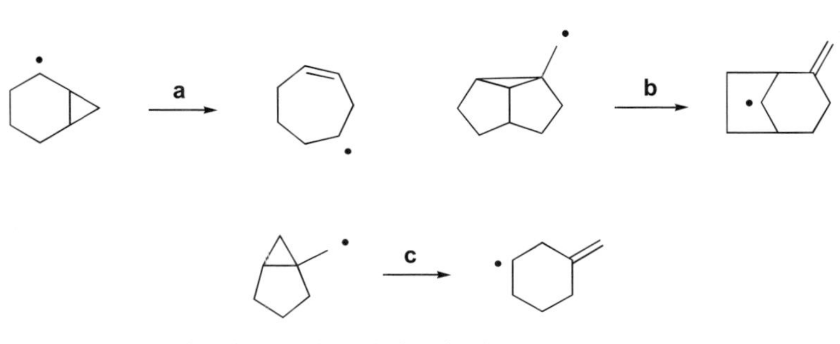

Scheme 21 Examples of cyclopropyl methyl radical rearrangement variants

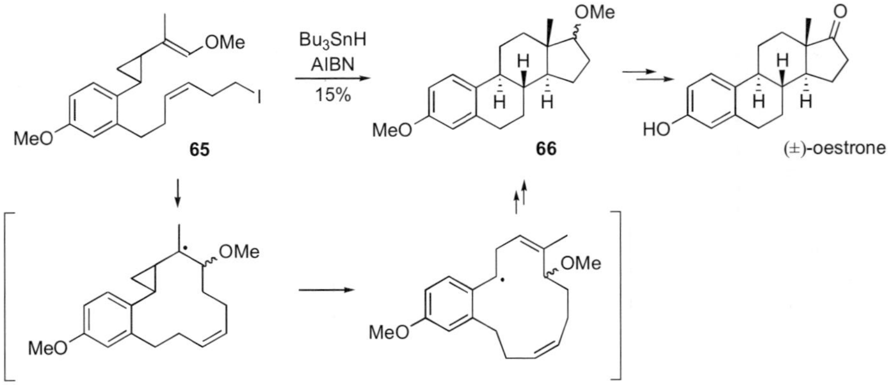

Scheme 22 Total synthesis of œstrone through radical fragmentation

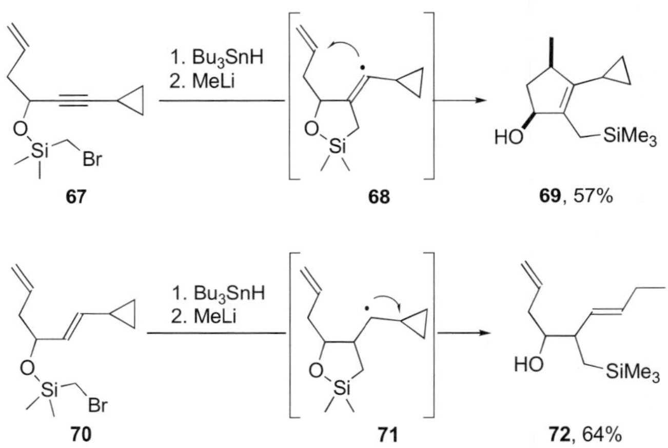

Scheme 23 Opening of vinylic α-cyclopropyl radicals

opening process. This result is especially significant when compared to the opposite outcome of the cyclization of the allyl precursor (**70**) that gave only 1,6-diene **72**. This latter adduct originates from the fast opening of the cyclopropyl ring on intermediate **71**.

Kilburn has extensively studied radical cascades centered on the use of methylenecyclopropane derivatives. An addition–fragmentation process apparented to pathway **c** opened new routes for the synthesis of carbocycles. In a recent work, a SmI_2-promoted cascade of propargyl ether **73** has been used to give bicyclic ether **78** with good diastereoselectivity (Scheme 24), thus providing a short route to the monoterpenoid paeonilactone B. The observed stereoselectivity in the **74** → **75** cyclization step was shown to be critically dependent on the presence of HMPA [84].

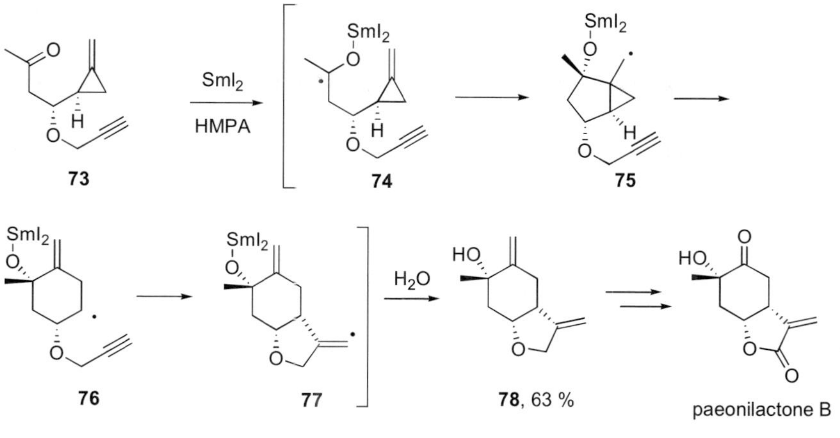

Scheme 24 Total synthesis of paeonilactone B

Along the same lines, Prévost and Shipman have developed an intramolecular radical rearrangement of 2-methyleneaziridines providing piperidines, decahydroquinolines, and octahydroindolizines [85].

Bacqué et al. have reported new approaches to cycloheptenes based on pathway **a** [86].

Based on strategy **b**, Lee has introduced a facile construction of the quadranoid skeleton from diyne **79** and eventually synthesized (±)-suberosenone [87]. In this process, the vinyl radical resulting from the initial addition of a stannyl radical onto the triple bond cyclized in a 5-*exo-trig* manner to give radical **81**. Intermediate **81** then underwent a homoallyl rearrangement to give **82** that finally cyclized in a 5-*exo-dig* manner (Scheme 25). A general

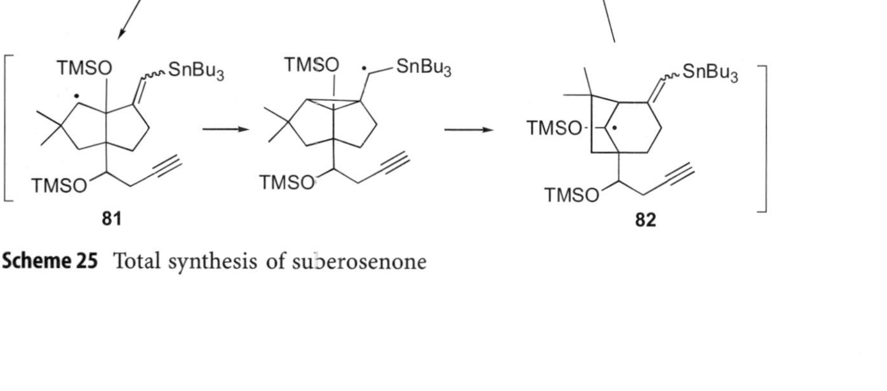

Scheme 25 Total synthesis of suberosenone

tandem radical cyclization route to tricyclo[$4.3.n.0^{1,5}$] alkanes has also been reported [88].

Toyota et al. have also used this rearrangement for the construction of bicyclo[2.2.2]octanes from enyne precursors [89]. This transformation proved to be highly dependent on the steric hindrance at C1 (Scheme 26). While compound **84** was almost exclusively formed from lactone precursor **83**, free alcohol **86** gave the rearranged bicyclo[2.2.2]octane product **87**. The increased steric hindrance at C1 on **83**—compared to **86**— presumably decreased the rate of the reduction of the intermediate radical originating from the 5-*exo-dig* cyclization. This favored the rearrangement. Gratifyingly, a formal synthesis of (±)-gibberellin A_{12} and a total synthesis of (±)-methyl gummiferolate could be achieved.

The same authors have proposed a similar approach for the total syntheses of serofendic acids [90] using tin-free conditions with the $NaBH_3CN/ZnCl_2$ reduction of a tosylhydrazone [91].

Also related to pathway **b**, Hodgson et al. were able to obtain 2-azabicyclo[2.2.1]hept-5-enes from 7-azabicyclo[2.2.1]heptadienes by a tandem featuring a thiyl intermolecular addition followed by a homoallylic radical rearrangement [92].

Mascareñas applied a strategy related to type **c** with an alkoxy radical to the chemistry of oxabicyclic systems (see **89** → **90**). He uncovered a very

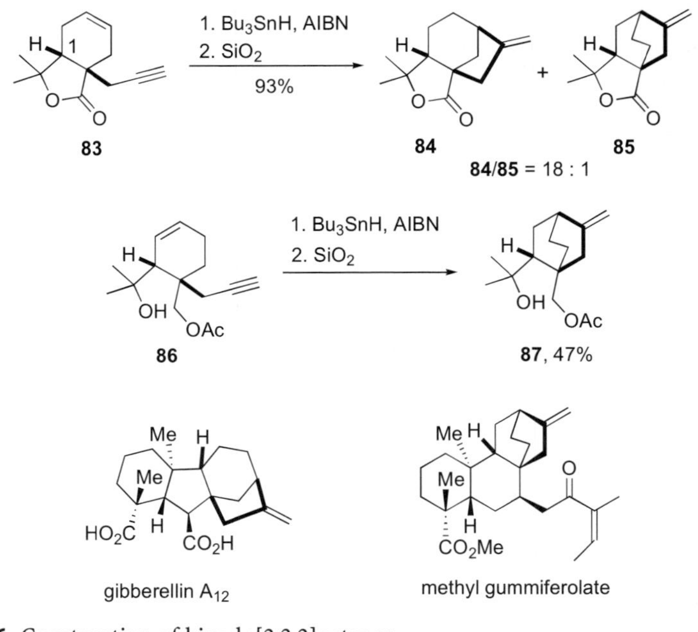

Scheme 26 Construction of bicyclo[2.2.2]octanes

efficient 7-*endo-trig*/Beckwith–Dowd rearrangement [93] cascade leading to bicyclo [5.3.1] undecane derivatives in high yield (81% for **92**, 3 : 1 ds, Scheme 27) [94]. This sequence could be extended to the preparation of bicyclo [4.3.1] decane skeletons, albeit in lower yields due to the higher strain. Such systems are of much interest, because they are present in a score of bioactive products.

Chuard et al. introduced a two-step radical alternative to the oxy-Cope reaction [95]. It relied on the opening of a strained α-oxetanyl radical to the allyloxy radical. Because of the favorable geometry, further fragmentation delivered enone radicals that could undergo a 6-*endo-trig* cyclization, which led to bicyclic ketones in 69% yield.

The radical synthesis of biaryls is a fascinating new endeavor that could be a nice alternative to transition-metal-catalyzed cross coupling reactions. In that context, Studer et al. have introduced an intramolecular radical aryl silicon to carbon migration (Scheme 28) [96]. This 1,5-aryl migration involves an initial *ipso* addition of an aryl radical, followed by a β-elimination of a silyl radical which restores aromaticity. Interestingly, the analogous 1,4-aryl migration reaction is less efficient. The same group further refined this chemistry and studied the stereoselectivity of this transformation [97].

Tokuyama et al. designed a highly elegant synthesis of indoles by reacting tributylstannyl radicals or hypophosphite salts with unsaturated

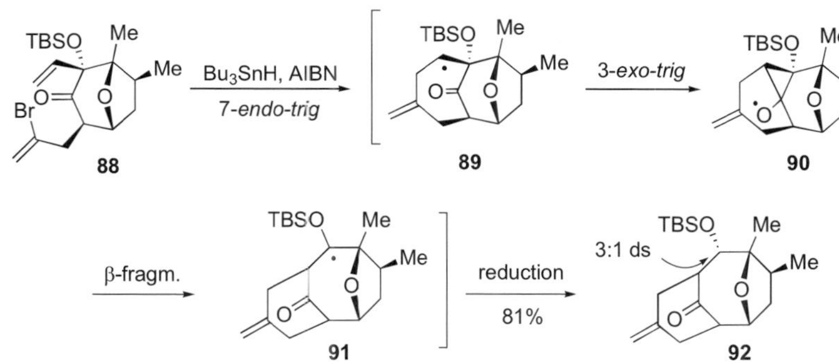

Scheme 27 Access to bicyclo [5.3.1] undecane derivatives

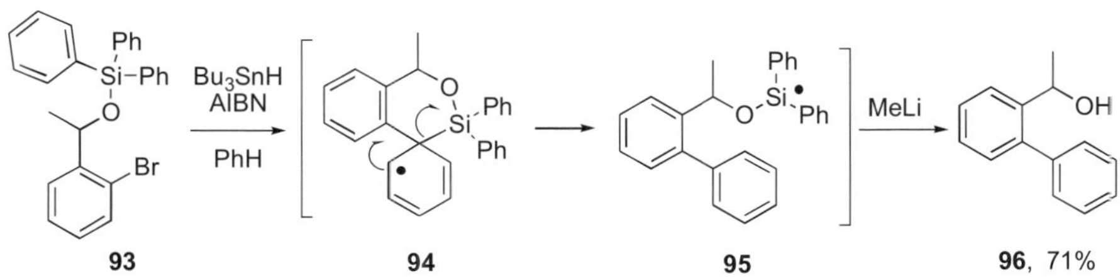

Scheme 28 1,5-Aryl silicon to carbon migration

thioanilides [98]. Initial regioselective addition of the P-centered radical onto the C = S bond generated a new stabilized carbon radical that could cyclize onto the double bond in the *ortho* position, thus giving rise to the carbon skeleton of indoles. Aromatization of the compound generated the desired 2,3-substituted indoles. The author used this reaction as the key step toward the total synthesis of (±)-catharantine, a presumed biological precursor of the antitumor alkaloids vinblastine and vincristine [99]. Following an initially similar methodology, Reynolds et al. have published an intramolecular carboxyarylation approach to Podophyllotoxin, as illustrated by the transformation of **97** into **98** (Scheme 29) [100]. Addition of a silyl radical onto a thiocarbonyl group gives birth to intermediate radical **99**. Rather than fragmenting, **99** underwent a 5-*exo-trig* cyclization leading to benzylic radical **100** which then adds in *ipso* fashion to give **101**. Rearomatization provides an alkoxy radical **102**, which suffers a rapid elimination of a new thiyl radical. This allowed regeneration of the radical mediator by hydride exchange. Based on the same strategy, the Sherburn group has published the total synthesis of (–)-arctigenin and (–)-matairesinol [101].

Other fragmentation approaches have involved the intramolecular addition of alkenyl radicals to furans. The resulting spiro radical intermediate then opens to a new α,δ-unsaturated carbonyl radical intermediate [102]. An interesting cascade process based on 2-cyanophenyl isothiocyanate, promoted by $Mn(OAc)_3$ and featuring the previously unknown addition of an iminyl radical to a thioureido group has been disclosed by Calestani et al. [103].

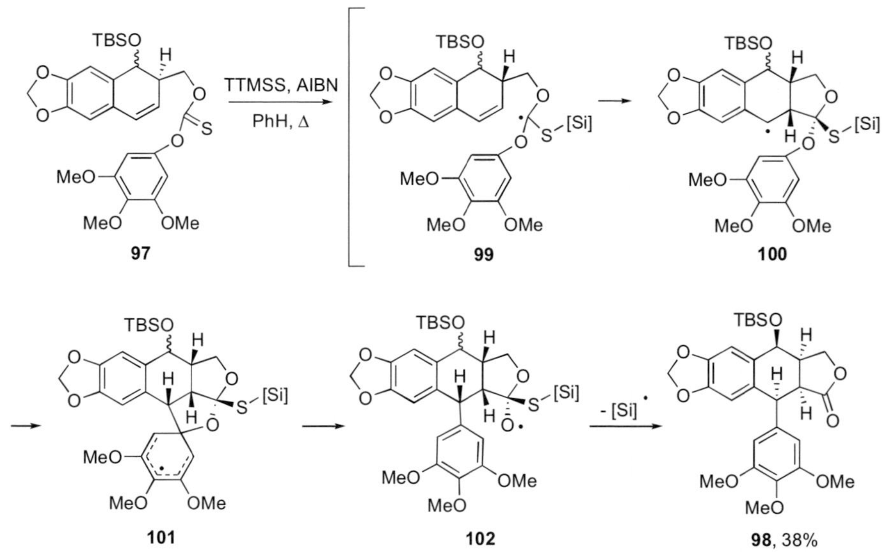

Scheme 29 Total synthesis of podophyllotoxin

2.5 Homolytic Substitution and Cyclization

Gansäuer et al. introduced a new titanocene-catalyzed domino process involving a final homolytic substitution (Scheme 30) [104]. The reaction of epoxide **103** with a titanocene derivative gave a ring-opened radical that underwent a *cis*-selective 5-*exo-trig* cyclization. Radical **104** possessed an ideal geometry for the final closure through homolytic substitution at the oxygen atom. Tricyclic ether **105** was thus obtained in 73% yield. The mechanism is reminiscent of the one involved in heme-mediated oxidations.

This reaction was not limited to alkyl-substituted olefins. Styrenes could also be involved, provided early reduction of the benzylic radical can be avoided [105]. We took advantage of this methodology to prepare densely

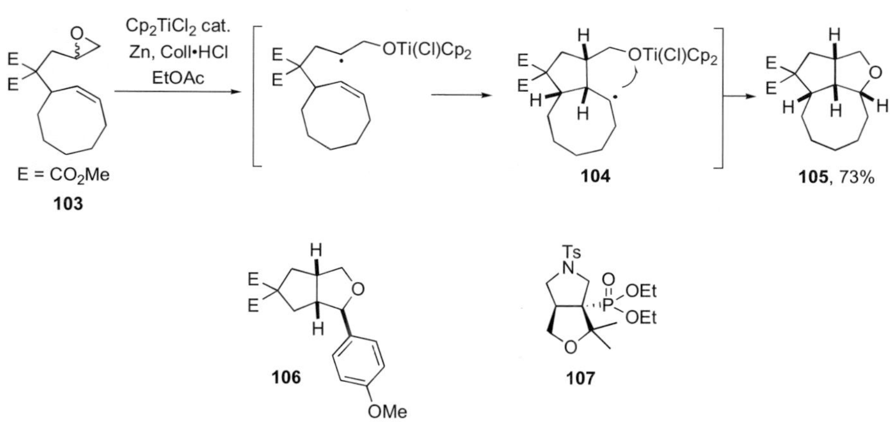

Scheme 30 Tetrahydrofurans through titanocene-mediated homolytic substitution

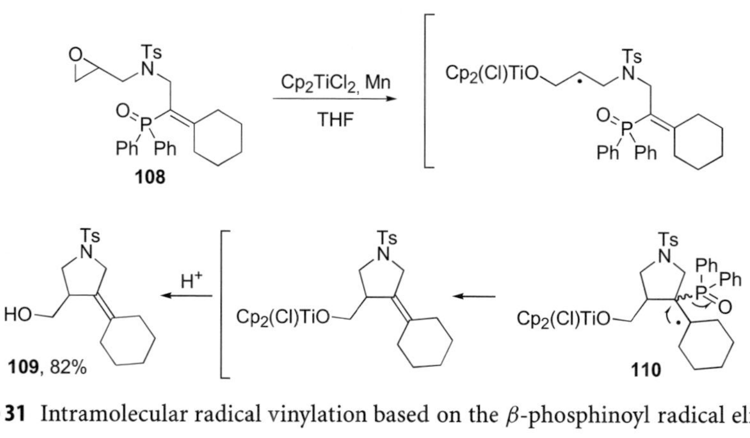

Scheme 31 Intramolecular radical vinylation based on the β-phosphinoyl radical elimination

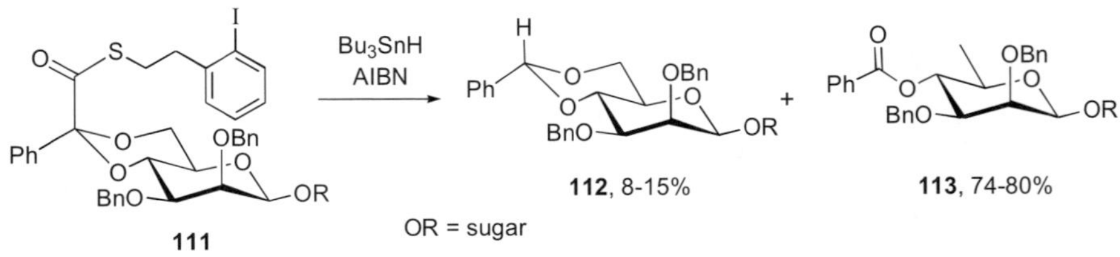

Scheme 32 Use of homolytic substitutions in carbohydrate chemistry

substituted phosphonate **107**. When a phosphine oxide was used, the reaction could be rerouted toward elimination of the heteroatom, and an original intramolecular vinylation sequence was observed (Scheme 31) [106]. The key elimination of β-phosphinoyl radicals is a reaction we discovered some time ago [64]. Thus, cyclohexylidene pyrrolidine **109** could be prepared in high yield starting from oxirane **108**. As stated above, the reaction transited via radical **110**. No tetrahydrofuran (THF) product was detected.

Crich and Yao have exploited a homolytic substitution at sulfur to trigger a radical cascade that includes a loss of carbon monoxide and a radical fragmentation of a 4,6-*O*-benzylidene moiety to give esters such as **113** after a final diastereoselective reduction (Scheme 32) [107]. This impressive outcome opens a direct avenue to β-D-rhamnopyranosides and other 6-deoxy sugars.

3 Tandem Reactions Involving Both Intra- and Intermolecular Steps

3.1 Tandem Reactions Leading to the Formation of Five-Membered Rings

3.1.1 Reactions Involving Radical Addition to 1,6-Diene or 1,5-Dienes Followed by 5-*exo-trig* Cyclization

3.1.1.1 Initial C-Heteroatom-Bond Formation

Transition-metal catalyzed tandem hydrosilylation/cyclization of dienes is a well known process. In order to render this reaction environmentally benign, Studer developed metal-free conditions using a silylated cyclohexadiene reagent **114** (Scheme 33) [108–110]. Thereby, the silyl radical released upon rearomatization of radical intermediate **115** reacts with a diene to form a β-silyl radical **116**, which undergoes cyclization into **117**. Reduction of **117** with cyclohexadiene **114** affords again **115**, as well as **118**, thus allowing for

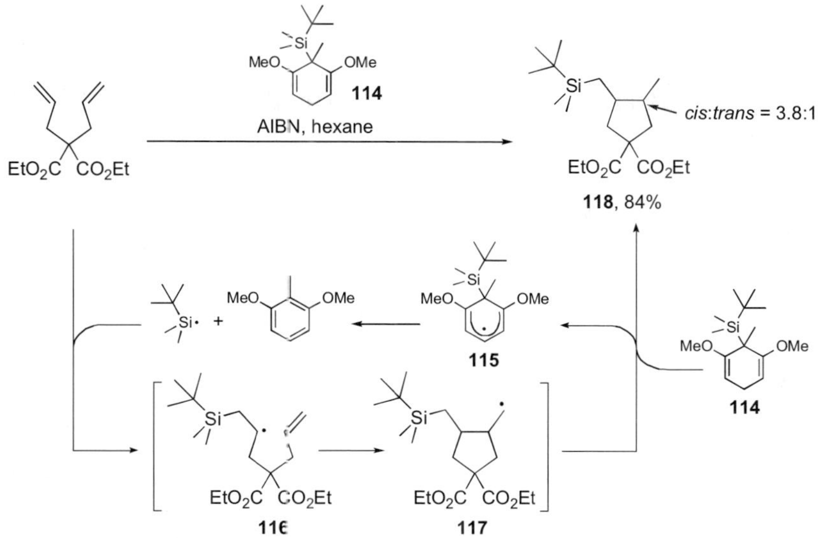

Scheme 33 Tin-free radical hydrosilylation processes

the propagation of the radical chain. This method enables the preparation of silylated cyclopentane and cyclohexane derivatives in good to reasonable yields, albeit in low diastereoselectivities.

3.1.1.2
Initial C–C-Bond Formation

Nitroxide radicals are known to easily trap alkyl radicals to form alkoxyamines, which in turn can undergo thermally induced homolysis to give back the starting carbonyl/nitroxide radical pair. Although this unique "go & return" propensity of alkoxyamines has been exploited for the design of highly efficient living radical polymerization processes, its utility for the purpose of organic synthesis has been recognized only recently [111]. In his quest for environmentally friendly radical reactions, Wetter et al. developed a thermal radical carboaminoxylation reaction with TEMPO-derived alkoxyamines, that could be combined with a cyclization process (Scheme 34) [112]. Thermal, reversible homolysis of alkoxyamine **119** generated TEMPO and a stabilized transient malonyl radical, which subsequently reacted with 1,6-diene **120**. After addition and 5-*exo-trig* cyclization, irreversible trapping of the radical adduct **121** with TEMPO afforded the carboaminoxylation product **122**. A limitation of this method is the slow addition rate of radicals issued from TEMPO derivatives other than malonates (although carboaminoxylation of simple olefins with TEMPO alkoxyamines derived from Weinreb amides, β-carbonyl phosphates and bisphosphates have recently been reported [111]). Further improvement of the tandem process by using microwave irradiation

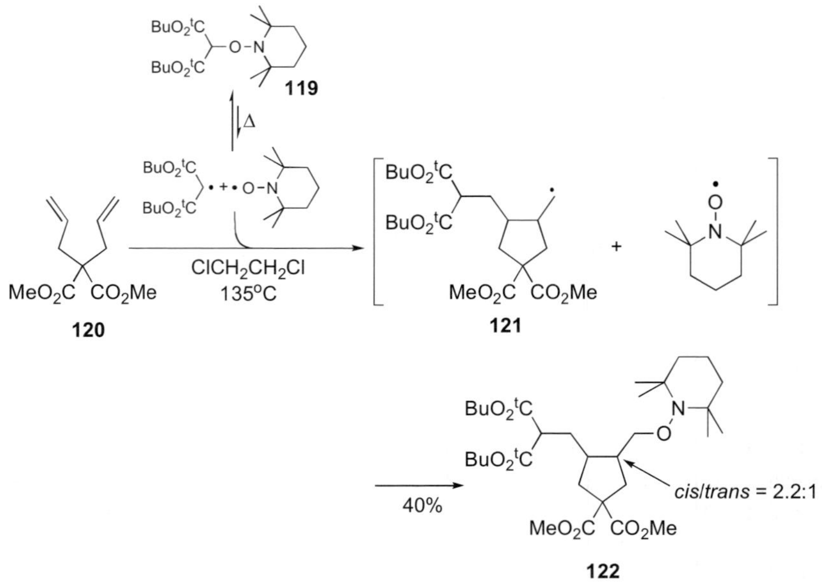

Scheme 34 Nitroxide-based carboaminoxylation

reduced the reaction time and enabled the synthesis of more difficult products such as γ-lactones and γ-lactams [113].

A similar process published by Leroi et al. using Tordo's alkoxyamine **123** allowed for new entries in the synthesis of indolinones and indolines. In this case, however, no trapping of the final radical by the nitroxyl radical occurred (Scheme 35) [114, 115].

Direct addition of an alkyl radical onto a carbonyl bond is difficult owing to its reversibility and the high π-bond strengths of the $C=O$ bond. However, the observation that phosphonate radicals constitute good leaving groups [116] triggered Kim et al. to propose acylphosphonates as acylating agents. He developed a tandem radical reaction involving the intramolecular addition of alkyl radicals to acylphosphonates and subsequent β-elimination of the corresponding phosphonyl radical from the resulting alkoxy radical (Scheme 36). This approach provided access to a wide range

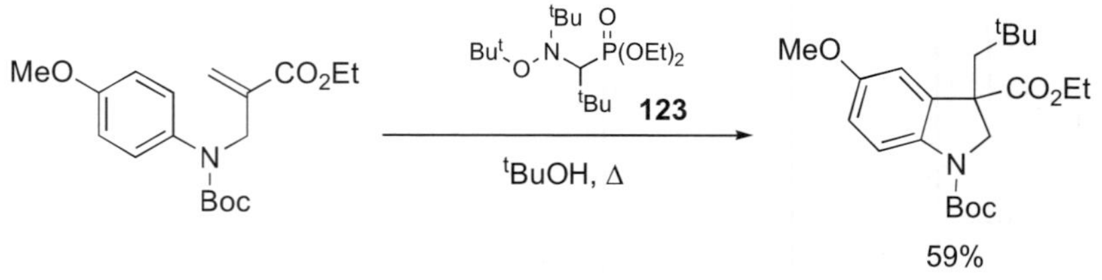

Scheme 35 Nitroxide-based access to nitrogen-containing heterocycles

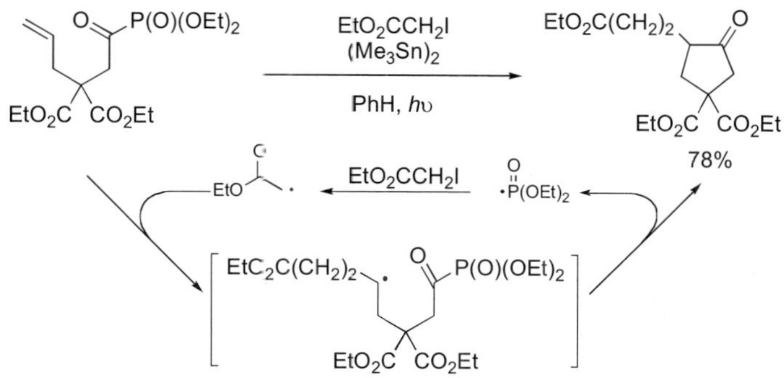

Scheme 36 Acylation using acylphosphonates

of β-functionalized cyclopentanone derivatives. In the case where alkyl radicals were obtained from alkyliodide precursors, the acylation reaction could be carried out with catalytic amounts of promoter, since the generated phosphonyl radical did promote the chain reaction by itself.

Miyabe et al. developed a tandem addition/cyclization reaction featuring an unprecedented addition of alkoxycarbonyl-stabilized radicals on oxime ethers [117], and leading to the diastereoselective formation of β-amino-γ-lactone derivatives [118, 119]. The reaction proceeds smoothly in the absence of toxic tin hydride and heavy metals via a route involving a triethylborane-mediated iodine atom-transfer process (Scheme 37). Decisive points for the success of this reaction are (1) the differentiation of the two electrophilic radical acceptors (the acrylate and the aldoxime ether moieties) towards the nucleophilic alkyl radical and (2) the high reactivity of triethylborane as a trapping reagent toward a key intermediate aminyl radical **125**. The presence of the bulky substituent R proved to be important not only for the

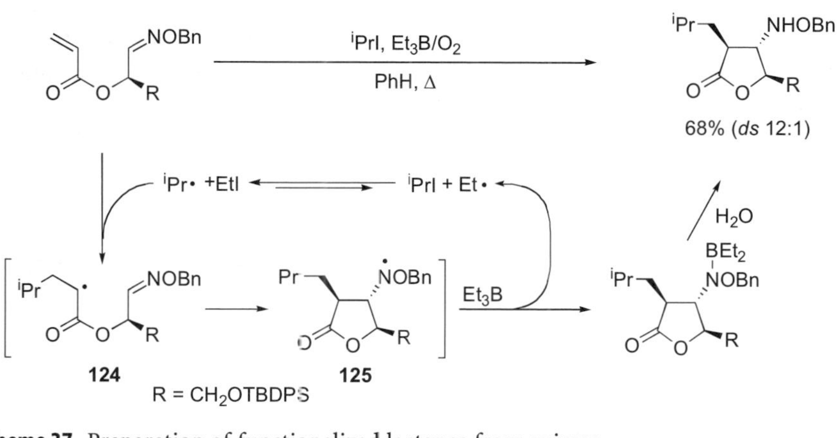

Scheme 37 Preparation of functionalized lactones from oximes

stereoselectivity but also for efficient cyclization of intermediate **124** into **125**. A further limitation is due to the impossibility of generating unstabilized radicals and primary alkyl radicals other than ethyl radicals under the iodine atom-transfer conditions. The reaction can also be conducted in a MeOH/water mixture, although with some loss in yield and diastereoselectivity. This strongly suggests that no water-unstable boryl enolate is involved in the reaction. Application of this tandem radical process for the synthesis of lactams and lactones on a solid support was also reported [120].

Briggs et al. proposed a new strategy for the synthesis of tricyclic structures using acyl xanthates as precursor for acyl radicals [121]. Irradiation with visible light of a solution of acyl xanthate in presence of 1,6-diene **126** afforded *cis*-fused bicyclic compound **127** in a good yield (Scheme 38). Radical reduction of xanthate and subsequent aldol condensation leads to the formation of [5.5.5]-fused ring systems similar to those of the triquinane terpene family.

Sibi et al. reported an elegant new tandem radical addition–cyclization process for the synthesis of oxacycles [122]. The method relies on the intermolecular addition of alkyl radical to a doubly activated Michael acceptor moiety, followed by a cyclization step (Scheme 39). The outcome of the re-

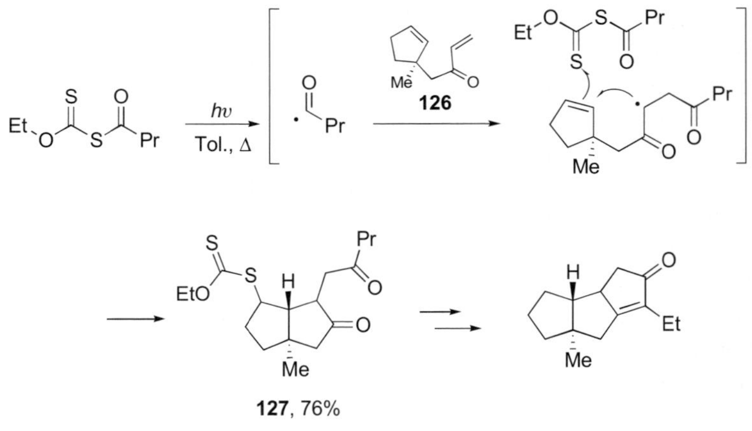

Scheme 38 Xanthate-based access to bicyclic ketones

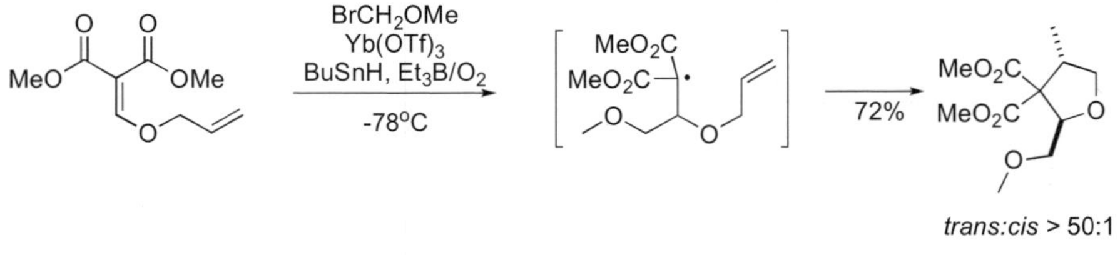

Scheme 39 Tandem synthesis of oxacycles

action is greatly enhanced by the use of $Yb(OTf)_3$ as Lewis acid. It exerts a beneficial effect on both the rate of intermolecular addition and the electrophilicity of the intermediate malonyl radical. The method proved to be highly versatile, as various oxacycles from size ranging from five to eight could be obtained in good yields and moderate to good diastereoselectivities via 5-*exo*, 6-*exo*, 6-*endo* or 7-*exo* cyclization steps respectively, depending on the ether type.

3.1.2 Internal Cyclization Followed by Intermolecular Addition

Kim et al. proposed a new successful strategy for the α-alkylation of unsaturated carboxylic acid derivatives relying on the addition of an alkyl radical to carboxylic imides (ketene *O,N*-acetals) **128** [123, 124]. In the course of his study, he also examined the sequential radical reaction involving cyclization and alkylation steps, which could not be achieved with other methods (Scheme 40). Although alkyl iodides and bromides bearing an α-electron withdrawing group underwent alkylation under tin-free conditions with 0.1 equiv. of AIBN or V-70—the chain reaction being maintained by silyl radicals issued from the conversion of the generated silyloxy radicals into silyl radicals—the tandem reaction needed to be run with equimolar amounts of hexamethylditin. Thus, treatment of iodide **129** with $(Me_3Sn)_2$ afforded **130** in reasonable yield as a 1 : 1 mixture of diastereoisomers.

Ollivier and Renaud built on his new ditin-mediated radical azidation with benzenesulfonyl azide [125] to introduce tandem intramolecular cyclization–azidation processes [126]. For example, iodoacetal **131** provides the corresponding tertiary azide **132** in high yield as a 3 : 2 mixture of *endo*/*exo* diastereoisomers (Scheme 41). The necessity of using ditin reagents still constitutes a drawback in this method, as DLP-mediated tin-free reaction conditions failed due to an inefficient chain process.

Organosulfone-mediated allylation, vinylation and azidation represent very effective tin-free radical processes. However, the reported methods do not work well with primary alkyl iodides and xanthate as radical precursor, owing to in-

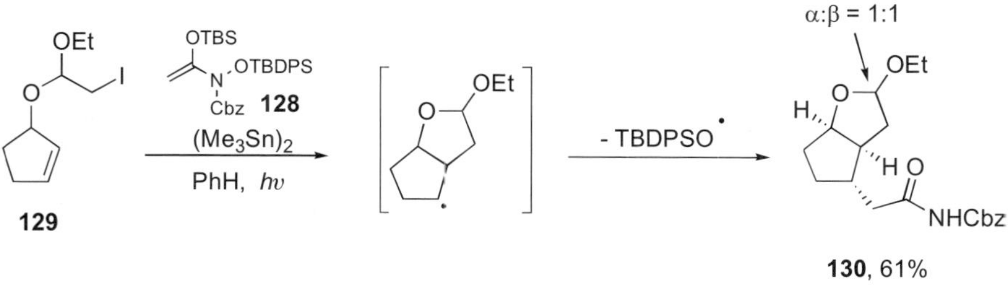

Scheme 40 γ-Alkylation of unsaturated carboxylic acid derivatives

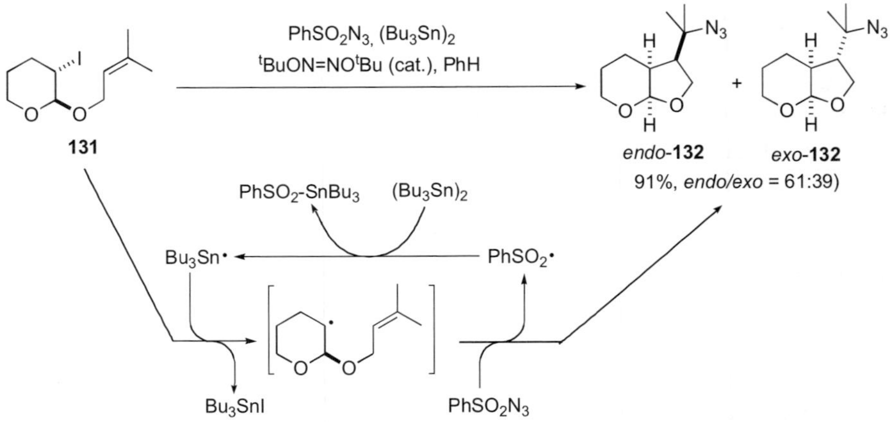

Scheme 41 Tandem cyclization/azidation

efficient xanthyl or iodide radicals transfer. Therefore, Kim and Lim developed a new reliable method and found that alkyl allyl sulfones represent highly efficient and reliable precursors for the generation of primary alkyl radicals under tin-free conditions [127]. This method can be applied to various C – C-bond forming reactions including cyanation, vinylation and allylation. Tandem reactions including a 5-*exo-trig* cyclization proved also to work well (Scheme 42). Thereby, the alkyl radical was obtained by thermal degradation of the alkyl sulfonyl radical generated by addition of *p*-toluenesulfonyl radical to alkyl allyl sulfone **133**. The key for the success is that the generated alkyl radical **136** should preferentially add to *p*-toluenesulfonyl cyanide and give **134**, rather than to the allyl sulfones **133** and **135**.

Kim further developed the alkyl allyl sulfone-based chemistry, and disclosed a carbonylation reaction to prepare thioesters (Scheme 43) [128, 129].

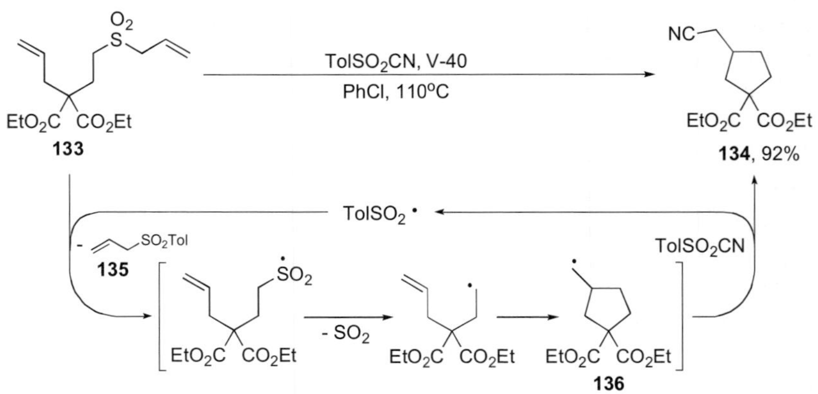

Scheme 42 Sulfone-based cyanations

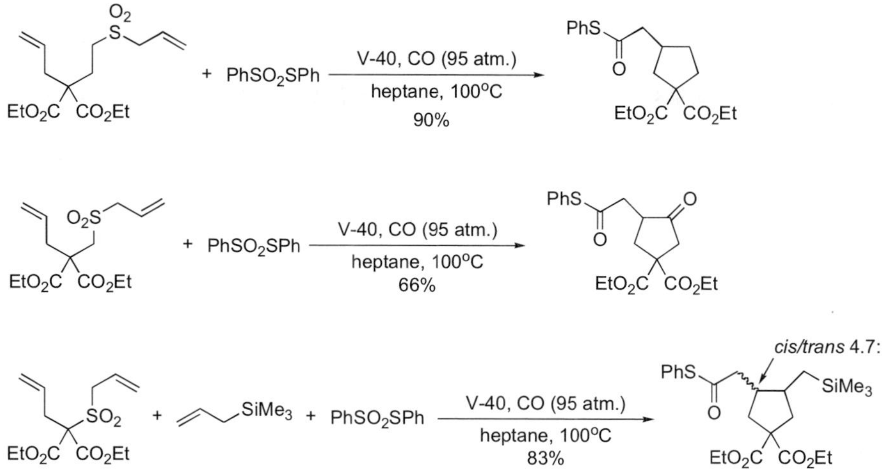

Scheme 43 Formation of thioesters through radical carbonylations

Depending on the length of the tether of the sulfone used, mono and di carbonylation products could be obtained through a multicomponent reaction. Benzenethiosulfonate thus constitutes a very efficient radical-trapping agent for acyl radicals.

The generation of amidyl radicals from *N*-allylsulfonamides and their subsequent cyclization was probed by Moutrille and Zard [130]. This strategy allowed the preparation of lactams such as **140** by treatment of acylsulfonamide **138** with lauryl peroxide and a xanthate in DCE (Scheme 44). However, when the stability of the generated amidyl radical (as with **137**) was not high enough, the extrusion of sulfur dioxide turned to be too slow, and premature cyclization of the *N*-amidosulfonyl radical intermediate took place, leading to **139**.

Uenoyama et al. also envisaged combining TEMPO-based radical cyclizations with carbonylations in order to yield precursors of the tetralone framework [131]. Surprisingly, treatment of **140** at 130 °C under a CO atmosphere directly yielded a cyclized carboxylic acid. This was shown to arise from intermediate **141** by an unknown mechanism. The free carboxylic acid derivative could be further cyclized into tricyclic ketone **142** using Otera's intramolecular Friedel-Crafts procedure.

Gansäuer et al. studied the tandem titanocene-mediated cyclization/intermolecular addition to Michael acceptors [132]. After treatment with *t*-butyl acrylate in the presence of Zn, collidine hydrochloride and catalytic amounts of titanocene [133], epoxide **143** yielded the desired bicyclic product **144** in reasonable yield and high diastereoselectivity (Scheme 46). Remarkably, alkyne **145** led to the first example of an intermolecular addition involving a vinylic radical. The success of this reaction was attributed by the authors

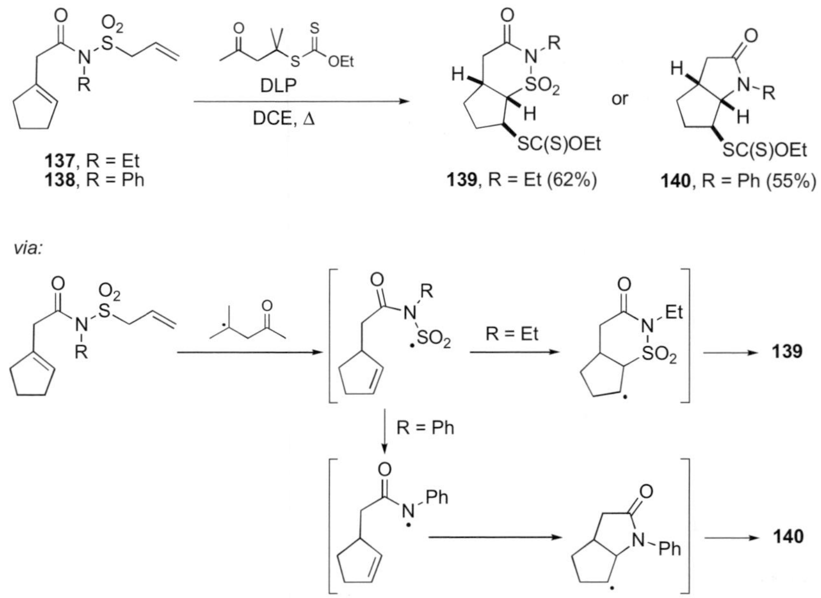

Scheme 44 Access to amidyl radicals through desulfonylation

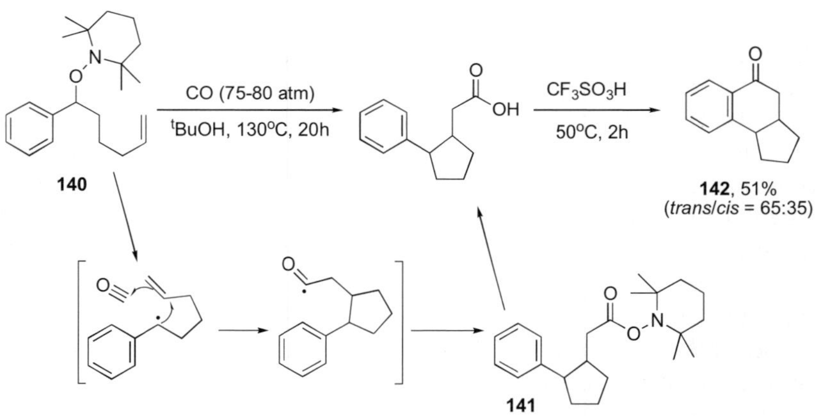

Scheme 45 Access to tetralones

to (1) the high grade of preorganization, (2) the electron deficiency of the β-titanoxy radicals and (3) the fact that THF —a solvent with high hydrogen donor propensity—could be advantageously replaced by AcOEt when needed.

Lee and Larock found that the characteristic tricyclic core of prostacyclines could be easily obtained by tandem radical reaction involving the trapping of a radical issued from a 5-*exo-trig* cyclization by an excess of stannyl enones [134].

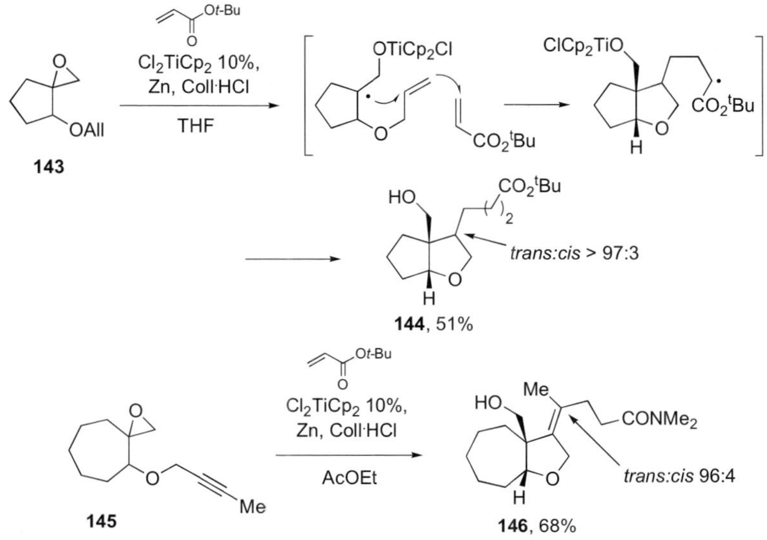

Scheme 46 Titanocene-mediated tandem reactions

3.1.3
Miscellaneous

In the course of his investigations on the radical synthesis of 2-acylindole alkaloids by addition of 2-indolylacyl radical onto Michael acceptors, Bennasar et al. noticed that when exchanging Bu_3SnH with a poorer hydrogen-atom donor such as TTMSSH, a tricyclic product was isolated in a noticeable amount, along with the desired linear addition product. It was therefore anticipated that carrying out the reaction in non-reductive conditions would allow access to the cyclopenta[b]indole core present in numerous indole alkaloids [135]. Heating a benzene solution of phenyl selenoester **147** and various unsaturated esters in the presence of $(Bu_3Sn)_2$ under light irradiation gave polycyclic compounds such as **148** (Scheme 47). However, when more reactive alkene acceptors such as acrylonitrile were tested, only the cyclohepta[b]indole product resulting from the tandem bis-addition/cyclization reaction could be isolated.

Tsuchii et al. reported a very interesting four-component domino process where an alkyne, two olefins and diphenyl diselenide sequentially react to form a highly functionalized cyclopentane derivative, after a linear addition sequence and 5-*exo-trig* cyclization [136]. This reaction can be seen as an interrupted polymerization process initiated by the addition of selenyl radical to an electron-deficient alkyne in the presence of a large excess of a Michael acceptor. The identity of each reaction partner is important for the outcome of the reaction. For instance, use of $(PhS)_2$ instead of $(PhSe)_2$ leads to the polymerization product rather than to the cyclization one, while $(PhTe)_2$ did

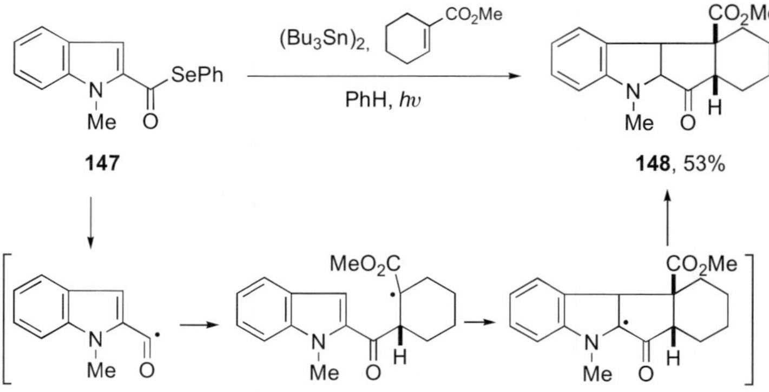

Scheme 47 Access to the cyclopenta[b]indole core

not initiate any sequential reaction, due to the low reactivity of the telluryl radical towards propiolates. Similarly, when vinyl ether was used as olefin, three-component coupled products were cleanly obtained. Finally, when the reaction was carried out in the presence of two different olefins, e.g., one olefin with an electron-withdrawing group and one with an electron-donating group, the mixed addition/cyclization product was obtained chemoselectively (Scheme 48).

In a thorough study on the homolytic substitution reaction at silicon, Studer and Steen reported a mild method for the formation of cyclic five-membered alkoxysilanes based on tandem intermolecular addition/intramolecular homolytic substitution [137], whereby the reaction of homoallylic stannylated silylether **149** with ethyl iodoacetate in the presence of $(Me_3Sn)_2$ led to the desired S_Hi product **150** in high yield as the sole *trans* diastereoisomer (Scheme 49). While the 1,2 induction proved to be very high, leading in some cases to a unique diastereoisomer, 1,3 stereoselectivities were

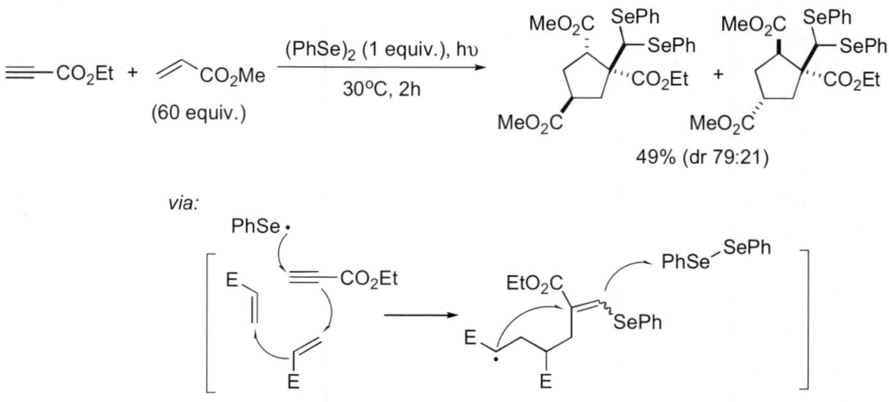

Scheme 48 Multicomponent diselenide-mediated annelation

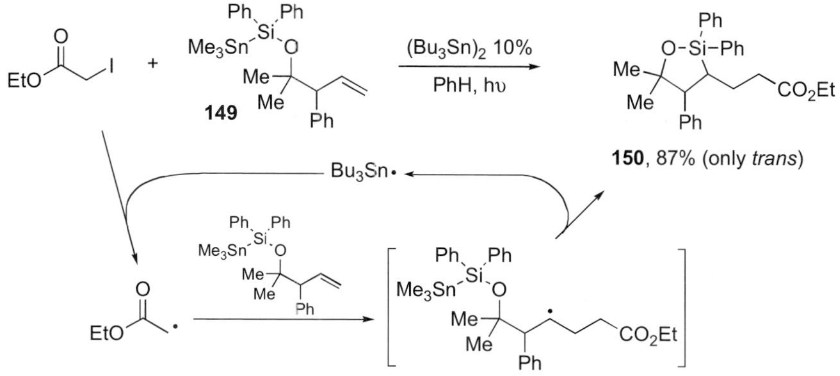

Scheme 49 S_Hi at silicon-based tandem reactions

rather poor. The products could easily be converted to the synthetically important 1,3 diol derivatives by Tamao-Fleming oxidation.

3.2 Reactions Leading to the Formation of Six-Membered Rings

3.2.1 [4 + 2] Annulation

Miranda et al. reported that benzolizidine and indolizidine systems can be obtained by tandem addition/cyclization/oxidation process starting from 1-(2-iodoethyl)indoles or pyrroles and methyl acrylates [138]. After having shown that this strategy worked successfully with hexabutylditin reagent as initiator, the authors reported a tin-free procedure relying on Fenton-type conditions [139]. This latter method offers environmentally benign conditions, and the advantage of solving the problem of the otherwise problematic oxidative rearomatization.

Hoffmann developed an original diastereoselective tandem addition/cyclization reaction of *N*,*N*-dimethylaniline with (5*R*)-menthyloxyfuran-2(5H)-one [140, 141]. The reaction was initiated by photochemical electron transfer (PET), using either aromatic ketones possessing electron-donating substituents or semiconductors such as TiO_2 or ZnS as sensitizer, thus allowing the generation of the nucleophilic α-aminoalkyl radical directly from the tertiary amine. Satisfactory results were obtained when the reaction was conducted in the presence of acetophenone derivatives as sensitizer (Scheme 50). Although it often offers the advantage of a simplified separation procedure, the implementation of the reaction under heterogeneous conditions using semiconductor particles as sensitizer proved to be less efficient [142].

Chlorodifluoromethylated derivatives represent good substrates for the preparation of *gem*-difluorinated heterocycles by means of electrochemical

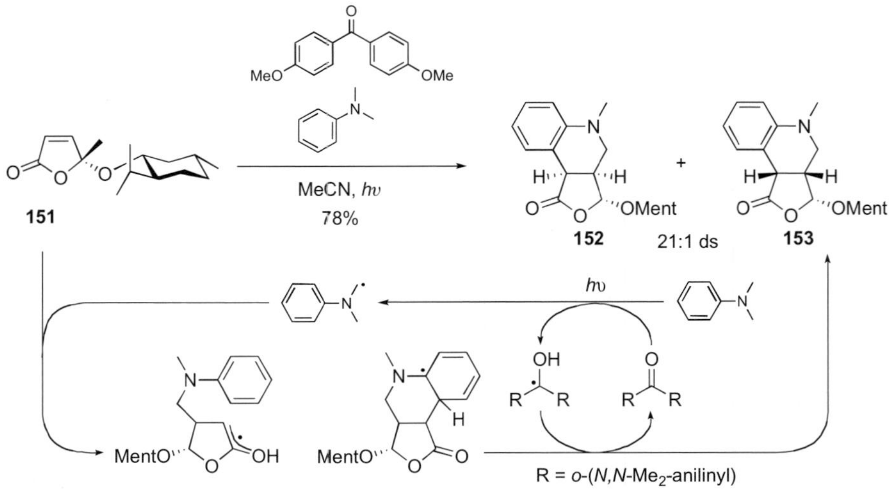

Scheme 50 Diastereoselective photoinduced tandem

free-radical cyclization reactions. The synthetic value of this method was illustrated by the elegant synthesis of biologically interesting compounds [143].

Pandey et al. discovered that silyl radical can be obtained trough mesolysis of the photoinduced electron transfer (PET) activation product of $PhSeSiMe_3$, and recognized that this process can be used for the design of catalytic phenylselenyl transfer reactions [144, 145]. In the course of his investigations, an interesting tandem reaction between allylsilane and malonate derivatives leading to the six-membered ring products was disclosed. The preference for the 6-*endo-trig* cyclization mode over the 5-*exo-trig* mode is believed to be due to a less bipolar transition state resulting from the presence of silyl group β to the radical.

Bachi has been active in the evaluation of several endoperoxides as potential antimalarial agents [146, 147]. The active compounds were prepared through a very elegant domino process involving molecular oxygen (Scheme 51). Thiyl radical addition onto diene **154** led to an adduct that was trapped by dioxygen to yield **155**. 5-*Exo-trig* cyclization of the hydroperoxyl radical delivered bicyclic radical **156**, which was reduced to hydroperoxide **157**. Triphenylphosphine-mediated reduction of **157** yielded alcohol **158** (70% over two steps) as an essentially 1 : 1 mixture of epimers at the benzylic position. This was due to low selectivity during the initial oxygen trapping, but it should be pointed out that all the other stereogenic centers were controlled.

Given the growing need for efficient drugs against malaria, O'Neill et al. examined the possible pharmacologic mechanisms that make artemisinin-related endoperoxides highly active molecules (Scheme 52) [146]. They showed that **159** acted as a Trojan-horse-type prodrug, which could enter the *Plasmodium falciparum* ferrous-rich food vacuole. The endoperoxide moiety

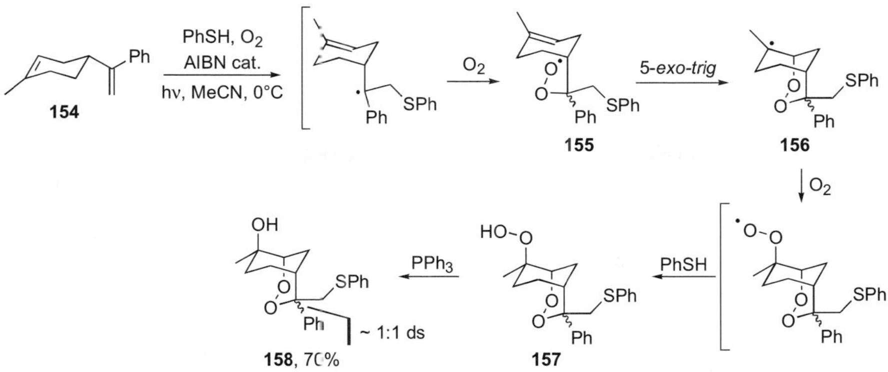

Scheme 51 Synthesis of antimalarial bicyclic peroxides

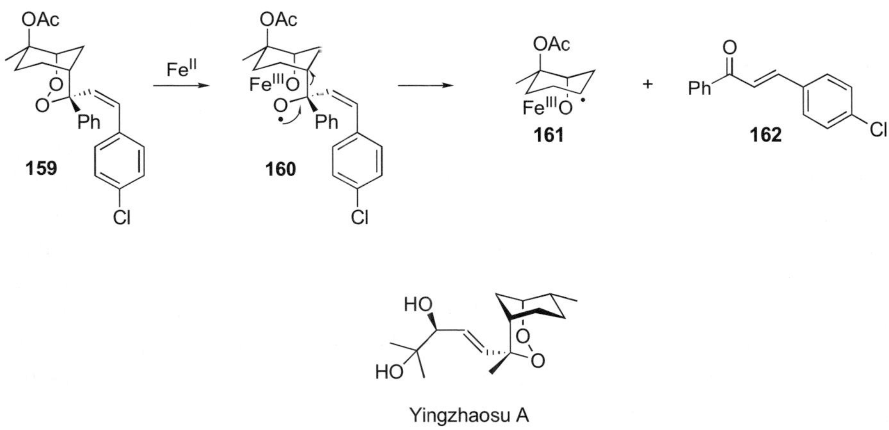

Scheme 52 Pharmaceutical model accounting for the antimalarial activity

is reduced by the iron(II) to deliver oxygen-centered radical **160**, which can collapse to generate radical **161** and toxic chalcone **162** in the living parasite, thus killing it. It is worthy of note that this biological pathway is also a radical tandem process!

Szpilman et al. used his methodology to prepare yingzhaosu A, a natural endoperoxide isolated from extracts of *Artabotrys uncinatus*, a traditional remedy for treatment of malaria in China [148].

3.2.2
1,6-Cyclizations

The unique reactivity of methylenecyclopropanes prompted Zhou to test on this substrate the manganese (III)-mediated free radical cyclization of alkenes with 1,3-dicarbonyl compounds [149]. It was found that benzyli-

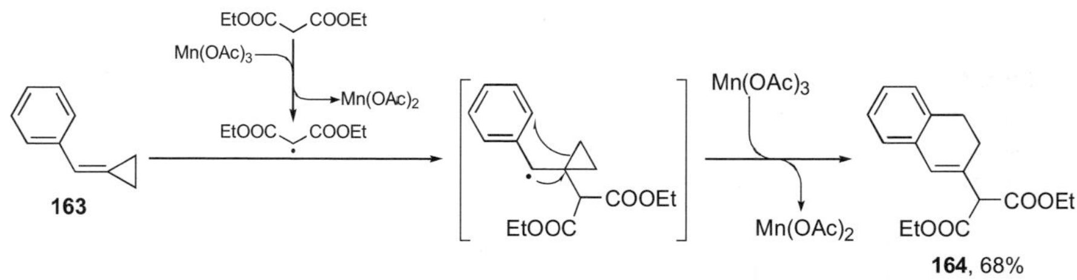

Scheme 53 Formation of dihydronaphthalenes

denecyclopropanes such as **163** react with diethylmalonate in the presence of $Mn(OAc)_3$ to generate 2-(3,4-dihydronaphthalen-2-yl)malonic acid derivatives (Scheme 53) in satisfactory yield. The reaction has been proposed to occur through formation of the malonyl radical, which adds to **163**. Opening of the resulting α-methylcyclopropane radical followed by internal cyclization onto the aromatic ring, and ensuing oxidation by a further $Mn(AcO)_3$ molecule results in the formation of product **164**.

3.3 Larger Cycles

In the course of his investigations toward the development of new strategies for the synthesis of bicyclic β-lactam systems alternative to classic β-lactam antibiotics, Alcaide et al. devised a new synthetic route relying on the tandem radical addition/cyclization of 2-azetidinone-tethered 1,7-, 1,8- and 1,9-enynes [150]. The method proved to be general and highly diastereoselective for the preparation of up to nine-membered fused rings (Scheme 54). Depending on the nature of the initial radical, two different products resulting from either an *endo-dig* (path A) or an *exo-trig* (path B) cyclization were obtained. The difference in reactivity can be explained by the nucleophilic character of benzylic radicals that would direct the initial addition to the Michael acceptor part of the substrate. The electrophilic sulfur- and tin-centered radicals on their part would preferentially add to the electron-rich triple bond. Further β-lactams were obtained by a similar strategy starting from 2-azetidinone-tethered allene-ynes [151].

3.4 Reactions Leading to the Formation of Polycyclic Compounds

In a pioneering work, Curran et al. demonstrated the practicability of radical [4 + 1] annulations with isonitriles for the synthesis of cyclopenta-fused quinolines, a structural motive constituent of camptopthecin family antitumor agents [152, 153]. This reaction proved to be highly efficient and of unrivalled flexibility, enabling the preparation of an unprecedented number

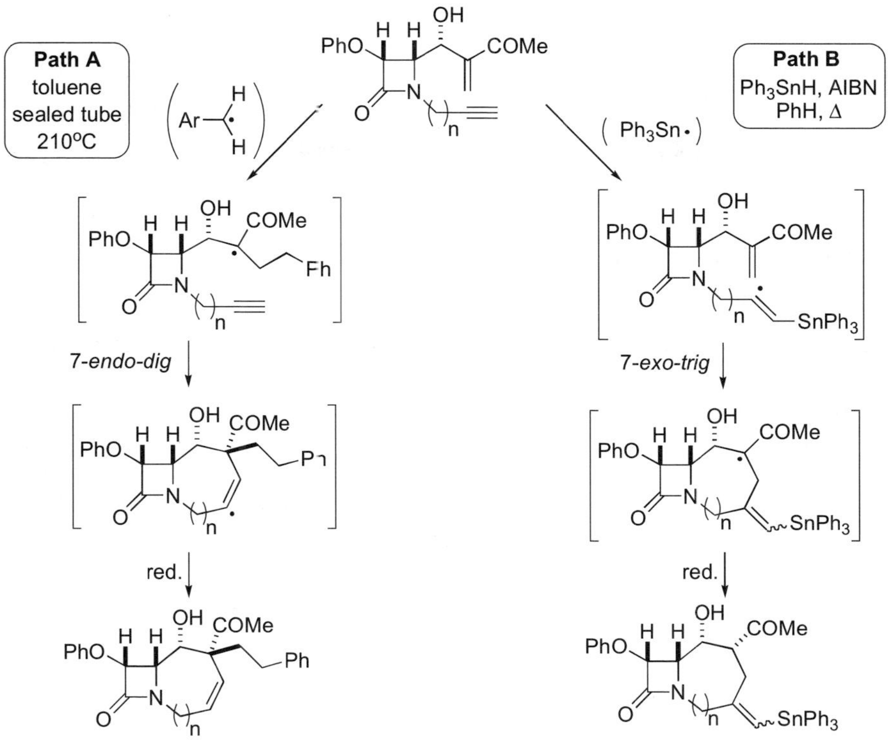

Scheme 54 Preparation of modified bicyclic β-lactams

of mappicine and camptothecin derivatives as well as whole combinatorial libraries [154, 155]. Use of the fluorous tag strategy [156] was even more beneficial. In two simultaneous reports, Zhang et al. disclosed the preparation of both enantiomers of mappicine through a quasi-racemic synthesis enabled by the fluorous method [157], and a solution-phase preparation of a 560-compound library of individually pure mappicine analogues (four-step approach; two one-pot and two parallel) [158]. Further substrate scope investigations showed that cascade reactions of *o,o′*-dialkyl phenylisonitriles with *N*-propargyl-6-iodopyridones provided mixtures of 9- and 12-substituted products, where the 9-substituted isomer **168** often predominates (Scheme 55) [159]. While the 12-substituted regioisomer **167** results from the β-fragmentation of an isopropyl group from the standard 1,6-cyclization product, regioisomer **168** is the product arising from an initial concurring 1,5 *ipso* cyclization of the vinyl radical intermediate. This reaction provides a route to 9-substituted products, which are otherwise difficult or impossible to form by radical cascade from monosubstituted phenyl isonitriles.

In a similar way, Zanardi and Nanni exploited [4 + 1] annulations leading to the synthesis of a series of benzothieno[2,3-b]quinoxalines [160, 161]. Add-

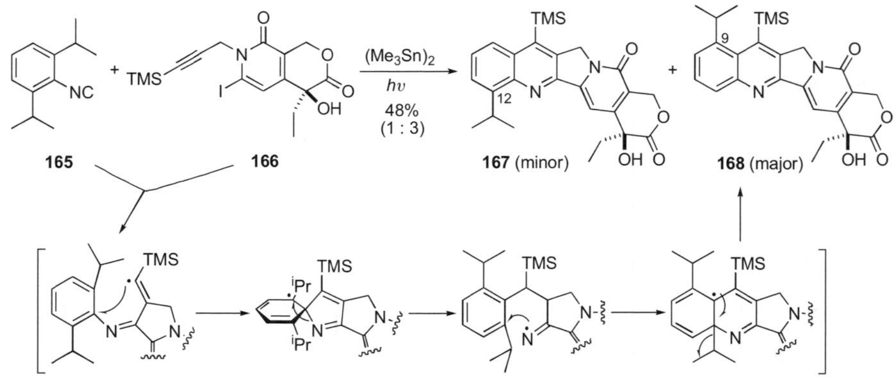

Scheme 55 Variations on the [4 + 1] radical cycloaddition

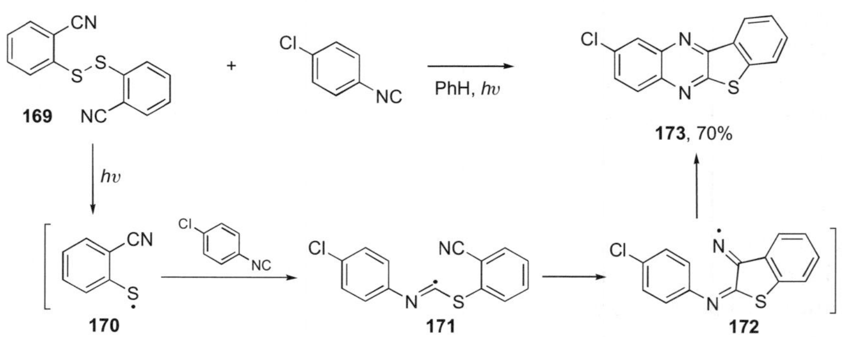

Scheme 56 Synthesis of sulfur-containing heterocycles through [4 + 1] radical cycloadditions

ition of 2-cyanoarylsulfanyl radical **170**—issued from the photolytic cleavage of disulfide **169**—to phenyl isonitrile leads to the formation of the imidoyl radical **171** (Scheme 56). 1,5-Cyclization to the cyano group and concomitant six-membered ring closure of iminyl radical **172** afforded benzothienoquinoxaline **173** in good yield. An alternative way to generate a-thio-substituted radical intermediates relying on the addition of C-centered radicals to isothiocyanates was also exploited by the authors. However, this strategy proved not to be as clean as the former [161–163].

3.5 Reaction Involving Hydrogen and Group Translocations

Cekovic has examined the addition of electron-deficient olefins to δ-hydroxy radicals [164]. Addition to acrylonitrile of the carbon-centered radicals resulting from the Barton photolysis of alkyl nitrite **173** and subsequent 1,5-hydrogen transfer, yielded α-oximino nitrile **174** (63%, Scheme 57) [165].

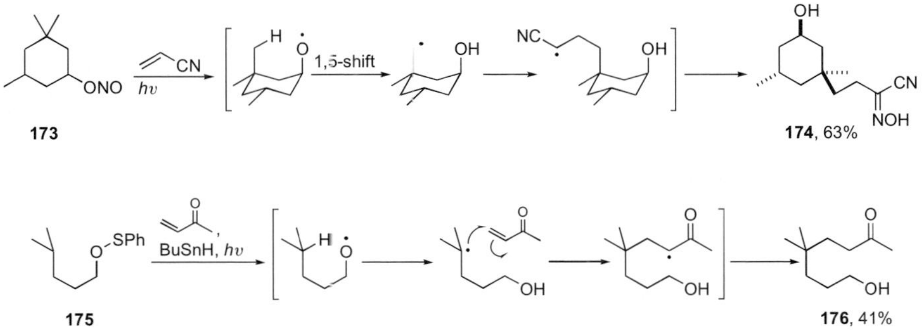

Scheme 57 Reaction of oxygen-centered radicals

However, a large excess (80 equiv.) of acrylonitrile had to be used in order to repress the normal Barton reaction (e.g., the coupling of the nitrosyl radical generated during the photolysis to the δ-hydroxy radical).

A reaction sequence where the alkoxyl radical is obtained from benzensulfonates such as **175**, leads to the formation of linear addition products [165, 166].

En route to the synthesis of modified oligonucleotides through radical addition of hypophosphorous substrate **177** to sugar **178**, Dubert et al. reported the unforeseen β-elimination of a phosphinoyl radical from a phosphinate (Scheme 58) [167]. This example came after a 1,5-H translocation giving a more stabilized α-oxy radical (from **179**), and it demonstrates further that β-phosphinoyl radicals are prone to undergo elimination (see Scheme 31), even when the substituents are all alkyl groups.

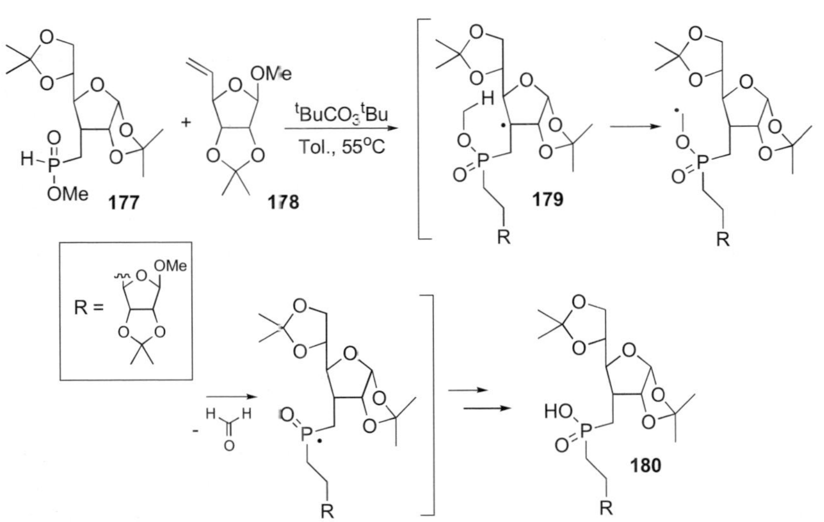

Scheme 58 Elimination of phosphinoyl radicals in carbohydrates

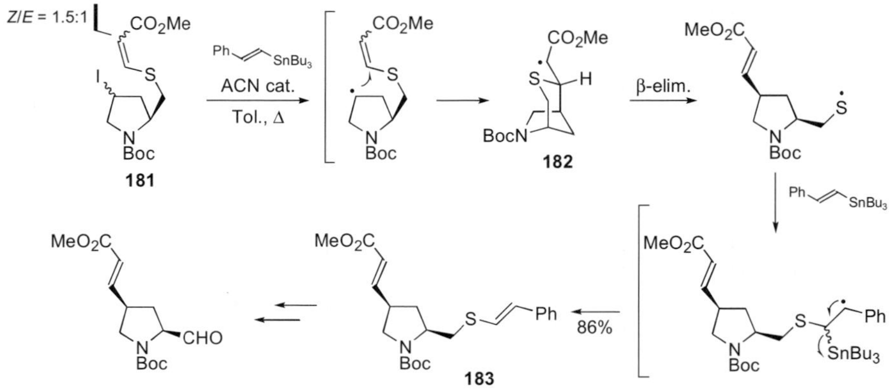

Scheme 59 Bachi's double vinylation

Not only hydrogen migration was exploited for the purpose of tandem radical reactions. Ouvry and Zard proposed a tin-free radical cascade involving a 1,2-aryl migration, and resulting in a net homoallylation process [168]. The neophyl rearrangement, on which the overall reaction relies, is a relatively slow and reversible process. However, the following elimination of a methanesulfonyl radical and its decomposition into SO_2 and the highly reactive methyl radical drives the reaction.

Korshin et al. reported an outstanding double stereoselective alkenylation through a 1,5 sulfur-to-carbon translocation on a proline-derived scaffold [169]. 5-*Exo-trig* cyclization of the radical derived from **181** led to bicyclic adduct **182**, which collapsed to the open thiyl radical (Scheme 59). The sole stereocontrol exerted by the cyclization step allows for the resulting vinyl translocation to occur entirely stereoselectively. Since the reaction was carried out in the presence of a styryltin derivative, consecutive intermolecular vinylation occurred, leading to bisvinyl compound **183** in very high yield (86%). The styrylsulfonyl moiety could be converted to a formyl group.

4
Intermolecular Tandem Reactions

4.1
Multicomponent Reactions

Miyazoe et al. introduced silyltellurides as partners for group-transfer three-component couplings with isocyanides (Scheme 60) [170]. The reaction probably started through initial addition of silyl radicals to a carbonyl derivative, yielding ketyl radical **185**, which underwent addition to phenylisocyanide (a C1 radical synthon). Final phenyltelluride group transfer to intermediate

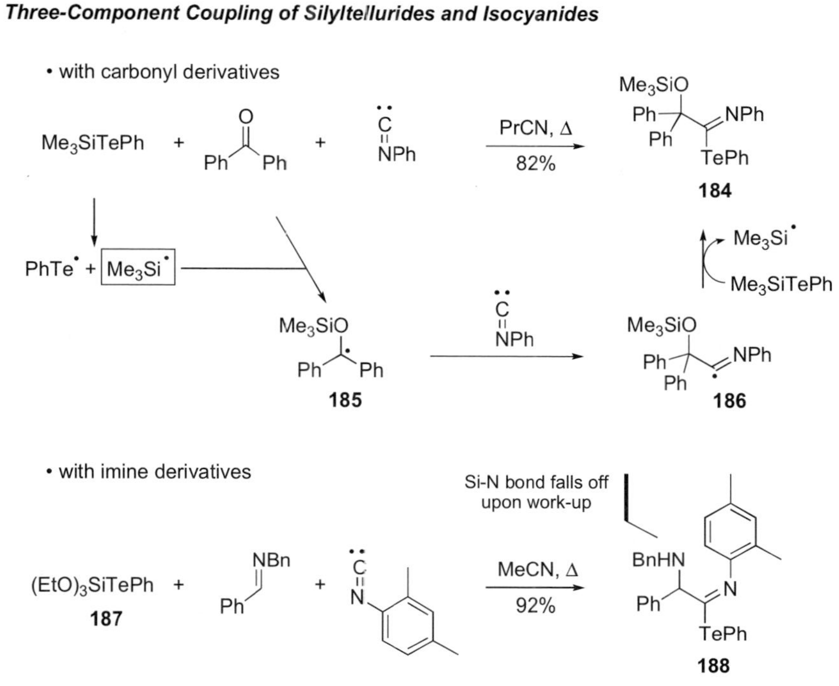

Scheme 60 Three-component coupling of silyltellurides

radical **186** yielded final product **184** in high yield. This method could not be directly extended to imines. Nonetheless, switching to triethoxysilylphenyltelluride **187** proved rewarding. α-Amino derivatives such as **188** could be prepared in excellent yields [171].

The same authors also reported a modified sequential reaction, replacing the isocyanide partner with an alkyne (Scheme 61) [172]. Varied functionalized TMS-protected *E*-allylic alcohols could be prepared. In addition, the phenyltellurium group final transfer to vinyl radical **191** afforded vinyl telluride **189**, which could be further elaborated. Once again, this reaction proved highly versatile, as basically all kind of carbonyl compounds were suitable (aromatic and aliphatic ketones and aldehydes). Electron-poor alkynes did not give the desired products.

Trimethylsilylphenyltelluride could also be used to efficiently bis-silylate quinones to the corresponding bis-protected hydroquinones (Scheme 62) [173]. The reaction required two equivalents of the silyltelluride and diphenylditelluride was also isolated. The proposed mechanism is slightly different from above, featuring an initial single electron-transfer to form the quinone radical-anion, which was presumably silylated to form phenoxyl radical **192**. Subsequent reaction with trimethylsilylphenyltelluride delivered **193** and diphenylditelluride.

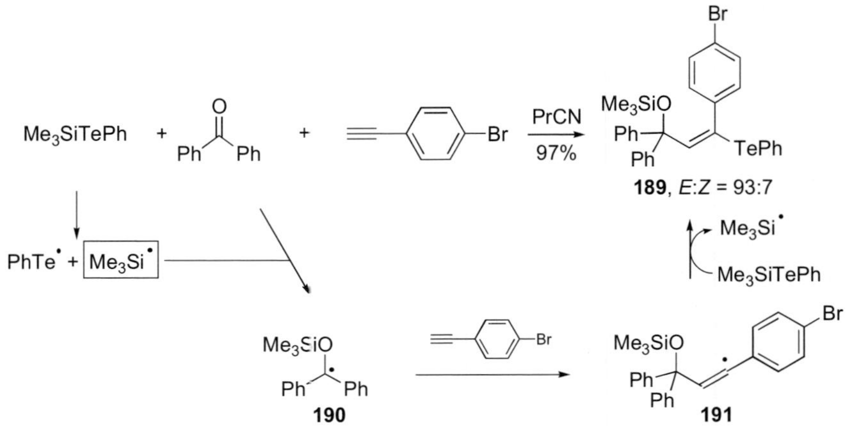

Scheme 61 Three-component coupling of silyltellurides with alkynes

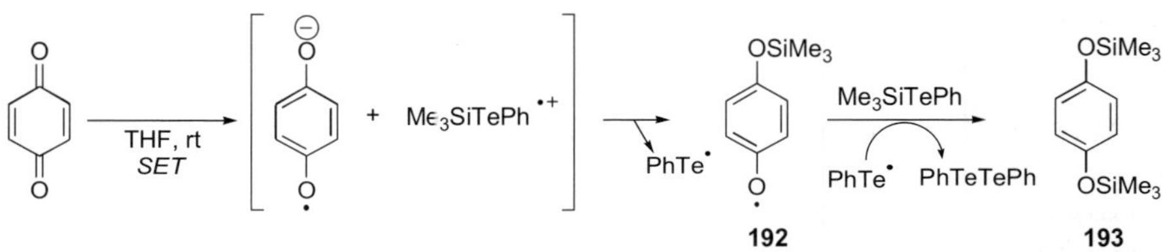

Scheme 62 Bis-silylation of quinones

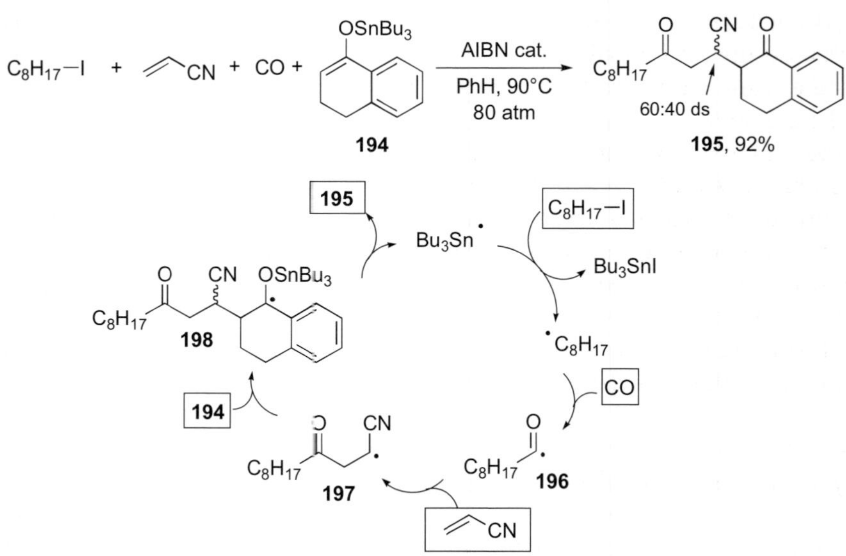

Scheme 63 Four-component radical cascades

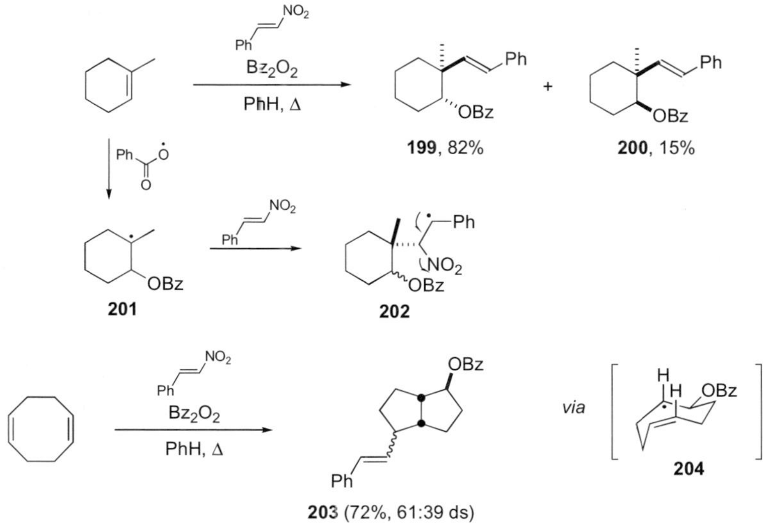

Scheme 64 Homoallylic alcohols from nitro-olefins

Miura et al. built on Ryu's previous work to achieve four-component radical cascades leading to diketones (Scheme 63) [174]. This outstanding result relies on initial carbonylation of alkyl radicals to form acyl radicals, such as **196**. The nucleophilicity of acyl radicals allowed them to react with electron-deficient olefins to form α-cyano radicals (**197**), whose philicity is now reversed. Thus, they were able to add onto stannyl enolates and led to ketyl radicals such as **198**. Those latter radicals underwent β-elimination of tributylstannyl radicals. This key elimination regenerated the mediator for the initial dehalogenation. This very fine tuning of the radical reactivities is the key element that makes the whole process work.

Jang et al. reported a highly diastereoselective tandem radical reaction to prepare *E*-polysubstituted homoallylic alcohols (Scheme 64) [175]. This new process relies on the initial addition of benzoyl radicals onto an olefin. The intermediate radicals such as **201** underwent a stereoselective vinylation (two elementary steps) to form the desired Bz-protected homoallylic alcohols in good yields. The stereochemical outcome of the reaction is strongly dependent on polar factors such as solvent polarity of Lewis acid additives. More sophisticated domino processes including cyclizations can be devised, as is the case for the formation of **203**.

4.2 Asymmetric Intermolecular Tandem Reactions

Starting from earlier data showing the positive role of Lewis acid for asymmetric tandem addition/allylation of unsaturated compounds derived from

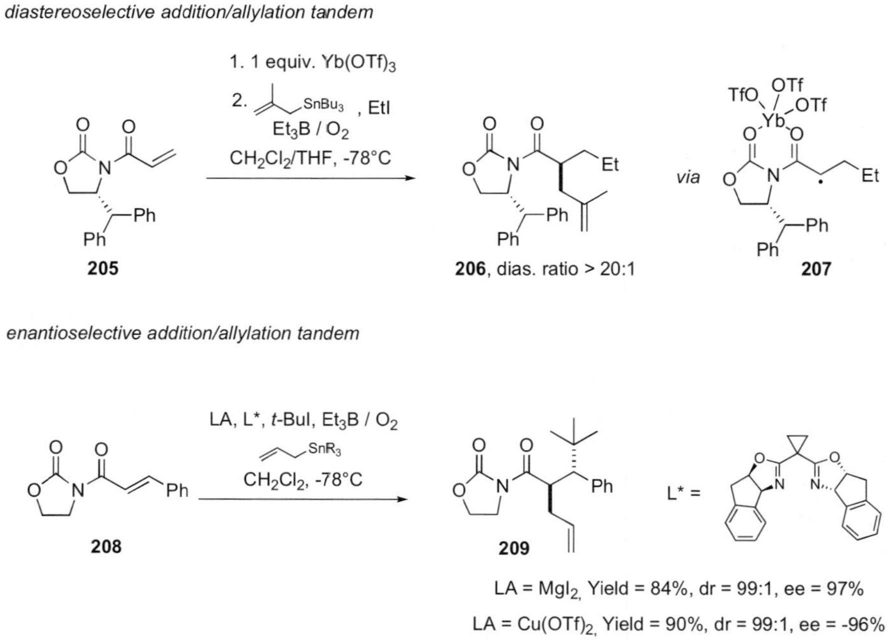

Scheme 65 Addition/allylation of enoates derived from oxazolidinones

chiral oxazolidinones such as **205**—which proceeded with high diastereoselectivity through chelate intermediates – [176], Sibi and Chen showed that two contiguous stereogenic centers could be established in one reaction, with excellent diastereo- and enantioselectivity (Scheme 65) [177]. This new tandem reaction required only substoichiometric amounts of chiral bisoxazoline-based Lewis acids. Depending on the metal used, either one or the other enantiomers could be obtained. The results suggest that the β-carbon stereochemistry is largely determinant of that of the α-carbon. Thus, the role of the ligand is to control the initial addition.

Watanabe et al. examined a similar addition/trapping tandem involving vinylic sulfones [178]. Acceptable ee's were obtained when bidentate chelation was made possible, as was the case when benzimidazolyl moities were attached to the sulfur atom (the ligand was again a box derivative). Yet only one stereogenic center was created during the reaction.

Enantioselective radical reactions have been reviewed by Sibi et al. [179].

5 Tandem Reactions Involving Radicals and Other Intermediates

5.1 Reactions Involving Reductive Steps

5.1.1 $S_{RN}1$ Mechanism

The radical nucleophilic substitution is perfectly suited for tandem reactions [180]. Recent examples have been reported by the Rossi group (Scheme 66). Dihydrobenzofuranes and dihydroindoles substituted at the 3-position were prepared from *ortho*-functionalized haloaromatic compounds in high yields [181]. The nucleophiles involved in the initial electron transfer and subsequent coupling are varied. In particular, starting form naphthyl derivative **210**, phosphinyl anions lead to tricyclic phosphine oxide **211** (after oxidation) in 98% yield.

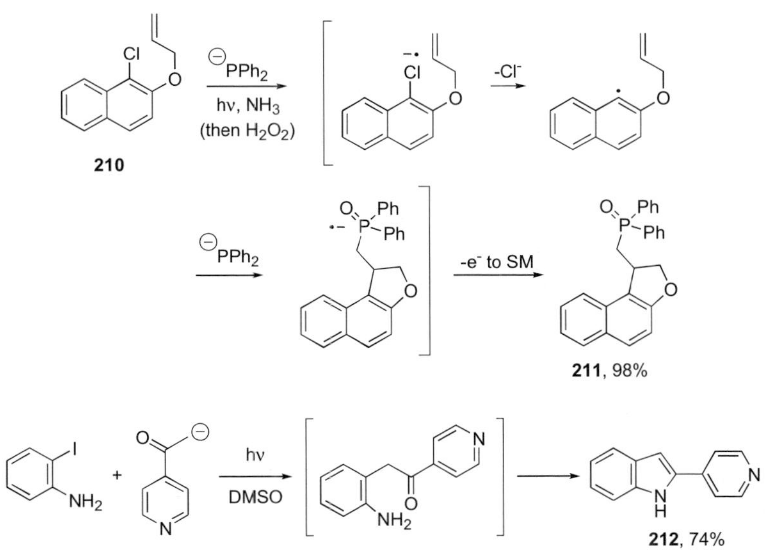

Scheme 66 $S_{RN}1$-based tandem reactions. *DMSO* Dimethyl sulfoxide

Alternatively, the $S_{RN}1$ mechanism can be coupled to nucleophilic additions as was the case during the formation of substituted indole **212** [182], an interesting variation on a reaction first reported by Beugelmans and Roussi [183].

5.1.2 Indium and Samarium-Mediated Tandem Reactions

Ueda et al. reported a tandem radical addition–cyclization reaction in aqueous media [184]. This reaction was initiated by single-electron transfer from indium to an alkyl iodide. Fragmentation of the *iso*-propyl iodide radical anion generated the *iso*-propyl radical, which triggered the addition/cyclization tandem. Final SET and in situ hydrolysis delivered cyclic sulfonamides in good yield but low stereoselectivity.

Rivkin et al. introduced a radical/polar crossover tandem reaction to gain entry to the BCD ring system of penitrem D (Scheme 67) [185]. Upon reaction with SmI_2/HMPA, aryl iodide **213** led to an aryl radical that underwent a 6-*exo-trig* cyclization on the cyclobutene moiety, thus delivering tricyclic adduct **214**. Further reduction by SmI_2 gave organosamarium derivative **215**, which added onto acetone to give alcohol **216** in 60% yield after hydrolysis. Thus, the left part of penitrem D, as well as the tertiary alcohol on the four-membered B ring were installed.

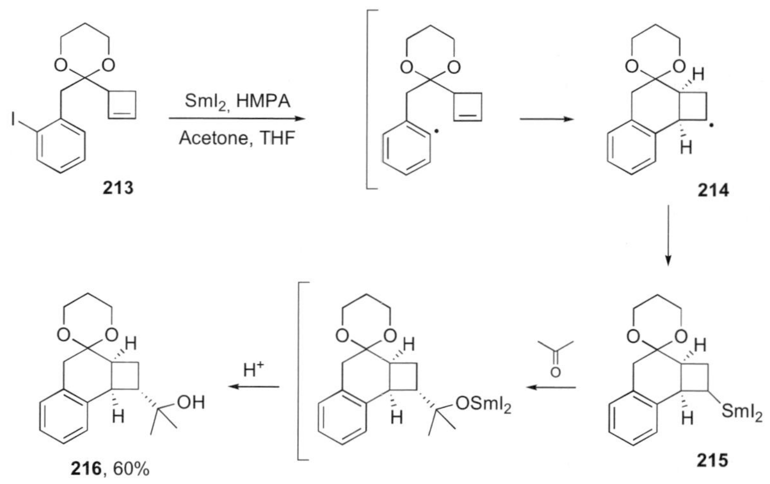

Scheme 67 SmI_2-mediated approach to Penitrem D

Howells et al. introduced a very sophisticated mixed domino process mediated by samarium diiodide (Scheme 68) [186]. Enone **217** reacted with SmI_2 to afford an allylic radical that is probably further reduced to the samarium enolate. This radical/polar crossover triggered an anionic cascade, during which enolate **218** first underwent an intramolecular aldol reaction on enone **217**. Adduct **219** further rearranged into tricyclic ketone **220**, which was ideally suited to undergo a SmI_2-mediated 5-*exo-trig* radical cyclization to install the THF ring. Overall yield of **221** was quite high (67%).

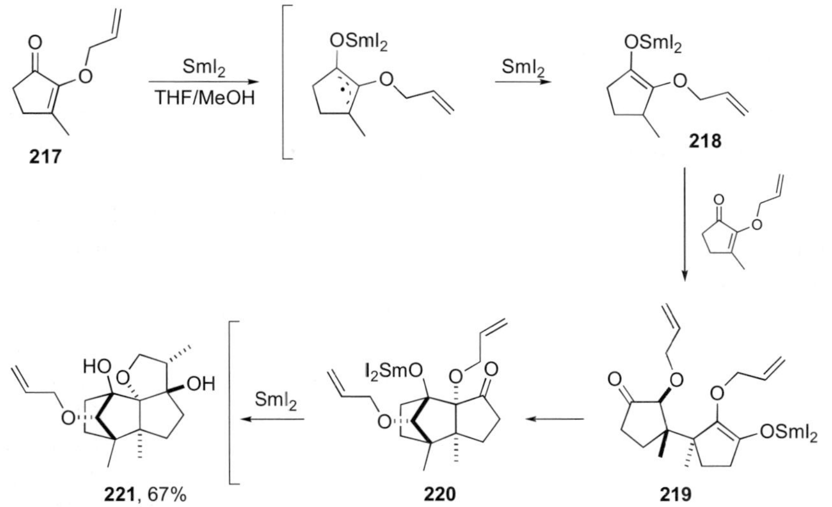

Scheme 68 SmI_2-mediated cascade from α-oxygenated enones

5.1.3 Tandem Reactions Involving Reductive Organometallic Steps

Wakabayashi et al. introduced a very powerful mixed cobalt-catalyzed tandem reaction based on a radical cyclization followed by cross-couplings (Scheme 69) [187]. The reactive species is believed to be a 17-electron cobalt ate-complex, which would undergo SET to yield an anion radical derived from **222**, and thus a radical upon loss of bromide. The cobalt(I) complex gen-

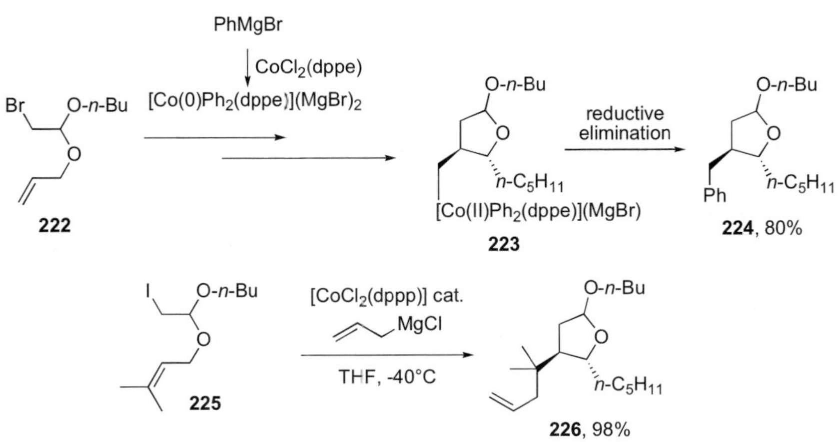

Scheme 69 Cobalt-catalyzed tandem radical reactions

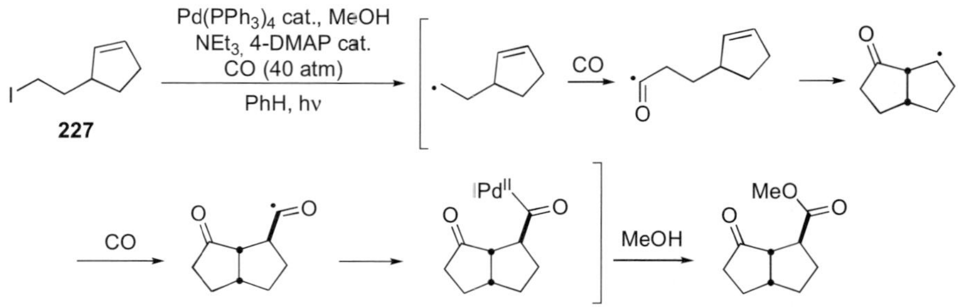

227

228, 78%

Scheme 70 Pd(0)/light-catalyzed multiple carbonylations of 4-alkenyl iodides

erated would recombine with the newly created cyclic radical to lead to **223**. Reductive elimination delivered acetal **224** in high yield. Further work by the same group extended this methodology to allylations and benzylations [188]. For example, reaction of acetal **225** in the presence of allyl magnesium chloride, yielded **226** in quantitative yield.

Ryu et al. showed that cyclizative multiple carbonylations of 4-alkenyl iodides could be catalyzed by a Pd(0)/light system (Scheme 70) [189]. The cascade reaction resulted from the interplay of radical and organopalladium species. Initial light-induced SET from the palladium atom to iodide **227** triggered the cascade. After several carbonylation/cyclization steps, the final acyl radical possibly recombined with the palladium(I) complex generated during the first electron transfer. Coupling of the acyl-palladium with methanol delivered bicyclic ketoester **228** in good overall yield (78%).

5.2
Reactions Involving Oxidative Steps

5.2.1
Use of Hypervalent Iodine Derivatives

Suarez has studied extensively the formation and reactivities of Oxygen- and nitrogen-centered radicals through oxidation by iodine (III) reagents coupled with iodine. In particular, sugar-based acids or α-amino acid lead to the formation of iminium cations such as **230** (Scheme 71), which can be trapped by various external nucleophiles in a one-pot, two-step reaction [190]. The reaction can be rerouted toward the formation of iodinated pyrrolidines through additional electrophilic substitution to **231** [191]. Compound **232** was obtained in fair yields (66%). Iglesias-Arteaga et al. reported a similar behavior of carboxyperhydropyrimidinones [192].

Oxidation at the anomeric center of carbohydrates of the **233**-type generates an oxyradical that undergoes β-fragmentation leading to open radical **234** (Scheme 72). Depending on the substituton pattern on **234**, two types of

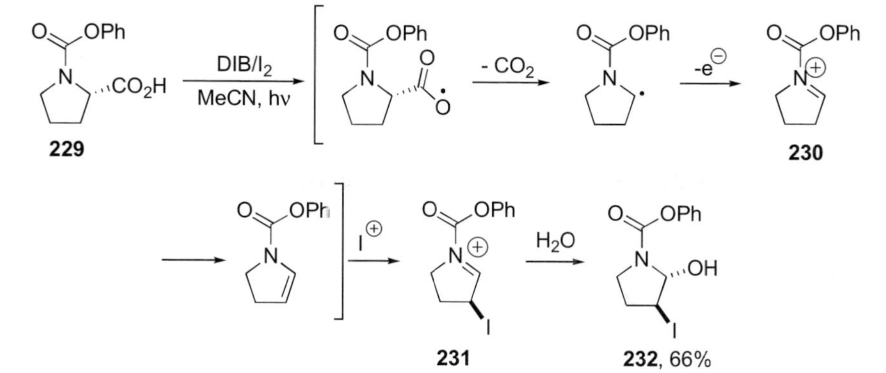

Scheme 71 Radicals from oxidation of carboxylic acids

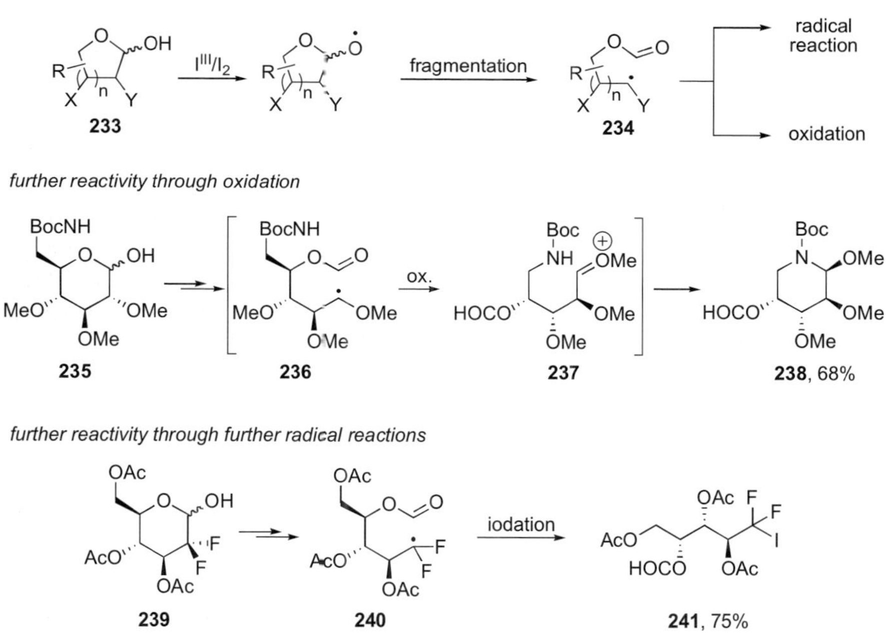

Scheme 72 Radicals from oxidation at the anomeric position

reactivities were evidenced by the Suarez group: radical **234** can be further oxidized or it can follow a radical pathway.

Examples of the former take place when radical **234** is stabilized by a heteroatom. When Y is an alkoxy moiety, an oxonium is formed and trapped by several external nucleophiles [193]. When an amide was present on the starting material, as was the case with **235**, amino sugars were obtained in good yields (68% for **238**) [194]. Finally, when Y is an azide, oxidation to the α-azido cation delivers nitriles upon loss of nitrogen [195].

The reaction follows the latter pathway when oxidation of radical **234** is not so easy. In those cases, an alternate course involving iodine atom transfer to the substrate is followed. This allows the formation of iodoalditols [196, 197], difluoroalditols (**241**, 75%) [198], vinyl azides [199], or vinyl sulfones (although in that particular case, the mechanism may be more sophisticated due to the possible β-elimination of the sulfur group) [200].

Hypervalent iodine derivatives can also oxidize alcohols or amides to the corresponding radicals. Those radicals can undergo hydrogen transfers with suitably located hydrogen atoms, leading to elegant domino processes involving oxidation and radical steps (Scheme 73) [201, 202]. For example, sugar derivative **242** leads to radical **244** after the aforementioned tandem process. Oxidation to the oxonium cation **245** was followed by intramolecular trapping by the regenerated alcohol and gave good yields of bicyclic product **246**. The 1,5- as well as 1,6-translocation can take place [201]. A rare example of 1,8-H

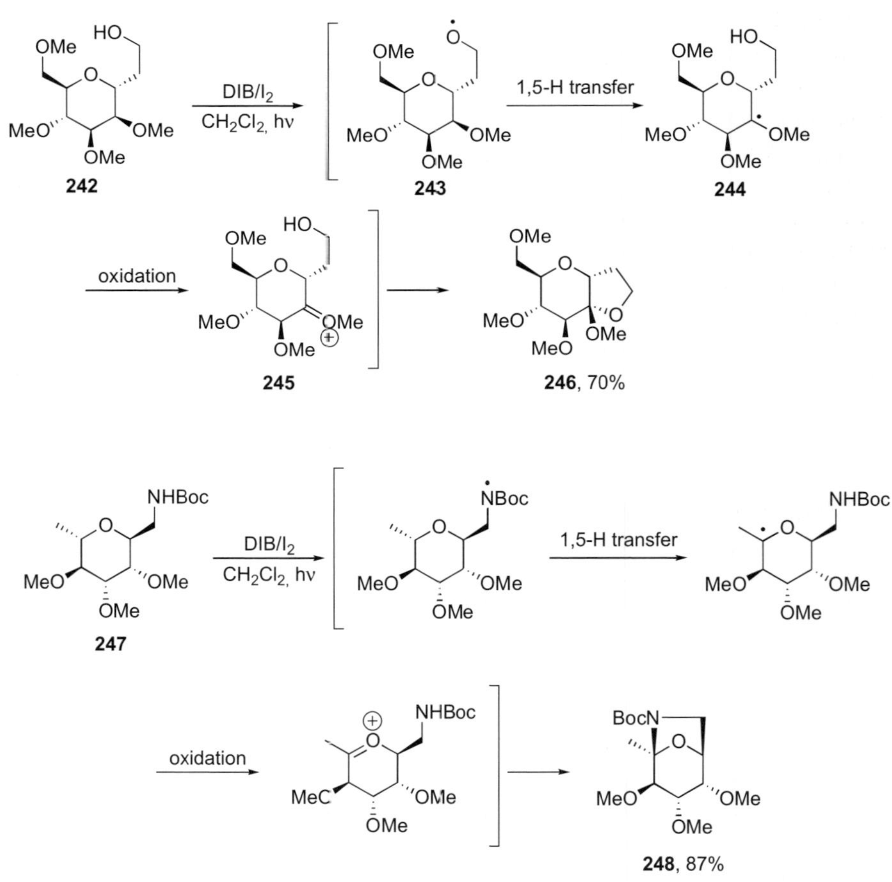

Scheme 73 Cascades involving 1,5-H transfer and oxidation

transfer was also evidenced [203]. The factors governing the reactivity of the oxyradicals (including the initial translocation/fragmentation competition) have been studied [201, 204]. As stated before, similar outcomes have been evidenced starting from nitrogen-centered radicals. For example, reaction of amino sugar **247** led to the substituted 8-oxa-6-aza bicyclo[3.2.1] octane **248** in very high yield (87%) [205]. In the particular case of amidyl radicals, it has been shown that electronic factors govern the nucleophilic addition of the ambident amides [206].

5.2.2 Oxidation of Anions

Jahn et al. reported an elegant new domino process based on the combination of anionic and radical reactions relying on a key oxidation of anion **249** followed by a radical cyclization leading to **250** (Scheme 74) [207]. Radical **250** can be trapped by TEMPO to yield functionalized pyrrolidine **251** in high yield and acceptable stereoselectivity: the 2–3 *cis* relationship was controlled in the initial anionic addition, but the relative configurations at C3 and C4 were largely non-controlled. All-carbon cyclopentanes [208], in particular prostanes [209] could also be prepared through this method. More recently, Jahn and Rudakov reported its extension to nitroalkenes [210].

The previous examples relied on a radical termination. Cationic termination could also be used, after suitable oxidation of the final radical (Scheme 75) [211]. Thus, ester **252** was treated with base and ethylchloroformate to give anion **253**, which was oxidized to malonyl radical **254** by the ferrocenium ion. Cyclization/oxidation gave cation **256**, which yielded 93% of malonate **257** by loss of the TMS group. This adduct was further elaborated upon and led to a cyclopentanoid monoterpene, dihydronepetalactone.

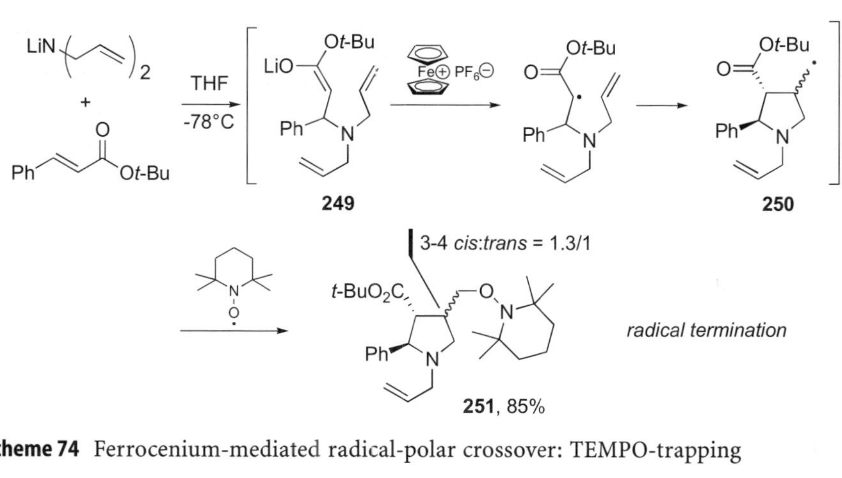

Scheme 74 Ferrocenium-mediated radical-polar crossover: TEMPO-trapping

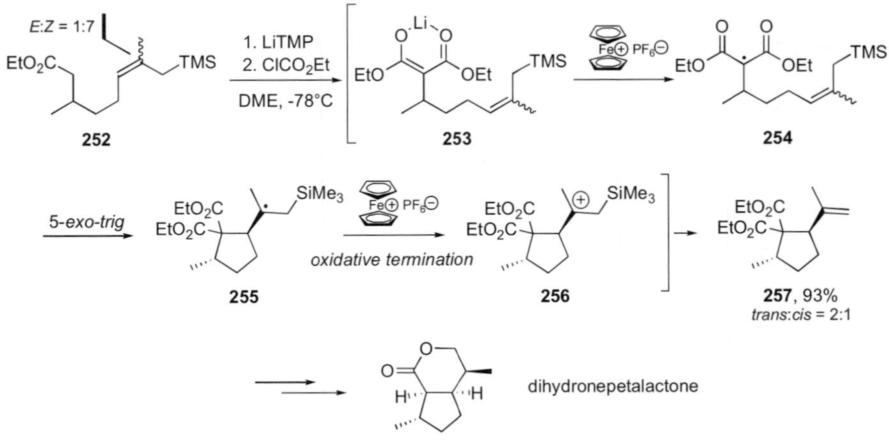

Scheme 75 Ferrocenium-mediated radical-polar crossover: intramolecular allylation

Work examining the factors directing the termination step has been reported by the Jahn group [212].

Denès et al. reported an interesting example of a radical tandem 1,4-addition/carbocyclization reaction initiated by oxidation of dialkylzinc by dissolved oxygen in the solvent [213]. Reduction of the final radical with dialkylzinc through a radical/polar crossover reaction afforded a new organozinc derivative that could be further functionalized.

5.2.3 Oxidative Opening of Cyclopropylsilyl Ethers

Booker-Milburn et al. used the oxidative ring-expansion/cyclization strategy they previously developed as an efficient way to stereoselectively build the bicyclo [5.3.0] framework of terpene-based natural products. For example, reaction of substrate **258** with iron (III) nitrate in DMF led to the formation of keto-alcohol **260** in fair yield as a single diastereomer (Scheme 76) [214]. Reduction of the final radical coming from the cyclization of **259** by 1,4-cyclohexadiene was smooth and the overall reaction proceeded without protection of the tertiary alcohol, opening the way to the total synthesis of pogostol and kessane.

Similar strategies were used by Booker-Milburn et al. in approaches toward the synthesis of dictyol C and α-eudesmol [215], and by Blake et al. in approaches toward the synthesis of ophiobolin F and fusicoccin A [216].

An elegant extension of this work has been contributed by Lee et al., who showed that aryl-cyclopropylamines easily obtained by a Kulinkovich cyclopropanation could also undergo the oxidative fragmentation/cyclization reaction (Scheme 76) [217]. Interestingly, the radical issued from the cyclization of **262** was trapped by molecular oxygen and afforded alcohol **263** after

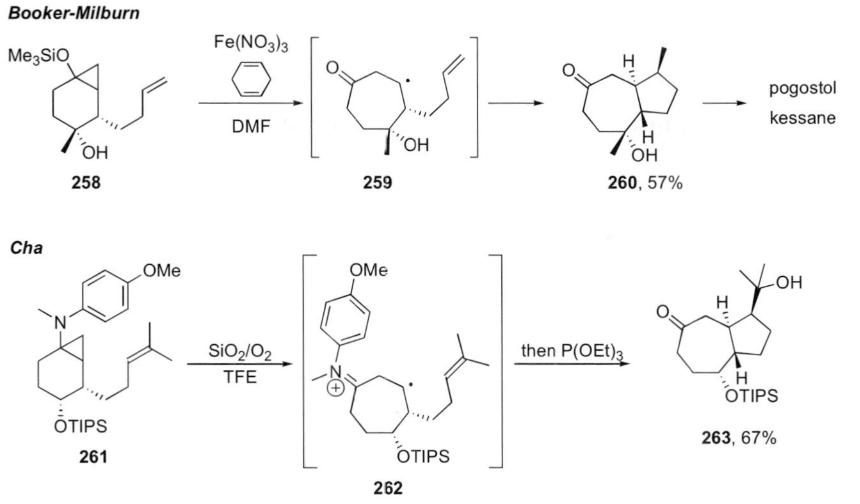

Scheme 76 Oxidative approaches to the bicyclo [5.3.0] framework

treatment by a phosphite. Optimized conditions involve the use of silica gel in a fluorous solvent, which acts as an oxygen stock.

Booker-Milburn et al. examined the reactivity of simpler substrates in which there is no ring attached to the cyclopropylether (Scheme 77) [218]. Gratifyingly, the reaction course was very different, leading to tricyclic product **266** from **264** in good overall yield. This sophisticated domino process featured an initial fragmentation followed by 5-*exo-trig* cyclization leading to **265**. Radical cyclization on the aldehyde was probably followed by intramolecular 1,5-H transfer to deliver a benzylic radical, then oxidation to the benzylic cation and intramolecular nucleophilic addition. Logically, the key step for this cascade is the cyclization onto the carbonyl moiety. When ketones were

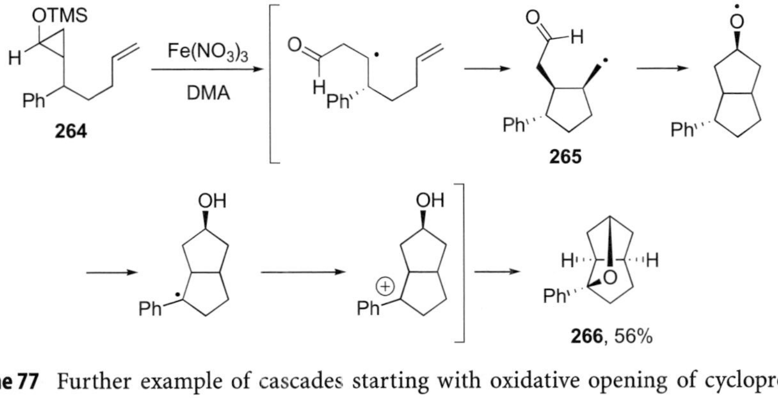

Scheme 77 Further example of cascades starting with oxidative opening of cyclopropylsilyl ethers

reacted (i.e., starting from trisubstituted cyclopropylethers), the yields in tricyclic adducts dropped, as one would expect considering the drop in the corresponding rate constants. Products arising from the early reduction of **265** were also isolated.

5.3 Other Reactions Involving Radical and Polar Steps

5.3.1 Radical-Cations from β-Phosphatoxyl Radicals

Crich and Newcomb have found that β-phosphatoxyl radicals underwent rapid fragmentation leading to the corresponding radical-cations and phosphatoxyl anions. When the reaction of **267** was carried out in the presence of allyl alcohol, a substituted tetrahydrofuran could be prepared (Scheme 78) [219]. The key step was the nucleophilic addition of allyl alcohol to radical-cation **268**, which generated radical **269**. Standard radical steps led to the desired heterocycle **270** in good yield.

When an amine is placed in an internal position, more complex frameworks could be assembled through the radical/polar crossover tandem sequence. Depending on the substitution pattern, pyrrolizidines [220], indolizidines [221], and azabicyclo[3.2.1] octanes [221] skeletons could be prepared [222].

Taking advantage of the tightness of the ion pair generated during the reactions, Crich et al. extended the method to enantiomerically controlled sequences [223]. In particular, substrate **271** yielded **272** in good yield and with a good ee (Scheme 79). The absolute configuration was explained by “an at-

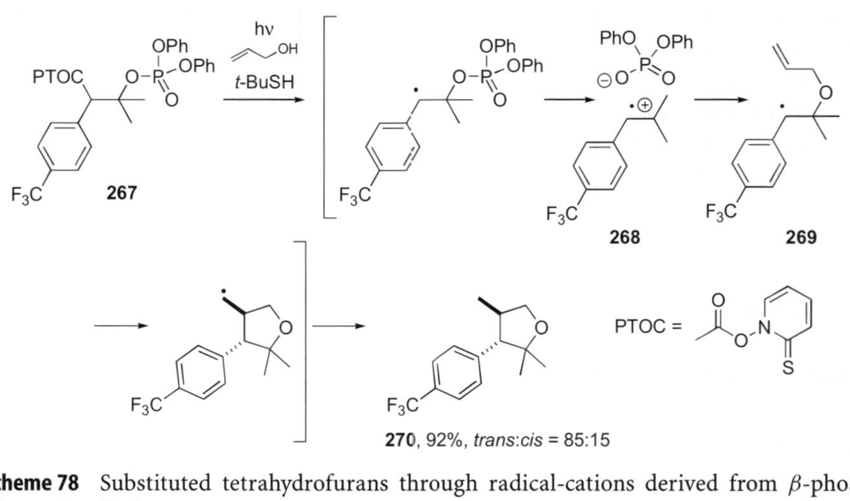

Scheme 78 Substituted tetrahydrofurans through radical-cations derived from β-phosphatoxyl radicals

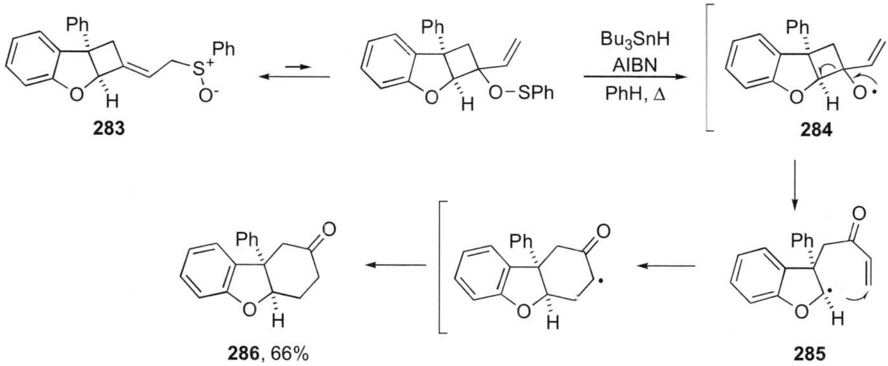

Scheme 83 Allylsufoxides as precursors of oxygen-centered radicals

ketone **286** was obtained in 66% yield with complete regioselectivity. This is due to the preferred formation of α-oxysubstituted radical **285** from **284**.

6 Conclusion

This review's purpose was to prove the relevance of radical reactions. Free radicals are uniquely suited for consecutive processes. The latter are important ways to efficiently increase the molecular complexity in a single step, and thus participate in the shift of synthetic chemistry toward green productions. No doubt the development of eco-compatible reagents, the ever-tighter control of the selectivity for each single step, and the incorporation of radical steps in mixed domino reactions will trigger a burgeoning of highly innovative approaches. This should keep organic chemists busy in the near future.

References

1. Devin P, Fensterbank L, Malacria M (1998) J Org Chem 63:6764
2. Renaud P, Sibi MP (2001) Radicals in organic synthesis. Wiley, Weinheim
3. Jasperse CP, Curran DP, Fevig TL (1991) Chem Rev 91:1237
4. Muratake H, Natsume M (2004) Angew Chem Int Ed 43:4646
5. Reddy LR, Fournier J-F, Reddy BVS, Corey EJ (2005) Org Lett 7:2699
6. Julia M (1967) Pure Appl Chem 15:167
7. Surzur J-M (1982) Radical Cyclizations by Intramolecular Additions. In: Abramovitch RA (ed) Reactive intermediates, vol 2. Plenum, New York, p 121
8. Butler GB, Angelo RJ (1957) J Am Chem Soc 79:3128
9. Marvel CS, Vest RD (1957) J Am Chem Soc 79:5771
10. Arai S, Sato S, Shida S (1960) J Chem Phys 33:1277
11. Gordon AS, Smith SR (1962) J Phys Chem 66:521

12. Julia M, Surzur J-M, Katz L (1960) CR Acad Sci 251:1031
13. Lamb RC, Ayers PW, Toney MK (1963) J Am Chem Soc 85:3483
14. Chatgilialoglu C, Ingold KU, Scaiano JC (1981) J Am Chem Soc 103:7739
15. Newcomb M (1993) Tetrahedron 49:1151
16. Beckwith ALJ, Schiesser CH (1985) Tetrahedron 41:3925
17. Spellmeyer DC, Houk KN (1987) J Org Chem 52:959
18. Curran DP, Porter NA, Giese B (1996) Stereochemistry of radical reactions. Wiley, Weinheim
19. McCarroll AJ, Walton JC (2001) Angew Chem Int Ed 40:2224
20. Dhimane A-L, Fensterbank L, Malacria M (2001) Polycyclic Compounds via Radical Cascade Reactions. In: Renaud P, Sibi MP (eds) Radicals in organic synthesis, vol 2. Wiley, Weinheim, p 350
21. Haney BP, Curran DP (2000) J Org Chem 65:2007
22. Takasu K, Maiti S, Katsumata A, Ihara M (2001) Tetrahedron Lett 42:2157
23. Tripp JC, Schiesser CH, Curran DP (2005) J Am Chem Soc 127:5518
24. Takasu K, Ohsato H, Kuroyanagi J-I, Ihara M (2002) J Org Chem 67:6001
25. Toyota M, Ilangovan A, Kashiwagi Y, Ihara M (2004) Org Lett 6:3629
26. Harrington-Frost NM, Pattenden G (2000) Tetrahedron Lett 41:403
27. Rychlet Elliott M, Dhimane A-L, Malacria M (1997) J Am Chem Soc 119:3427
28. Aïssa C, Dhimane A-L, Malacria M (2000) Synlett 2000(11):1585
29. Dhimane A-L, Aïssa C, Malacria M (2002) Angew Chem Int Ed 41:3284
30. Handa S, Pattenden G (1999) J Chem Soc Perkin Trans 1:843
31. Boehm HM, Handa S, Pattenden G, Roberts L, Blake AJ, Li W-S (2000) J Chem Soc Perkin Trans 1:3522
32. Harrowven DC, Nunn MIT, Fenwick DR (2002) Tetrahedron Lett 43:7345
33. Joshi SN, Phalgune UD, Bhawal BM, Deshmukh ARAS (2003) Tetrahedron Lett 44:1827
34. Takayama H, Watanabe F, Kuroda A, Kitajima M, Aimi N (2000) Tetrahedron Lett 56:6547
35. Fensterbank L, Malacria M, Sieburth SM (1997) Synthesis 2000(8):813
36. Lee E, Song HY, Kang JW, Kim D-S, Jung C-K, Joo JM (2002) J Am Chem Soc 124:384
37. Flisinska-Luczak J, Lesniak S, Nazarski RB (2004) Tetrahedron 60:8181
38. Du W, Curran DP (2003) Org Lett 5:1765
39. Marion F, Courillon C, Malacria M (2003) Org Lett 5:5095
40. Shen L, Hsung RP (2005) Org Lett 7:775
41. Taniguchi T, Iwasaki K, Uchiyama M, Tamura O, Ishibashi H (2005) Org Lett 7:4389
42. Lizos DE, Murphy JA (2003) Org Biomol Chem 117
43. Zhou S-Z, Bommezijn S, Murphy JA (2002) Org Lett 4:443
44. Patro B, Murphy JA (2000) Org Lett 2:3599
45. Bowman WR, Bridge CF, Brookes P (2000) Tetrahedron Lett 41:8989
46. Benati L, Bencivenni G, Leardini R, Minozzi M, Nanni D, Scialpi R, Spagnolo P, Zanardi G, Rizzoli C (2004) Org Lett 6:417
47. Senboku H, Hasegawa E, Orito K, Tokuda M (2000) Tetrahedron Lett 41:5699
48. Hasegawa H, Senboku H, Kajizuka Y, Orito K, Tokuda M (2003) Tetrahedron 59:827
49. Cassayre J, Gagosz F, Zard SZ (2002) Angew Chem Int Ed 41:1783
50. Gagosz F, Moutrille C, Zard SZ (2002) Org Lett 4:2707
51. Barclay GL, Quiclet-Sire B, Sanchez-Jimenez G, Zard SZ (2005) Org Biomol Chem 823
52. Yang D, Gao Q, Lee O-Y (2002) Org Lett 4:1239
53. Yang D, Gu S, Yan Y-L, Zhao H-W, Zhu N-Y (2002) Angew Chem Int Ed 41:3014

54. Snider BB (1996) J Org Chem 61:7832
55. Yang D, Xu M, Bian M-Y (2001) Org Lett 3:111
56. Yang D, Xu M (2001) Org Lett 3:1785
57. Gansäuer A, Bluhm H (2000) Chem Rev 100:2771
58. Justicia J, Oller-Lopez JL, Campana AG, Oltra JE, Cuerva JM, Bunuel E, Cardenas DJ (2005) J Am Chem Soc 127:14911
59. Barrero AF, Cuerva JN, Herrador MM, Valdivia MV (2001) J Org Chem 66:4074
60. Justicia J, Oltra JE, Cuerva JN (2005) J Org Chem 70:8265
61. Curran DP, Xu J (1996) J Am Chem Soc 118:3142
62. Huang XL, Dannenberg JJ (1991) J Org Chem 56:5421
63. Dorigo AE, Houk KN (1987) J Am Chem Soc 109:2195
64. Bogen S, Gulea M, Fensterbank L, Malacria M (1999) J Org Chem 64:4920
65. Bogen S, Fensterbank L, Malacria M (1999) J Org Chem 64:819
66. Clive DLJ, Yang W, MacDonald AC, Wang Z, Cantin M (2001) J Org Chem 1966
67. Beaufils F, Dénès F, Renaud P (2004) Org Lett 6:2563
68. Curran DP, Shen W (1993) J Am Chem Soc 115:6051
69. Lachia M, Dénès F, Beaufils F, Renaud P (2005) Org Lett 7:4103
70. Renaud P, Beaufils F, Feray L, Schenk K (2003) Angew Chem Int Ed 2003:4230
71. Beaufils F, Dénès F, Renaud P (2005) Angew Chem Int Ed 44:5273
72. Leca D, Fensterbank L, Lacôte E, Malacria M (2005) Chem Soc Rev 34:858
73. Marque S, Tordo P (2005) Top Curr Chem 250:43
74. Khan TA, Tripoli R, Crawford JJ, Martin CG, Murphy JA (2003) Org Lett 5:2971
75. Hilton ST, Ho TCT, Pljevaljcic G, Schulte M, Jones K (2001) Chem Commun 209
76. Devin P, Fensterbank L, Malacria M (1999) Tetrahedron Lett 40:5511
77. Devin P, Fensterbank L, Malacria M (1998) Tetrahedron Lett 39:833
78. Fensterbank L, Mainetti E, Devin P, Malacria M (2000) Synlett :1342
79. Gulea M, Lopez-Romero JM, Fensterbank L, Malacria M (2000) Org Lett 2:2591
80. Dupuis J, Giese R, Rüegge D, Fischer H, Korth H-G, Sustmann R (1984) Angew Chem Int Ed 23:896
81. Pattenden G, Reddy LK, Walter A (2004) Tetrahedron Lett 45:4027
82. Nonhebel DC (1993) Chem Soc Rev 22:347
83. Mainetti E, Fensterbank L, Malacria M (2002) Synlett :923
84. Boffey RJ, Whittingham WG, Kilburn JD (2001) J Chem Soc, Perkin Trans 1:487
85. Prévost N, Shipman M (2001) Org Lett 3:2383
86. Bacqué E, Pautrat F, Zard SZ (2003) Org Lett 5:325
87. Lee H-Y, Kim BG (2000) Org Lett 2:1951
88. Lee H-Y, Lee S, Kim BG, Bahn JS (2004) Tetrahedron Lett 45:7225
89. Toyota M, Yokota M, Ihara M (2001) J Am Chem Soc 123:1856
90. Toyota M, Asano T, Ihara M (2005) Org Lett 7:3929
91. Taber DF, Joshi PV (2001) CR Acad Sci 4:557
92. Hodgson DM, Bebbington MWP, Willis P (2001) Chem Commun 2001(10):889
93. Dowd P, Zhang W (1993) Chem Rev 93:2091
94. Rodriguez JR, Castedo L, Mascareñas JL (2001) Org Lett 3:1181
95. Chuard R, Giraud A, Renaud P (2002) Angew Chem Int Ed 41:4321
96. Studer A, Bossart M, Vasella T (2000) Org Lett 2:985
97. Amrein S, Bossart M, Vasella T, Studer A (2000) J Org Chem 65:4281
98. Tokuyama H, Yamashita T, Reding MT, Kaburagi Y, Fukuyama T (1999) J Am Chem Soc 121:3791
99. Reding MT, Fukuyama T (1999) Org Lett 1:973
100. Reynolds AJ, Scott AJ, Turner CI, Sherburn MS (2003) J Am Chem Soc 125:12108

101. Fischer J, Reynolds AJ, Sharp LA, Sherburn MS (2004) Org Lett 6:1345
102. Demircan A, Parsons PJ (2003) Eur J Org Chem :1729
103. Calestani G, Capella L, Leardini R, Minozzi M, Nanni D, Papa R, Zanardi G (2001) Tetrahedron 57:7221
104. Gansäuer A, Rinker B, Pierobon M, Grimme S, Gerenkamp M, Mück-Lichtenfeld C (2003) Angew Chem Int Ed 42:3687
105. Gansäuer A, Rinker B, Ndene-Schiffer N, Pierobon M, Grimme S, Gerenkamp M, Mück-Lichtenfeld C (2004) Eur J Org Chem :2337
106. Leca D, Fensterbank L, Lacôte E, Malacria M (2004) Angew Chem Int Ed 43:4220
107. Crich D, Yao Q (2003) Org Lett 5:2189
108. Amrein S, Studer A (2002) Helv Chim Acta 85:3559
109. Amrein S, Timmermann A, Studer A (2001) Org Lett 3:2357
110. Studer A, Amrein S (2000) Angew Chem Int Ed 39:3080
111. Studer A, Schulte T (2005) Chem Rec 5:27
112. Wetter C, Jantos K, Woithe K, Studer A (2003) Org Lett 5:2899
113. Wetter C, Studer A (2004) Chem Commun :174
114. Leroi C, Bertin D, Dufils P-E, Gigmes D, Marque S, Tordo P, Couturier J-L, Guerret O, Ciufolini MA (2003) Org Lett 5:4943
115. Leroi C, Fenet B, Couturier J-L, Guerret O, Ciufolini MA (2003) Org Lett 5:1079
116. Kim S, Cho CH, Lim CJ (2003) J Am Chem Soc 125:9574
117. Miyabe H, Ueda M, Naito T (2004) Synlett :1140
118. Miyabe H, Fujii K, Goto T, Naito T (2000) Org Lett 2:4071
119. Miyabe H, Ueda M, Fujii K, Nishimura A, Naito T (2003) J Org Chem 68:5618
120. Miyabe H, Fujii K, Tanaka H, Naito T (2001) Chem Commun 2001(9):831
121. Briggs ME, El Qacemi M, Kalai C, Zard SZ (2004) Tetrahedron Lett 45:6017
122. Sibi MP, Patil K, Rheault TR (2004) Eur J Org Chem 2004(2):372
123. Kim S, Lim CJ (2004) Angew Chem Int Ed 43:5378
124. Kim S, Lim CJ, Song C, Chung W-J (2002) J Am Chem Soc 124:14306
125. Ollivier C, Renaud P (2000) J Am Chem Soc 122:6496
126. Ollivier C, Renaud P (2001) J Am Chem Soc 123:4717
127. Kim S, Lim CJ (2002) Angew Chem Int Ed 41:3265
128. Kim S (2004) Adv Synth Catal 346:19
129. Kim S, Kim S, Otsuka N, Ryu I (2005) Angew Chem Int Ed 44:6183
130. Moutrille C, Zard SZ (2004) Chem Commun 2004(16):1848
131. Uenoyama Y, Tsukida M, Doi T, Ryu I, Studer A (2005) Org Lett 7:2985
132. Gansäuer A, Pierobon M, Bluhm H (2002) Angew Chem Int Ed 41:3206
133. Gansäuer A, Lauterbach T, Narayan S (2003) Angew Chem Int Ed 42:5556
134. Lee NH, Larock RC (2001) Bull Korean Chem Soc 22:857
135. Bennasar ML, Roca T, Griera R, Bosch J (2001) J Org Chem 66:7547
136. Tsuchii K, Doi M, Hirao T, Ogawa A (2003) Angew Chem Int Ed 42:3490
137. Studer A, Steen H (1999) Chem Eur J 5:759
138. Miranda LD, Cruz-Almanza R, Pavon M, Romero Y, Muchowski JM (2000) Tetrahedron Lett 41:10181
139. Miranda LD, Cruz-Almanza R, Pavon M (2002) Arkivoc 2002(XII):15
140. Bertrand S, Hoffmann N, Humbel S, Pete JP (2000) J Org Chem 65:8690
141. Marinkovic S, Brule C, Hoffmann N, Prost E, Nuzillard J-M, Bulach V (2004) J Org Chem 69:1646
142. Marinkovic S, Hoffmann N (2004) Eur J Org Chem 2004(14):3102
143. Hapiot P, Medebielle M (2001) J Fluorine Chem 107:285
144. Pandey G, Gadre SR (2004) Acc Chem Res 37:201

145. Pandey G, Rao KSSP, Rao KVN (2000) J Org Chem 65:4309
146. O'Neill PM, Stocks PA, Pugh MD, Araujo NC, Korshin EE, Bickley JF, Ward SA, Bray PG, Pasini E, Davies J, Verissimo E, Bachi M (2004) Angew Chem Int Ed 43:4193
147. Bachi MD, Korshin EE, Hoos R, Szpilman AM, Ploypradith P, Xie S, Shapiro TA, Posner GH (2003) J Med Chem 46:2516
148. Szpilman AM, Korshin EE, Rozenberg H, Bachi MD (2005) J Org Chem 70:3618
149. Zhou H, Huang X, Chen W (2004) J Org Chem 69:5471
150. Alcaide B, Almendros P, Aragoncillo C (2001) J Org Chem 66:1612
151. Alcaide B, Almendros P, Aragoncillo C (2003) Org Lett 5:3795
152. Curran DP, Liu H (1991) J Am Chem Soc 113:2127
153. Josien H, Ko S-B, Bom D, Curran DP (1998) Chem Eur J 4:67
154. Gabarda AE, Curran DP (2003) J Comb Chem 5:617
155. de Frutos O, Curran DP (2000) J Comb Chem 2:639
156. Luo Z, Zhang Q, Oderaotoshi Y, Curran DP (2001) Science 291:1766
157. Zhang Q, Rivkin A, Curran DP (2002) J Am Chem Soc 124:5774
158. Zhang W, Luo Z, Chen CH-T, Curran DP (2002) J Am Chem Soc 124:10443
159. Du W, Curran DP (2003) Synlett :1299
160. Camaggi CM, Leardini R, Nanni D, Zanardi G (1998) Tetrahedron 54:5587
161. Benati L, Leardini R, Minozzi M, Nanni D, Spagnolo P, Zanardi G (2000) J Org Chem 65:8669
162. Benati L, Calestani G, Leardini R, Minozzi M, Nanni D, Spagnolo P, Strazzari S, Zanardi G (2003) J Org Chem 68:3454
163. Leardini R, Nanni D, Pareschi P, Tundo A, Zanardi G (1997) J Org Chem 62:8394
164. Cekovic Z (2003) Tetrahedron 59:8073
165. Petrovic G, Cekovic Z (2000) Org Lett 2:3769
166. Petrovic G, Cekovic Z (2004) Synthesis 2004(10):1671
167. Dubert O, Gautier A, Condamine E, Piettre SR (2002) Org Lett 4:359
168. Ouvry G, Zard SZ (2003) Synlett 2003(11):1627
169. Korshin EE, Bilokin YV, Zheng H, Bachi MD (2004) J Am Chem Soc 126:2708
170. Miyazoe H, Yamago S, Yoshida J-I (2000) Angew Chem Int Ed 39:3669
171. Yamago S, Miyazoe H, Nakayama T, Miyoshi M, Yoshida J (2003) Angew Chem Int Ed 42:117
172. Yamago S, Miyoshi M, Miyazoe H, Yoshida J (2002) Angew Chem Int Ed 41:1407
173. Yamago S, Miyazoe H, Iida K, Yoshida J-I (2000) Org Lett 2:3671
174. Miura K, Tojino M, Fujisawa N, Hosomi A, Ryu I (2004) Angew Chem Int Ed 43:2423
175. Jang Y-J, Wu J, Lin Y-F, Yao C-F (2004) Tetrahedron 60:6565
176. Sibi MP, Rheault TR (2000) J Am Chem Soc 122:8873
177. Sibi MP, Chen J (2001) J Am Chem Soc 123:9472
178. Watanabe Y, Mase N, Furue R, Toru T (2001) Tetrahedron Lett 42:2981
179. Sibi MP, Manyem S, Zimmerman J (2003) Chem Rev 103:3263
180. Rossi RA, Pierini AB, Peñeñory AB (2003) Chem Rev 103:71
181. Vaillard SE, Postigo A, Rossi RA (2002) J Org Chem 67:8500
182. Barolo SM, Lukach AE, Rossi RA (2003) J Org Chem 68:2807
183. Beugelmans R, Roussi G (1979) J Chem Soc, Chem Commun 1979(21):950
184. Ueda M, Miyabe H, Nishimura A, Miyata O, Takemoto Y, Naito T (2003) Org Lett 5:3835
185. Rivkin A, Gonzalez-Lopez de Turiso F, Nagashima T, Curran DP (2004) J Org Chem 69:3719
186. Howells DM, Barker SM, Watson FC, Light ME, Hursthouse MB, Kilburn JD (2004) Org Lett 6:1943
187. Wakabayashi K, Yorimitsu H, Oshima K (2001) J Am Chem Soc 123:5374

188. Ohmiya H, Tsuji T, Yorimitsu H, Oshima K (2004) Chem Eur J 10:5640
189. Ryu I, Kreimerman S, Araki F, Nishitani S, Oderaotoshi Y, Minakata S, Komatsu M (2002) J Am Chem Soc 124:3812
190. Boto A, Hernandez R, Suarez E (2000) J Org Chem 65:4930
191. Boto A, Hernandez R, Suarez E (2000) Tetrahedron Lett 41:2495
192. Iglesias-Arteaga MA, Avila-Ortiz CG, Juaristi E (2002) Tetrahedron Lett 43:5297
193. Boto A, Hernandez R, Suarez E (2001) Tetrahedron Lett 42:9167
194. Francisco CG, Freire R, Gonzalez CC, Leon EI, Riesco-Fagundo C, Suarez E (2001) J Org Chem 66:1861
195. Hernandez R, Leon EI, Moreno P, Riesco-Fagundo C, Suarez E (2004) J Org Chem 69:8437
196. Gonzalez CC, Kennedy AR, Leon EI, Riesco-Fagundo C, Suarez E (2001) Angew Chem Int Ed 40:2326
197. Gonzalez CC, Kennedy AR, Leon EI, Riesco-Fagundo C, Suarez E (2003) Chem Eur J 9:5800
198. Francisco CG, Gonzalez CC, Paz NR, Suarez E (2003) Org Lett 5:4171
199. Alonso-Cruz CR, Kennedy AR, Rodriguez MS, Suarez E (2003) Org Lett 5:3729
200. Alonso-Cruz CR, Leon EI, Ortiz-Lopez FJ, Rodriguez MS, Suarez E (2005) Tetrahedron Lett 46:5265
201. Francisco CG, Freire R, Herrera AJ, Perez-Martin I, Suarez E (2002) Org Lett 4:1959
202. Francisco CG, Herrera AJ, Suarez E (2002) J Org Chem 67:7439
203. Francisco CG, Herrera AJ, Kennedy AR, Melian D, Suarez E (2002) Angew Chem Int Ed 41:856
204. Boto A, Hernandez D, Hernandez R, Suarez E (2003) J Org Chem 68:5310
205. Francisco CG, Herrera AJ, Suarez E (2003) J Org Chem 68:1012
206. Martin A, Perez-Martin I, Suarez E (2005) Org Lett 7:2027
207. Jahn U, Müller M, Aussieker S (2000) J Am Chem Soc 122:5212
208. Jahn U (2001) Chem Commun 2001(17):1600
209. Jahn U, Hartmann P, Dix I, Jones PG (2002) Eur J Org Chem 2002(4):718
210. Jahn U, Rudakov D (2004) Synlett 2004(7):1207
211. Jahn U, Hartmann P, Kaasalainen E (2004) Org Lett 6:257
212. Jahn U, Hartmann P, Dix I, Jones PG (2001) Eur J Org Chem 2001(17):3333
213. Denès F, Chemla F, Normant JF (2003) Angew Chem Int Ed 42:4043
214. Booker-Milburn KI, Jenkins H, Charmant JPH, Mohr P (2003) Org Lett 5:3309
215. Booker-Milburn KI, Cox B, Grady M, Halley F, Marrison S (2000) Tetrahedron Lett 41:4651
216. Blake AJ, Highton AJ, Majid TN, Simpkins NS (1999) Org Lett 1:1787
217. Lee HB, Sung MJ, Blackstock SC, Cha JK (2001) J Am Chem Soc 123:11322
218. Booker-Milburn KI, Jones JL, Sibley GEM, Cox R, Meadows J (2003) Org Lett 5:1107
219. Crich D, Huang X, Newcomb M (2000) J Org Chem 65:523
220. Crich D, Ranganathan K, Huang X (2001) Org Lett 3:1917
221. Crich D, Neelamkavil S (2002) Org Lett 4:2573
222. Crich D, Ranganathan K, Neelamkavil S, Huang X (2003) J Am Chem Soc 125:7942
223. Crich D, Shirai M, Rumthao S (2003) Org Lett 5:3767
224. Crich D, Ranganathan K (2005) J Am Chem Soc 127:9924
225. Crich D, Ranganathan K (2002) J Am Chem Soc 124:12422
226. Sartillo-Piscil F, Vargas M, Anaya de Parrodi C, Quintero L (2003) Tetrahedron Lett 44:3919
227. Tojino M, Uenoyama Y, Fukuyama T, Ryu I (2004) Chem Commun 2004(21):2482
228. Booker-Milburn KI, Baker JR, Bruce I (2004) Org Lett 6:1481
229. Chuard R, Giraud A, Renaud P (2002) Angew Chem Int Ed 41:4323

Top Curr Chem (2006) 264: 63–91
DOI 10.1007/128_025

Published online: 20 December 2005

Cp_2TiCl in Natural Product Synthesis

Juan M. Cuerva · José Justicia · Juan L. Oller-López · J. Enrique Oltra (✉)

Department of Organic Chemistry, Faculty of Sciences, University of Granada, Campus Fuentenueva s/n, 18071 Granada, Spain
joltra@ugr.es

Abstract The synthesis of complex natural products constitutes one of the most exigent tests to prove the usefulness of a new reagent or catalyst. During the past 10 years, methods based on the Cp_2TiCl-promoted and/or -catalyzed radical epoxide opening have been used by several authors as the key step for the synthesis of more than 50 natural products and advanced synthons. At first, owing to its selectivity and the extremely mild experimental conditions employed, stoichiometric Cp_2TiCl was chosen for deoxygenation and reductive epoxide opening on densely functionalized substrates, but during the last 5 years catalytic Cp_2TiCl has been increasingly employed to achieve radical cyclizations of different epoxyalkenes and epoxyalkynes, affording straightforward strategies for the synthesis of complex natural products. In these fields Cp_2TiCl-based methods have already largely proved their synthetic usefulness but other Cp_2TiCl-mediated reactions have as yet been scarcely applied. This review focuses on the increasing scope of applications that employ the Cp_2TiCl-mediated procedures in the synthesis of natural products. We also present the basic concepts of these methods to facilitate and encourage further synthetic applications.

Keywords Natural products · Radicals · Synthesis · Titanocene

Abbreviations

col	2,4,6-Collidine
Cp	Cyclopentadienyl
DMP	Dess–Martin periodinane
m-CPBA	3-Chloroperbenzoic acid
min	Minutes
Py	Pyridine
THF	Tetrahydrofuran

1 Introduction

The synthesis of natural products constitutes one of the most exigent tests to prove the usefulness of a new reagent or catalyst. (For an excellent overview of the most exciting examples of natural product synthesis reported during the past century, see [1].) Methods for synthesizing complex natural products require selectivity, mild experimental conditions, and wide functional group tolerance. Radical reactions mediated by bis(cyclopentadienyl)titanium(III) chloride have shown themselves to conform to these requirements and, therefore, in the last 10 years they have been used by several authors as the key step for the synthesis of more than 50 natural products and advanced synthons. Curiously, the question concerning the true nature of the active form of this $[Ti^{III}]$ single-electron-transfer reagent has been the subject of controversy for quite some time [2–4]. It is known that the complexes obtained by the reduction of commercial Cp_2TiCl_2 with metals such as Mn, Zn, or Mg crystallize as trinuclear species $(Cp_2TiCl)_2MCl_2$ (M = Mn, Zn, Mg) [5, 6]. Nevertheless, cyclic voltammetry and kinetic measurements carried out by the Skrydstrup and Daasbjerg group have finally confirmed that the complex commonly generated in situ by stirring Cp_2TiCl_2 with Mn, Zn, or other reductive metals in THF solution actually takes the form of an equilibrium mixture of the mononuclear species Cp_2TiCl and its dimer $(Cp_2TiCl)_2$, whatever the reductive metal used [4]. Therefore, in this review we have represented the titanocene(III) complex either as Cp_2TiCl or simply as $[Ti^{III}]$ for the sake of clarity.

In 1988, Nugent and RajanBabu introduced the Cp_2TiCl-induced radical cyclization of epoxyolefins and subsequently Gansäuer et al. developed the catalytic version of the reaction, two major advances in the application of transition-metal-centered radicals for organic synthesis [7–9]. Since then both the stoichiometric and catalytic versions have been objects of active research, and several excellent reviews have covered the results reported [10–15]. Over the last 5 years or so, however, relevant new findings have been published and the method has been applied to the synthesis of a wide array of natural products. This review, therefore, is not meant to be exhaustive but rather focuses on the increasing scope of applications that the Cp_2TiCl-based procedures are being put to in the synthesis of natural products. We also present the basic concepts of these methods to facilitate and encourage further natural product syntheses.

2 Basic Transformations Mediated by Cp_2TiCl

2.1 Reactions Based on Cp_2TiCl-Promoted Radical Epoxide Openings

Epoxides (oxiranes) constitute one of the most versatile intermediates in organic synthesis as they can be prepared without difficulty from alkenes, carbonyl compounds, and other easily accessible starting materials and subsequently transformed into many other functional groups [16]. Moreover, the successful methods for asymmetric epoxidation developed by Sharpless, Katsuki, Jacobsen, Shi, and others have resulted in epoxides becoming crucial intermediates for enantioselective synthesis [17]. It has been known for a long time that when treated with acidic reagents or carbon nucleophiles epoxides undergo heterolytic ring opening, thus facilitating C – C bond formation via carbocation- or carbanion-type chemistry, respectively [18, 19].

Within this context, reports published by RajanBabu and Nugent between 1988 and 1994 introduced a novel concept into epoxide chemistry: homolytic ring opening mediated by Cp_2TiCl [7, 20–22]. On the basis of an analogy to the rearrangement of cyclopropylmethyl radicals to homoallyl radicals, these authors reported a fundamental hypothesis: that a σ complex of an epoxide with a transition metal possessing a half-filled d orbital might undergo C – O bond cleavage to release ring strain (Scheme 1) [7]. Experimental evidence provided by the authors confirmed their hypothesis and proved that the treatment of epoxides with an excess of Cp_2TiCl can be used not only for selective reduction and deoxygenation reactions, but also for C – C bond forming

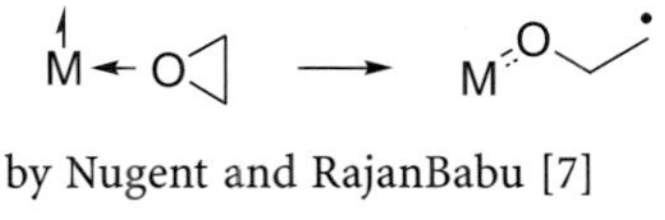

Scheme 1 Analogy realized by Nugent and RajanBabu [7]

processes such as the cyclization of 6,7-epoxyalkenes and 6,7-epoxyalkynes, and intermolecular additions to activated olefins (Scheme 2) [7, 20–22]. It is worth noting that as the reaction mechanism generally proceeds via the most substituted (i.e., energetically most stable) radical, the regiochemistry is the opposite (and complementary) to that which is normally to be expected for conventional S_N2-type epoxide openings with carbon nucleophiles or hydride reagents [19, 23].

Apart from the report by RajanBabu, Nugent, and Beattie [21], among the processes employing stoichiometric proportions of Cp_2TiCl there have been numerous studies dealing with reductions and deoxygenations of different epoxides, including epoxy alcohols, epoxy ketones, and epoxy carvone derivatives [24–37]. Six years ago, Fernandez-Mateos et al. published the titanocene-promoted intramolecular addition of epoxides to carbonyl groups, a new radical reaction affording compounds ranging from

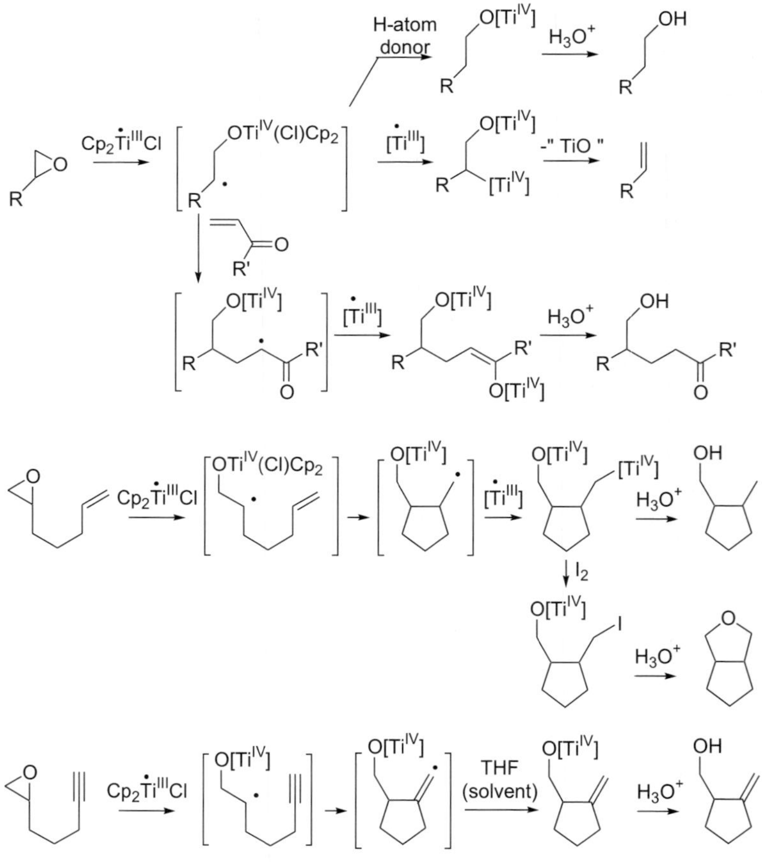

Scheme 2 Basic Cp_2TiCl-promoted transformations reported by RajanBabu and Nugent [7, 20–22]

cyclopropanols to cyclohexanols (Scheme 3) [38]. This procedure has subsequently been extended to the cyclization of epoxynitriles to give β-hydroxycycloalkanones [39].

Some descriptions of titanocene-promoted intermolecular additions of epoxides to α,β-unsaturated chromium and tungsten carbene complexes and other activated alkenes have been published by different authors [40–43]. More recently, the intramolecular version of the addition of epoxides to activated olefins has been independently developed by the groups of Fernandez-Mateos and Gansäuer, employing stoichiometric and catalytic quantities of titanocene, respectively [44–46]. This method allowed the construction of three- to six- and eight-membered carbocycles at medium to high yields, although those reported for the synthesis of seven-membered rings were considerably lower (Scheme 4).

To study the effects of water and other solvents on titanocene(III)-mediated processes we used the transannular cyclization of epoxygermacrolides as a model reaction [47]. Thus, we found that in anhydrous, non-halogenated solvents such as THF the reaction led selectively to decalins with an exocyclic double bond (Scheme 5). In an aqueous medium (THF/H_2O), however, the characteristic lime green color of Cp_2TiCl turned deep blue and the main product was a reduced decalin (Scheme 5). Under these conditions, water (either H_2O or D_2O) proved to be more effective than the toxic and expensive hydrogen-atom donor 1,4-cyclohexadiene for the reduction of tertiary radicals [47]. This is an unusual phenomenon in free-radical chemistry [48–50], subsequently exploited by us for the selective reduction of aromatic ketones as we shall see later [51, 52].

Some controversy remains concerning contemporary radical chemistry as to whether radical cascade cyclizations take place in a concerted or stepwise fashion [48]. Within this context we have recently reported theoretical and

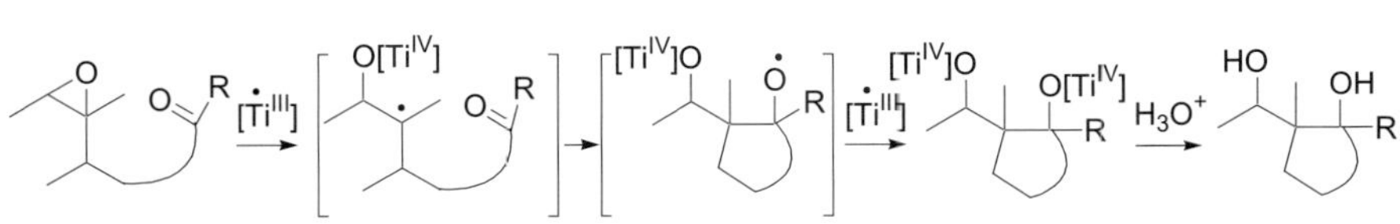

Scheme 3 1,3-Cycloalkanediols from epoxy carbonyl compounds [38]

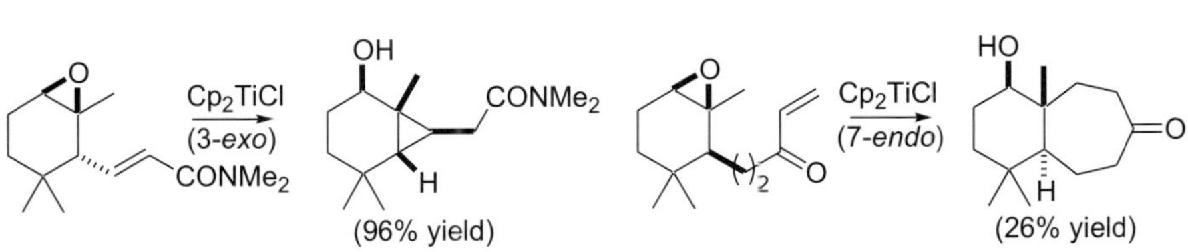

Scheme 4 Intramolecular additions of epoxides to activated olefins [44–46]

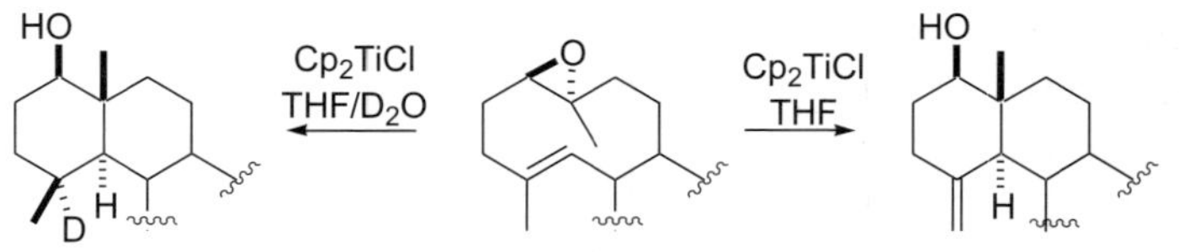

Scheme 5 Cp_2TiCl-mediated transannular cyclization of epoxygermacrolides [47]

experimental evidence to support the idea that titanocene(III)-promoted cascade cyclization of epoxypolyprenes takes place in a nonconcerted fashion via discrete carbon-centered radicals [53]. The 6-*endo* consecutive cyclizations proceed stereoselectively, and the end step can easily be controlled to give exocyclic alkenes or reduction products by simply excluding or adding water to the medium (Scheme 6). Cascade cyclizations promoted by stoichiometric quantities of Cp_2TiCl, however, require high dilution conditions to avoid the premature trapping of intermediate radicals by an excess of titanocene, which results in decreased yields [54]. As we shall see later, this drawback is circumvented and the yields are improved considerably by using the catalytic version of the process [53].

Cp_2TiCl-mediated intramolecular radical vinylations based on the elimination of β-phosphinoyl radicals have recently been reported by Leca

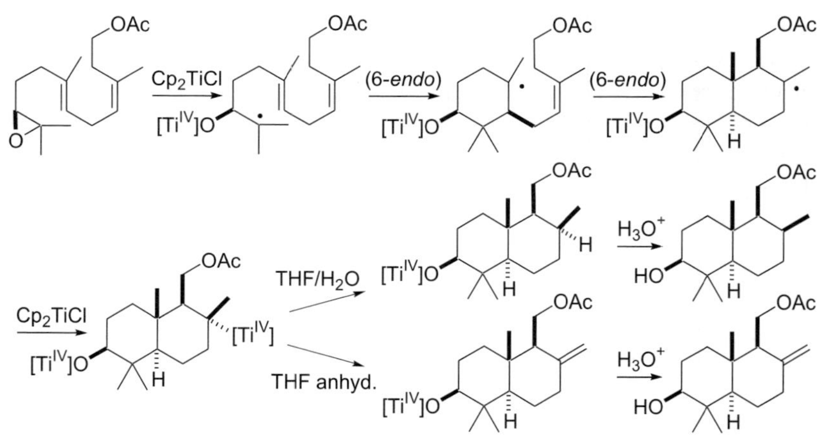

Scheme 6 Cp_2TiCl-promoted cascade cyclization of epoxypolyprenes [53, 54]

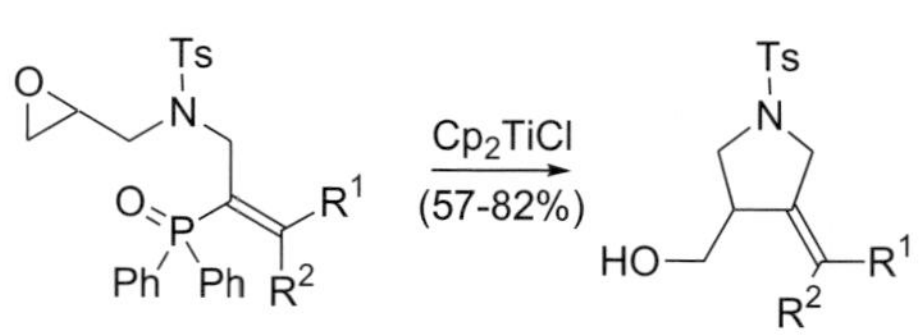

Scheme 7 Cp_2TiCl-mediated preparation of alkylidenepyrrolidines [55, 56]

et al. [55, 56]. The procedure, which is new to organic synthesis, has proved to be useful in synthesizing alkylidenepyrrolidines (Scheme 7).

2.2 Processes Based on Cp_2TiCl-Catalyzed Epoxide Openings

In 1998, 10 years after the pioneering report by Nugent and RajanBabu [7], Gansäuer et al. described the first catalytic procedure to achieve homolytic epoxide openings [8, 9, 57]. Their method was based on regenerating the precatalyst Cp_2TiCl_2 by protonating the titanium-bound oxygen atom. Subsequent in situ reduction by a stoichiometric reductant (Mn or Zn) once more gives the active titanium(III) complex, and closes the catalytic cycle (Scheme 8). The acid employed in this process must be weak enough not to open the epoxide in a heterolytic form, but strong enough to protonate $O-Ti^{IV}$ bonds [9]. The authors anticipated that Brønsted acids, with an interval of pK_a values between 5.25 and 12.5, would be suitable. In fact, both pyridine hydrochloride and 2,6-lutidine hydrochloride proved to be capable of maintaining the catalytic cycle, but it was 2,4,6-collidine hydrochloride (col·HCl; pK_a = 7.43) which gave the best results [8, 9, 57].

This titanocene-catalyzed procedure was immediately extended by Gansäuer et al. to the enantioselective opening of *meso*-epoxides by employing substoichiometric quantities of titanocene complexes with chiral ligands [58–60]. It has also been applied by this group in racemic form not only for reductive epoxide openings and intermolecular additions to α,β-unsaturated carbonyl compounds, but also to achieve 3-*exo*, 4-*exo*, and 5-*exo* cyclizations, as well as tandem cyclization addition reactions featuring vinyl radicals (Scheme 9) [8, 9, 44, 46, 57, 61–65].

Another novel titanocene-catalyzed tandem reaction reported by Gansäuer et al. provides a straightforward way of synthesizing structurally complex

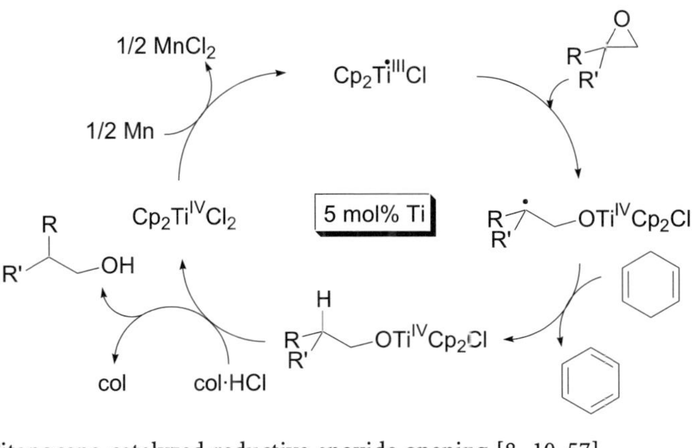

Scheme 8 Titanocene-catalyzed reductive epoxide opening [8–10, 57]

tetrahydrofurans in only one step (Scheme 10) [66, 67]. In the final step of the process a homolytic, concerted substitution reaction (S_H2) closes the tetrahydrofuran ring and regenerates the [Ti^{III}] catalyst. A closely related S_H2 reaction has been more recently observed by Leca et al. during the Cp_2TiCl-mediated cyclization of a vinylic phosphonate derivative [56].

When we assessed col·HCl as the titanocene-regenerating agent for achieving a catalytic version of the transannular cyclization of epoxygermacrolides, instead of the desired exocyclic alkene **3**, we obtained a high proportion of the reduction product **2**, presumably deriving from the protonation of an intermediate $C-Ti^{IV}$ bond (Scheme 11) [68]. To avoid this undesired background we introduced a new titanocene-regenerating agent, the Me_3SiCl/col combination, which presumably generates in situ the nonprotic reagent 2,4,6-trimethyl-1-trimethylsilylpyridinium chloride. This compound is both compatible with epoxides and capable of regenerating Cp_2TiCl_2 not only from oxygen-bound titanium atoms, including $Cp_2Ti(Cl)OAc$, but also from $Cp_2Ti(Cl)H$ formed by β-hydride elimination reactions from alkyl-Ti^{IV} complexes toward alkenes. Therefore, as we expected, the titanocene-catalyzed cyclization of **1** with this novel reagent gave improved proportions of exocyclic alkene **3** (Scheme 11) [68].

The Me_3SiCl/col couple has also proved to be a suitable regenerating agent for Cp_2TiCl-catalyzed cascade cyclizations of epoxypolyprenes [53]. Thus, the catalytic cyclization of epoxyfarnesyl acetate (**4**) (Scheme 12), for example, provided substantially increased yields of alkene **5** (40% versus the 25% obtained by the stoichiometric version) whilst employing lower titanocene proportions and dilution levels by one and two orders of magnitude, respectively.

We have also achieved unusual 7-*endo*-dig and 7-*endo*-trig radical cyclizations employing the Me_3SiCl/col combination as titanocene-regenerating agent [69]. Moreover, we have provided theoretical and experimental evi-

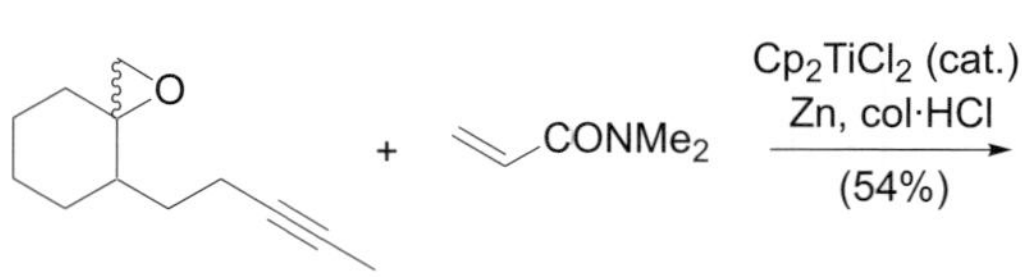

Scheme 9 Titanocene-catalyzed tandem reactions with alkynes in EtOAc [65]

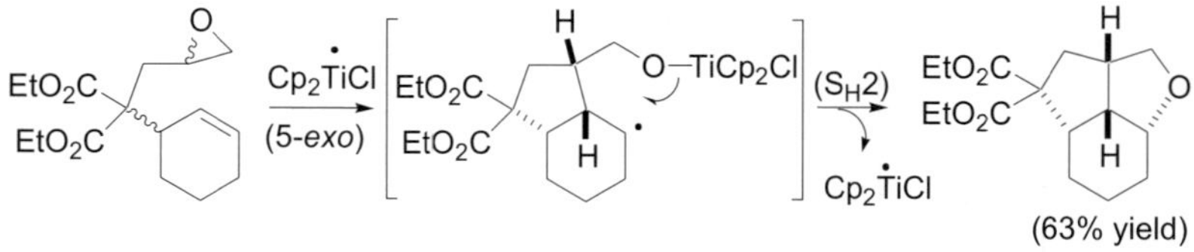

Scheme 10 Formation of tetrahydrofurans by a novel radical tandem cyclization [66, 67]

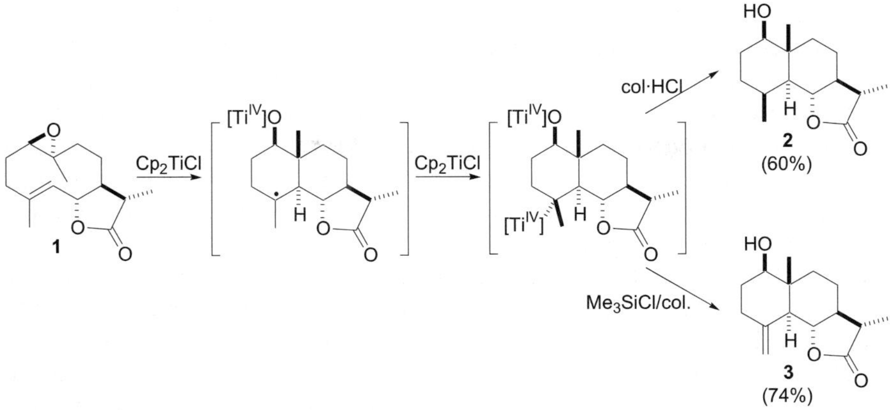

Scheme 11 Titanocene-catalyzed transannular cyclization of epoxygermacrolides [68]

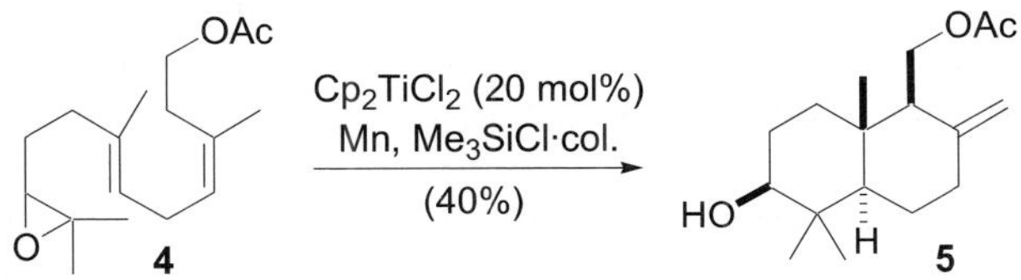

Scheme 12 Titanocene-catalyzed cyclization of epoxyfarnesyl acetate [53]

dence in support of a plausible mechanism which may rationalize the preference for the unusual 7-*endo* cyclization mode shown by radicals with substitution patterns characteristic of the linalyl, nerolidyl (Scheme 13), and geranyl linalyl systems [69]. The Me_3SiCl/col couple has also been successfully used by Leca et al. to achieve the titanocene-catalyzed tandem 5-*exo*/S_H2 cyclization of a phosphonate derivative [56].

Finally, Fuse et al. recently proposed the use of Et_3B together with 2,6-lutidine hydrochloride or 2,4,6-collidine hydrochloride to improve the capacity of the system for regenerating titanocene(III) from $Cp_2Ti(Cl)H$ in the $[Ti^{III}]$-catalyzed cyclization of 6,7-epoxygeranyl acetate [70].

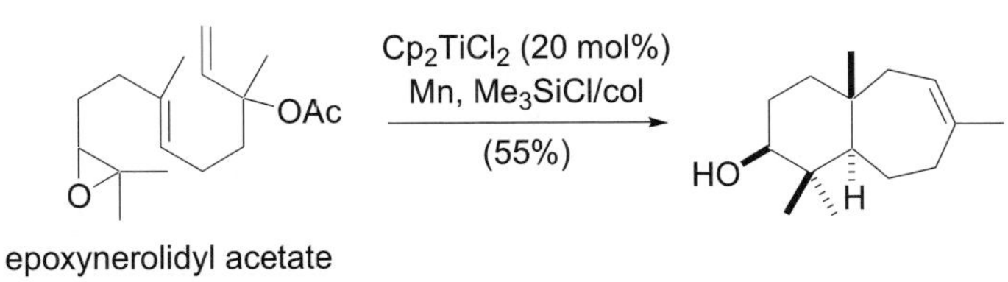

Scheme 13 Titanocene-catalyzed tandem 6-*endo*/7-*endo* cyclization of epoxynerolidyl acetate [69]

2.3 Other Reactions Mediated by Cp_2TiCl

Apart from the processes based on the homolytic epoxide opening summarized above, other synthetic transformations employing Cp_2TiCl have been reported by several authors. These reactions, however, have received scarce application for natural product synthesis so far and, therefore, we only present a short resume of them.

In 1987, Handa and Inanaga reported the reductive homocoupling of aromatic and α,β-unsaturated aldehydes promoted by titanocene(III) to yield symmetrical 1,2-diol (pinacols) with high *threo* selectivity [71]. The trinuclear species $(Cp_2TiCl)_2MgCl_2$ was claimed by the authors to be responsible for the diastereoselectivity observed in these pinacolizations [71]. Nevertheless, recent studies carried out by the Skrydstrup and Daasbjerg group have cast serious doubt on this hypothesis [4, 72]. The catalytic versions developed by Gansäuer and Bauer employed $Me_3SiCl/MgBr_2$ or different pyridine hydrochlorides as titanocene-regenerating agents [73, 74]. Hirao et al. have reported the titanocene-catalyzed pinacol coupling of aliphatic aldehydes with the assistance of zinc powder and Me_3SiCl, and Dunlap and Nicholas have analyzed the catalyst structural effects in titanocene-catalyzed pinacolizations [3, 75]. In the pinacol coupling of benzaldehyde in aqueous medium (THF/H_2O, 4 : 1) reported by Barden and Schwartz, the assistance of a considerable excess of NaCl (62 equiv per Ti) was required to recover the green color characteristic of Cp_2TiCl (from the blue species formed under these conditions) and complete the pinacolization process with high stereoselectivity [76]. In this sense, we have recently observed that the "blue titanocene species" formed in THF/H_2O can catalyze the selective reduction of aromatic ketones by employing water as the proton source instead of the toxic and expensive hydrogen-atom donors generally used in free-radical chemistry [51]. In a similar way, Cp_2TiCl-mediated selective reduction of α,β-unsaturated ketones in THF/MeOH employed methanol as proton source [77]. Moreover, blue titanocene generated in simple water has proved to be capable of promoting the pinacol coupling of benzaldehyde, but with complete loss of stereoselectivity [52].

In the past few years, several authors have shown the capacity of Cp_2TiCl to promote and/or catalyze C – C bond forming reactions between activated alkyl halides and carbonyl compounds. Thus, in 2003 Parrish et al. reported Cp_2TiCl-promoted Reformatsky-type additions [78]. In 2004, Roy's group reported the allylation of aldehydes mediated by stoichiometric Cp_2TiCl [79]. Simultaneously, we developed the first Barbier-type allylations, benzylations, and propargylations of aldehydes and ketones catalyzed by Cp_2TiCl [80]. Moreover, we demonstrated that chiral titanocene complexes are capable of catalyzing enantioselective C – C bond forming processes between allylic halides and carbonyl compounds [80]. More recently, Roy's group has reported the Cp_2TiCl-promoted conjugate addition of activated bromo compounds to α,β-unsaturated carbonyl derivatives [81]. Some results presented in this last report are intriguing because we previously observed that reactions between citral or β-ionone with allyl bromide, catalyzed by Cp_2TiCl, mainly gave the products from carbonyl carbon addition [80].

In the absence of carbonyl compounds, allylic bromides treated with Cp_2TiCl underwent the expected dimerization process, which has been exploited for the synthesis of symmetrical terpenoids such as squalene and β-onocerin [82]. In contrast, non-allylic olefinic iodoethers and α-bromo carbonyl compounds underwent 5-*exo* cyclization to the corresponding tetrahydrofuran derivatives [83, 84].

Titanocene(III) reagents for the synthesis of glycals and C-glycosides have been reviewed by Spencer and Schwartz [85]. Moreover, a catalytic cycle for the generation of glycosyl carbanions with Cp_2TiCl has been developed by the Skrydstrup and Daasbjerg group and applied to glycal synthesis [86]. Cp_2TiCl has also been the subject of electrochemical and photochemical studies [87–90].

3
Stoichiometric Cp_2TiCl in Natural Product Synthesis

The following syntheses of natural products were based on the use of stoichiometric quantities of Cp_2TiCl (often interpreted as a considerable excess) to promote different chemical processes initiated by homolytic epoxide opening, such as deoxygenations and reductive epoxide openings, simple 5-*exo* and 6-*endo* cyclizations, and so on. They have been classified according to the chemical process promoted by Cp_2TiCl employed.

3.1
Deoxygenations and Reductive Epoxide Openings

There are two especially relevant cases where Cp_2TiCl-promoted epoxide deoxygenations have been demonstrated to conform to the requirements of selectivity, mildness, and wide functional group tolerance desirable in natural product synthesis: the chemical correlation between cryptophycin-23 and cryptophycin-45 and the synthesis of anhydrovinblastine, the key intermediate in the synthesis of the anticancer drug Navelbine, from leurosine (Scheme 14) [91–93].

Cp_2TiCl-mediated reductive epoxide openings, using hydrogen-atom donors such as 1,4-cyclohexadiene or *t*-BuSH, have been the key step in the total synthesis of natural δ-lactones found in *Cryptocarya latifolia* (Scheme 15) [31]. Cp_2TiCl-mediated reductive epoxide openings have also been employed for the preparation of oxygenated segments of rhizoxin, taxol,

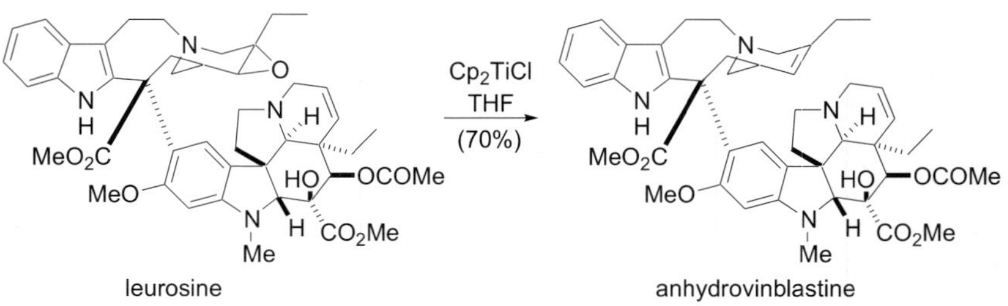

Scheme 14 Synthesis of anhydrovinblastine from leurosine [92, 93]

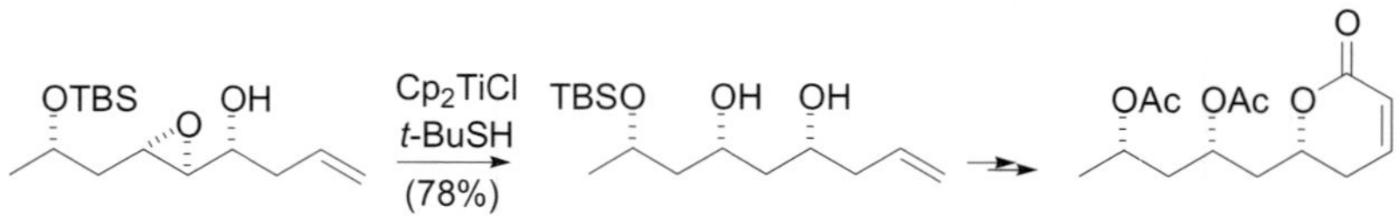

Scheme 15 Cp_2TiCl-mediated synthesis of a lactone from *C. latifolia* [31]

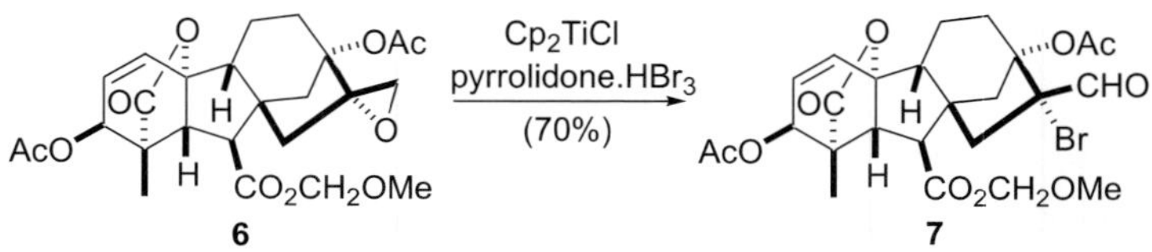

Scheme 16 Cp_2TiCl-promoted synthesis of bromoaldehyde **7** [94]

and epothilones as well as for the synthesis of (+)-prelactone C and (+)-prelactone B [26, 27, 30, 34, 36].

Bhaskar and Mander used the Cp_2TiCl-promoted opening of epoxide **6** in the presence of pyrrolidone hydrotribromide to obtain α-bromoaldehyde **7**, a crucial intermediate in the synthesis of the biologically potent gibberellin GA_{32} (Scheme 16) [94].

3.2 5-*exo* Cyclizations

Cp_2TiCl-induced 5-*exo*-dig cyclizations have been used by Clive's group in the preparation of the sesquiterpenoid (±)-ceratopicanol (Scheme 17) [95,

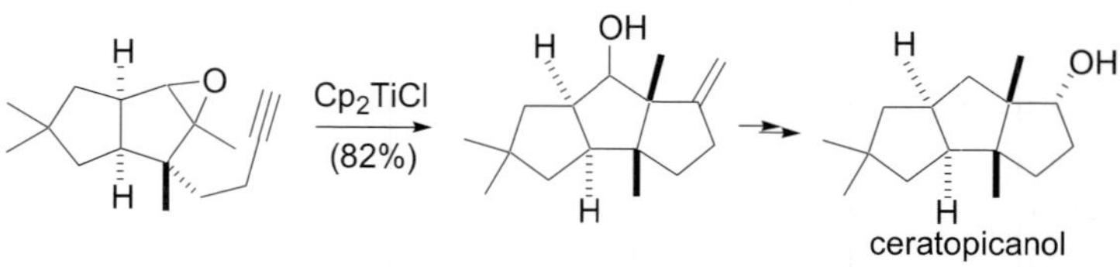

Scheme 17 Cp_2TiCl-mediated synthesis of (±)-ceratopicanol [95, 96]

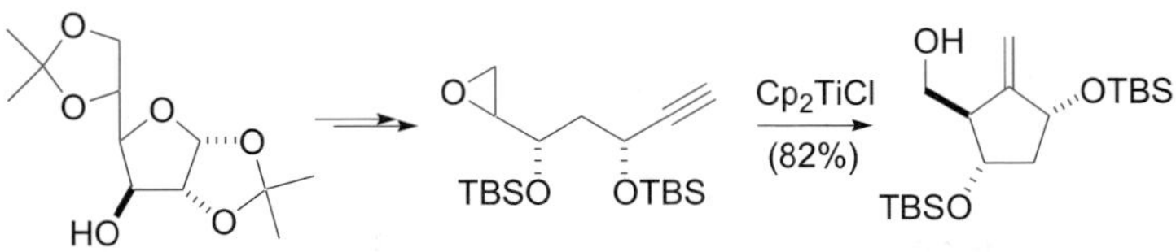

Scheme 18 Synthesis of a protected carbocyclic core of entecavir from D-diacetone glucose [99]

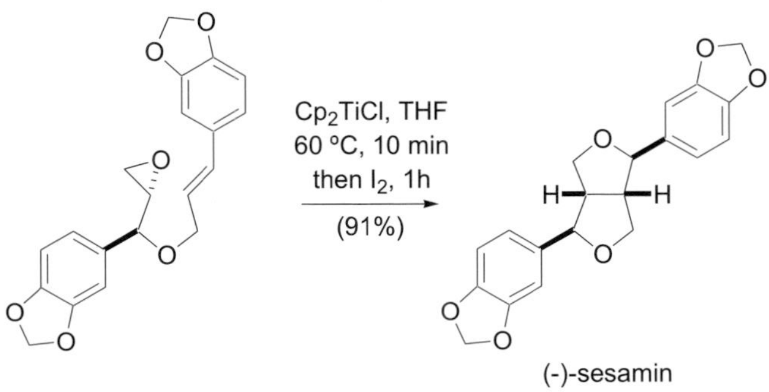

Scheme 19 Cp_2TiCl-mediated synthesis of (–)-sesamin [104]

96]. Cp_2TiCl-promoted 5-*exo*-dig cyclizations have also been employed by Roy's group for the synthesis of (±)-methylenolactocin and (±)-protolichesterinic acid [97, 98]. More recently Ziegler and Sarpong used a Cp_2TiCl-promoted 5-*exo*-dig cyclization for the synthesis of a protected carbocyclic core of BMS-200475 (entecavir) (Scheme 18), a nucleoside active at the nanomolar level against the hepatitis B virus [99].

Cp_2TiCl-mediated 5-*exo*-trig cyclizations have been intensively exploited by Roy et al. for the total synthesis of (±)-dihydroprotolichesterinic and (±)-roccellaric acids, (±)-sesamin, (±)-dihydrosesamin, (±)-acuminatin, (±)-eudesmin, (±)-lariciresinol, (±)-pinoresinol, (±)-piperitol, (±)-acuminatin methyl ether, (±)-sanshodiol methyl ether, (±)-piperitol methyl ether, (±)-pinoresinol monomethyl ether, (±)-lariciresinol monomethyl ether, and (±)-lariciresinol dimethyl ether [100–103]. Moreover, this group has very recently reported the enantioselective synthesis of (–)-sesamin (Scheme 19), (–)-dihydrosesamin, (–)-acuminatin, and (–)-methyl piperitol by radical cyclization of chiral epoxides initiated by Cp_2TiCl [104].

3.3 Simple 6-*endo* Cyclizations

The Cp_2TiCl-mediated simple 6-*endo* cyclization of epoxygeranyl acetate was used by Nakai et al. in the preparation of synthons for building the A and C rings of paclitaxel [105], and a similar cyclization of epoxygeranylacetone derivative **8** by some of us for the first synthesis of achilleol A (**9**) (Scheme 20) [106]. Thus, it was possible to confirm the chemical structure of this unusual monocyclic triterpenoid more than 10 years after its discovery [107]. Interestingly, achilleol A has recently been found among the products obtained from the incubation of (3*S*)-2,3-oxidosqualene with a Gly600-deletion mutant of the enzyme squalene cyclase from *Alicyclobacillus acidocaldarius* [108].

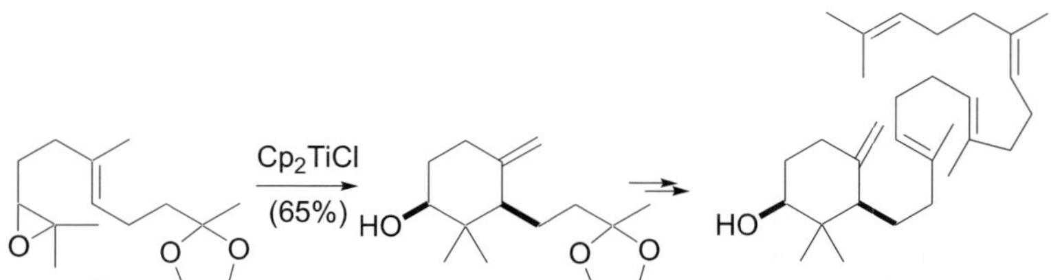

Scheme 20 Cp_2TiCl-promoted synthesis of achilleol A (**9**) [106]

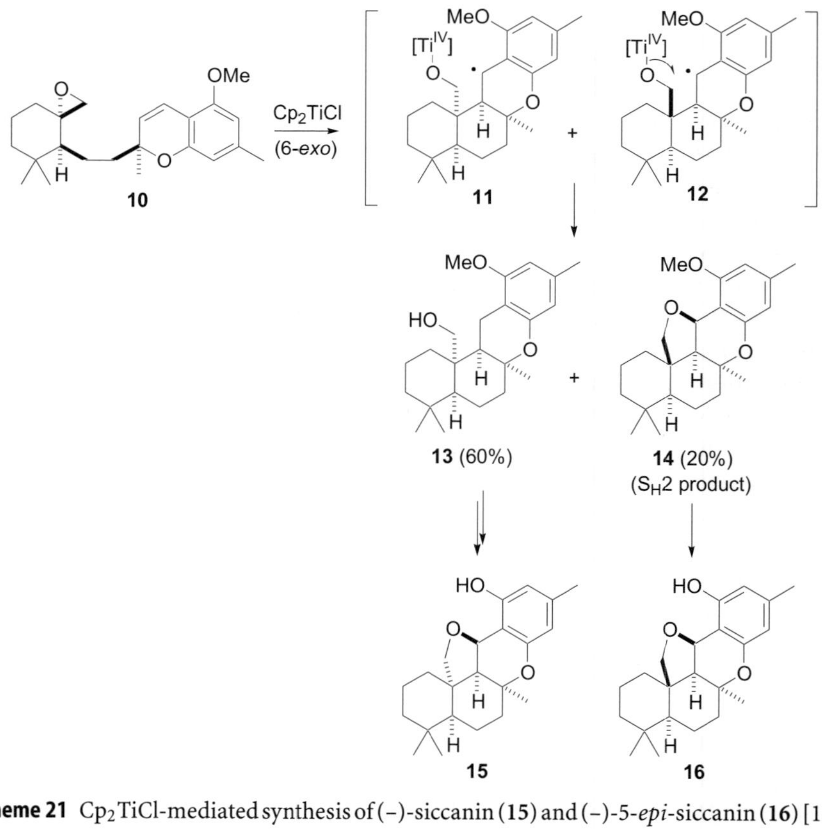

Scheme 21 Cp_2TiCl-mediated synthesis of (–)-siccanin (**15**) and (–)-5-*epi*-siccanin (**16**) [109]

Trost et al. have exploited the Cp_2TiCl-promoted 6-*exo* cyclization of **10** to **13** for the first enantioselective biomimetic total synthesis of the antifungal metabolite (–)-siccanin (**15**) (Scheme 21) [109]. These authors also observed a minor amount (20%) of by-product **14**, presumably derived from the radical intermediate **12** by a process reminiscent of the S_H2 reaction described by Gansäuer's group [66]. They took advantage of **14** for the preparation of (–)-5-*epi*-siccanin (**16**) (Scheme 21) [109].

Grande's group has studied the titanocene-promoted intramolecular addition of epoxides to activated alkenes in both 5-*exo* and 6-*endo* cyclization modes as a new way of preparing polycyclic β-lactam antibiotics [110–112].

3.4 Tandem 6-*endo*/6-*exo* Cyclizations

Stoichiometric quantities of Cp_2TiCl were used by Haruo et al. to promote a stereoselective tandem 6-*endo*-trig/6-*exo*-dig cyclization employed for the total synthesis of the sesquiterpenoid (±)-smenospondiol (**17**) (Scheme 22) [113].

3.5 Transannular Cyclizations

Transannular cyclizations generally contribute to the enhancement of molecular rigidity and structural complexity, two properties often associated with biological activity of small molecules [114]. In this scenario, we found that the Cp_2TiCl-mediated transannular cyclization of 1,10-epoxycostunolide, a homochiral material accessible in (multi)gram quantities, provides a straightforward way for the enantiospecific synthesis of eudesmanolides such as (+)-reynosin (**18**) (Scheme 23) and (+)-3α-hydroxyreynosin [47].

The conventional acid-induced transannular cyclization of 1,10-epoxycostunolide, via carbocationic chemistry, results in a eudesmanolide mix-

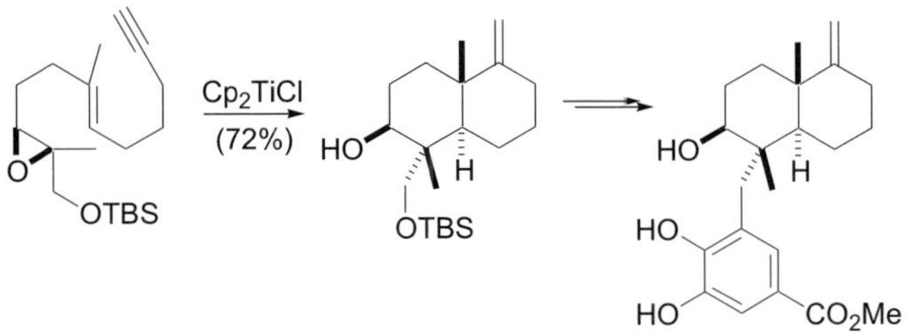

17

Scheme 22 Cp_2TiCl-promoted tandem cyclization in the synthesis of smenospondiol [113]

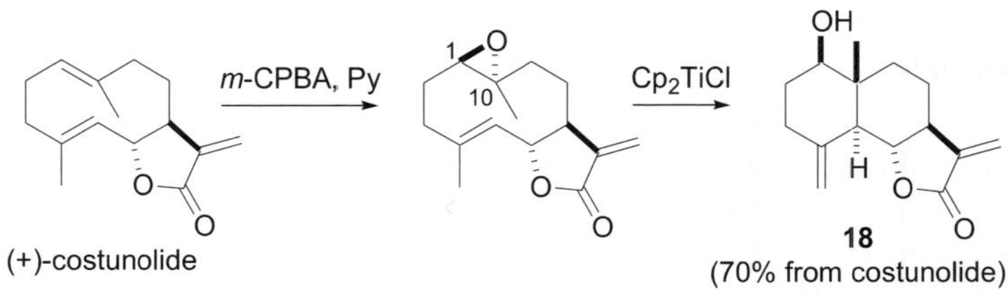

Scheme 23 Cp_2TiCl-mediated enantiospecific synthesis of (+)-reynosin (**18**) [47]

ture in which the desired exocyclic alkene **18** is one of the minor components. Furthermore, the palladium(II)-promoted rearrangement of germacrolides leads to similar results, probably also via carbocationic chemistry [115]. In contrast, the Cp_2TiCl-promoted radical cyclization of 1,10-epoxycostunolide regio- and stereoselectively leads to (+)-reynosin (**18**) in good yield (Scheme 23) [47].

4 Cp_2TiCl-Catalyzed Synthesis of Natural Products

The following syntheses of natural products were based on the use of substoichiometric quantities of Cp_2TiCl to catalyze different cyclization processes initiated by homolytic epoxide opening. They have been classified according to the type of cyclization process employed. Section 4.4 has been reserved for the Cp_2TiCl-catalyzed radical cyclization of 2,3-oxidosqualene due to the biological importance of this terpenoid and steroid precursor.

4.1 Simple 6-*endo* Cyclizations

Employing the Me_3SiCl/col couple as titanocene-regenerating agent, we have prepared both monocyclic sesquiterpenoids, such as the metabolite **19** found in the fragrant plant *Artemisia chamaemelifolia* (Scheme 24), and monocyclic triterpenoids, such as achilleol A (**9**), starting from epoxygeranylacetone derivative **8** and using only substoichiometric proportions (0.2 equiv) of the titanocene catalyst [53].

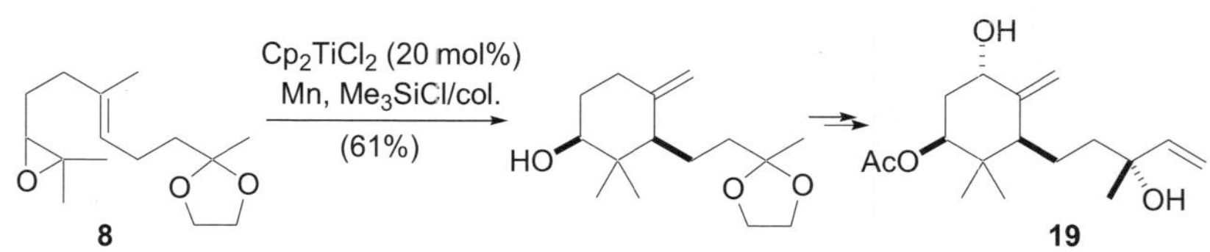

Scheme 24 Titanocene-catalyzed synthesis of **19** [53]

4.2 Tandem 6-*endo*/6-*endo* Cyclizations

Drimanes constitute a family of bicyclic sesquiterpenoids with interesting biological properties [116, 117]. The titanocene-catalyzed 6-*endo*/6-*endo* tandem cyclization of epoxyfarnesyl acetate (**4**) stereoselectively gave bicyclic intermediate **5** (Scheme 12), which proved to be a suitable precursor for the total synthesis of 3-hydroxydrimanes with different functionalization patterns, including (±)-isodrimenediol (**20**), triol **21**, (±)-3β-hydroxydihydroconfertifolin (**22**), (±)-3β-hydroxycinnamolide (**23**), and (±)-3β-acetoxydrimenin (**24**) (Scheme 25) [53, 118]. Interestingly, synthetic **22** showed antifeedant activity against the insect *Leptinotarsa decemlineata*, whereas the conjugated γ-lactone **23** was active against *Myzus persicae* [118].

Furthermore, the titanocene-catalyzed 6-*endo*/6-*endo* cyclization of epoxyfarnesylacetone derivative **25** led to the first total synthesis of (±)-3β-hydroxymanool (**27**) (Scheme 26), a bicyclic diterpenoid with a labdane skeleton from the fern *Gleichenia japonica* [53]. The subsequent palladium-mediated, long-range functionalization of the C-18 methyl group of **26** allowed the first synthesis of (±)-rostratone (**28**) (Scheme 26), a labdane diterpenoid with a characteristic γ-dioxygenated system on ring A, found in the Chilean plant *Nolana rostrata* [119, 120]. In our laboratory, the combination of [Pd^{II}] and [Ti^{III}] chemistries has also been useful for advanced synthetic approaches toward the pharmacologically active products aphidicolin and pyripyropene A [120].

The term "meroterpenoids" is generally used to denote a wide range of natural products of mixed (polyketide–terpenoid) biogenesis [121]. In our laboratory, the titanocene-catalyzed 6-*endo*/6-*endo* cyclization of **31**, ob-

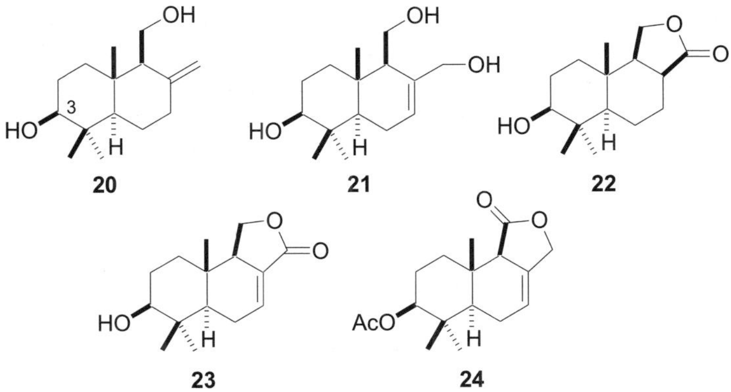

Scheme 25 Drimanes synthesized by titanocene-catalyzed 6-*endo*/6-*endo* cyclization of **4** [53, 118]

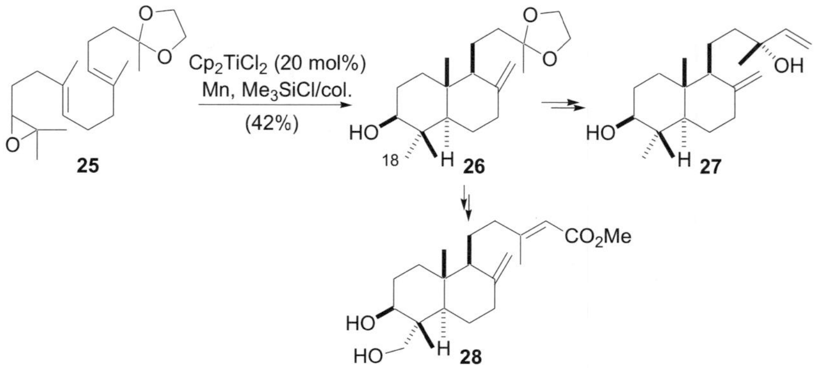

25 26 27 28

Scheme 26 Titanocene-catalyzed synthesis of (±)-3β-hydroxymanool (**27**) and (±)-rostratone (**28**) [53, 119, 120]

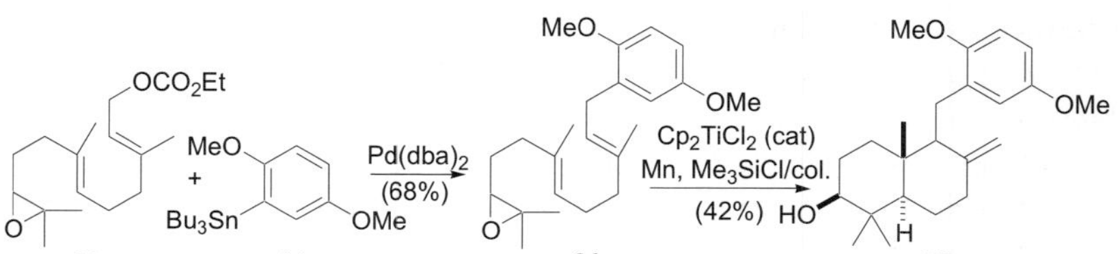

29 30 31 32

Scheme 27 Synthesis of **32** based on Stille couplings and titanocene catalysis [122]

tained by Stille coupling between carbonate **29** and stannane **30**, stereoselectively gave **32** (Scheme 27), an intermediate closely related to the marine meroterpenoid zonarol [122]. Interestingly, the aromatic ring of **31** and its oxygenated groups proved to be inert against the carbon-centered radicals formed during the cyclization process [122]. This is in contrast to what occurred in previously described cascade cyclizations toward meroterpenoids via carbocationic chemistry [123, 124].

Very recently, Gansäuer's group has described an alternative route based on copper-catalyzed allylic substitutions (instead of Stille coupling) for the preparation of precursors which might be useful for the synthesis of different meroterpenoids by means of Cp_2TiCl-promoted epoxypolyene cyclizations [125].

4.3
Cascade 6-*endo*/6-*endo*/6-*endo* Cyclizations

The marine metabolite stypoldione (**36**) has attracted the attention of chemists owing to both its pharmacological properties and its challenging chemical structure. Xing and Demuth have recently reported an elegant synthesis of **36** via the tricyclic intermediate **35** [126, 127]. In our laboratory

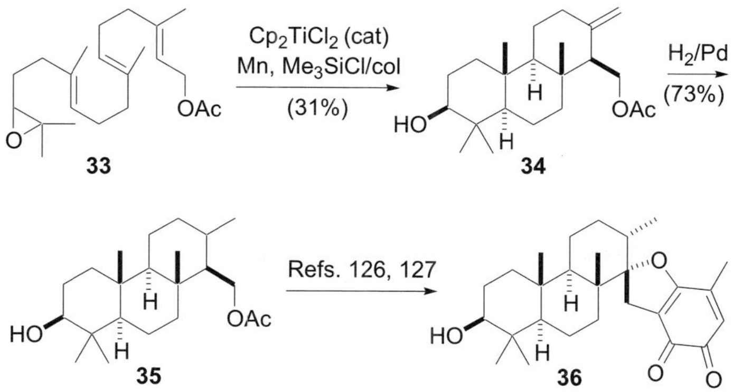

Scheme 28 Titanocene-catalyzed formal synthesis of stypoldione (**36**) [53]

the titanocene-catalyzed 6-*endo*/6-*endo*/6-*endo* cascade cyclization of epoxypolyene **33**, prepared from commercial geranylgeraniol, gave tricyclic alkene **34** (Scheme 28) [53].

The moderate yield of **34** (31%) can nevertheless be regarded as satisfactory when bearing in mind that the reaction selectively afforded a product (**34**) containing three fused (*trans*/*anti*/*trans*) six-membered rings, an exocyclic double bond, and six stereogenic centers among more than 190 potential regio- and stereoisomers. Finally, the catalytic hydrogenation of **34** gave **35** and thus the formal synthesis of stypoldione (**36**) was completed [53].

4.4 Oxidosqualene Radical Cyclization

The increasing demand for selectivity and atom and step economy in organic synthesis will presumably have a decisive influence on the strategies employed by chemists in the coming years [128, 129]. The biosynthesis of lanosterol from squalene fits these requirements admirably, taking place as it does in only two steps: the enantioselective epoxidation of squalene followed by the stereoselective cascade cyclization of 2,3-oxidosqualene, only one proton being lost during the process (to form the double bond at Δ^8) [130–132]. Mimicking this natural transformation, Goldsmith, van Tamelen, and Corey, among others, have exploited the acid-induced cascade cyclization of epoxypolyprenes as a powerful procedure in the construction of polycyclic terpenoids via carbocationic chemistry [133–140]. To the best of our knowledge, however, free-radical chemistry was never applied to the cyclization of 2,3-oxidosqualene during the last century, probably due to the lack of a suitable protocol for the radical opening of epoxides. The Cp_2TiCl-based procedure and its catalytic versions have catered for this need, thus opening up the possibility of mimicking the enzyme lanosterol synthase via radical chemistry.

These observations encouraged us to treat 2,3-oxidosqualene (**37**) with a catalytic quantity of titanocene. In this case, the cascade cyclization proceeded in a 6-*endo*/6-*endo*/5-*exo* manner, giving a mixture of malabaricane **38** and its C-13 epimer **39** (39% combined yield), together with minor amounts of achilleol A (**9**) and the acyclic alcohol **40** (Scheme 29) [53].

In contrast, the acid-induced carbocationic cyclization of **37**, previously reported by van Tamelen, Sharpless, and coworkers, gave a mixture containing bicyclic (**41**) and tricyclic products (**42** and **43**, stereochemistry not specified in the original paper) as the major components (Scheme 30) [141, 142].

The above results suggest that in the carbocationic process the second 6-*endo* cyclization is relatively fast but the third 5-*exo* cyclization is quite slow. Thus, formation of the bicyclic rearranged compound **41** is allowed whereas that of monocyclic products is avoided. In the radical process, however, the second 6-*endo* cyclization seems to be relatively slow and the third 5-*exo* cyclization fairly fast, thus allowing the formation of monocyclic achilleol A (**9**) and avoiding bicyclic products. In other words, there are subtle

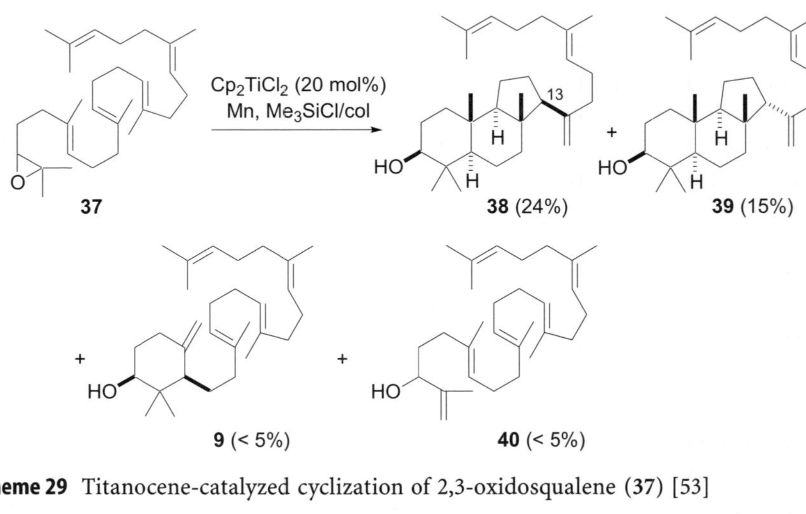

Scheme 29 Titanocene-catalyzed cyclization of 2,3-oxidosqualene (**37**) [53]

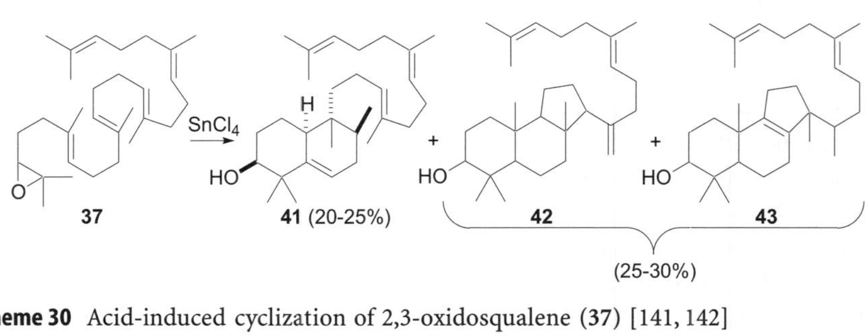

Scheme 30 Acid-induced cyclization of 2,3-oxidosqualene (**37**) [141, 142]

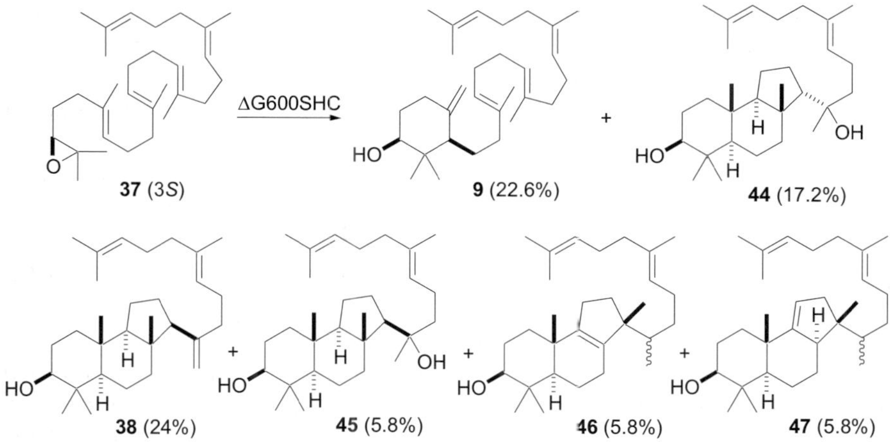

Scheme 31 2,3-Oxidosqualene cyclization catalyzed by a mutated enzyme [108]

but significant differences between the kinetics of radical and carbocationic cyclizations of 2,3-oxidosqualene that are reflected in the skeletal profile of the products obtained (monocyclic and tricyclic skeletons in the former case versus bicyclic and tricyclic ones in the latter).

In this context, Hoshino et al. have recently reported the cyclization of (3*S*)-2,3-oxidosqualene catalyzed by a Gly600-deletion mutant (ΔG600SHC) of the enzyme squalene-hopene cyclase (SHC) from *Alicyclobacillus acidocaldarius* (Scheme 31) [108]. The enzymatic biotransformation gave monocyclic (**9**) and tricyclic triterpenoids (**38** and **44–47**), but no detectable bicyclic products [108]. Despite the authors' biogenetic proposal of a conventional carbocationic pathway, the skeletal profile of products reported closely resembles that expected for a radical-type cyclization.

4.5 Simple 7-*endo* Cyclizations

Seven-membered carbocycles (as well as five- and six-membered ones) are often classified as "common rings" owing to their relatively low ring strain [143]. It is not surprising, therefore, that they are quite widespread in nature where they can be found in the carbon skeleton of numerous terpenoids, including monocyclic substances such as the monoterpenoid karahanaenone [144]. Synthetic methods based on simple 7-*endo* cyclizations have received scarce attention, however, possibly because of the general assumption that entropic factors do not favor cyclizations leading to seven-membered rings. We have found, however, that treatment of 6,7-epoxylinalyl acetate with a substoichiometric quantity of Cp_2TiCl_2 (0.2 equiv), Mn dust, and the Me_3SiCl/col mixture in dry THF gave the synthetic aroma chemical

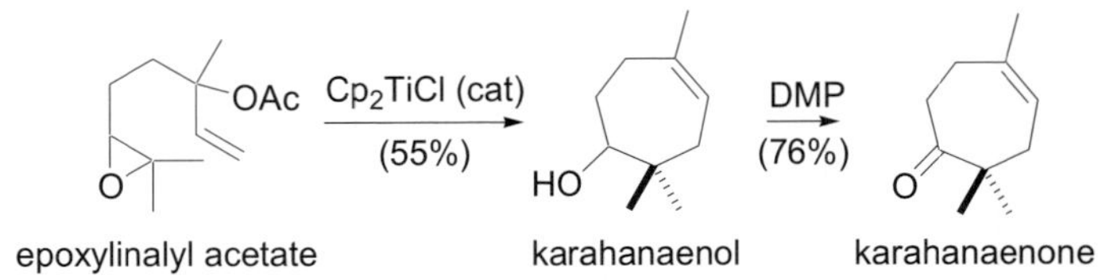

Scheme 32 Titanocene-catalyzed synthesis of karahanaenone [69, 120]

karahanaenol, which was easily oxidized to karahanaenone by Dess–Martin periodinane (DMP) (Scheme 32) [69, 120].

4.6 Tandem 6-*endo*/7-*endo* Cyclizations

Among the sesquiterpenoids with a reported daucane skeleton, one can find some confusing structural assignments. Thus, the daucadiene structure **48** was originally assigned to a metabolite from the liverwort *Bazzania trilobata* but, two years later, its mirror image was proposed for another metabolite with different NMR data found in the hybrid cypress species *Cupressocyparis leylandii* [145, 146]. In this context we deemed that the chemical synthesis of **48** from **49** (prepared by titanocene-catalyzed tandem 6-*endo*/7-*endo* cyclization of epoxynerolidyl acetate, as depicted in Scheme 13) might help to resolve this discrepancy and establish an unambiguous synthetic reference for facilitating further elucidation of the structure of related natural products. Thus, we treated **49** with PCl_5 and obtained a 75% yield of **48** (Scheme 33) [69]. NMR data for this synthetic product (**48**) matched those reported for the metabolite found in the cypress species, thus confirming the structure and relative stereochemistry proposed by Cool for this natural product [146].

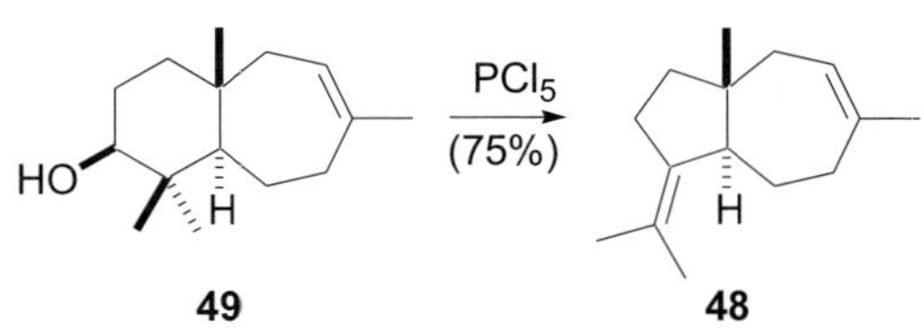

49 **48**

Scheme 33 Synthesis of daucadiene sesquiterpenoid **48** [69]

4.7 Cascade 6-*endo*/6-*endo*/7-*endo* Cyclizations

Marine natural products constitute one of the most exciting examples of chemical diversity. Many of them are biosynthesized by metabolic pathways exclusive to marine organisms and have shown an astonishing array of biological properties [147, 148]. In 1992, while studying pharmacologically active metabolites from sponges, Rudi and Kashman isolated barekoxide (**50**) (Scheme 34), a marine diterpenoid with an unprecedented carbon skeleton, the structure of which was established with the aid of X-ray analysis of its 3-bromo derivative [149–151]. Three years later Su et al. isolated the closely related terpenoid laukarlaol from the red alga *Laurencia karlae* and proposed the structure **51** (Scheme 34) on the basis of spectral data [152].

Barekoxide and laukarlaol attracted our attention not only because of the synthetic challenge they posed but also because of their intriguing biogenesis. Therefore, we attempted their chemical synthesis to corroborate structures **50** and **51**, provide chemical evidence concerning their biogenesis, and facilitate further biological analysis. Thus, Cp_2TiCl-catalyzed 6-*endo*/6-*endo*/7-*endo* cyclization of epoxygeranyllinalyl acetate (**52**) gave tricyclic alcohol **53** which, via tosylhydrazone **54**, provided synthetic barekoxide (**50**) (Scheme 35) [69]. Its physical and spectroscopic data matched those of the marine natural product, thus confirming the structure proposed by Kuniyoshi et al. [150, 151].

Tosylhydrazone **54** also served as a branch point to start a divergent route toward authentic laukarlaol (Scheme 36) [69]. Thus, the treatment of **54** with sodium hydride gave diene **55**, which underwent selective epoxidation at the α face of the Δ^{13} double bond. A nuclear Overhauser effect observed between H-14 and H_3-17 confirmed the α disposition of the oxirane ring of **56**. Finally, the perchloric acid opening of epoxide **56** completed the total synthesis of alcohol **57** (Scheme 36), which showed spectroscopic data matching those reported for natural laukarlaol [69]. Therefore, the relative 14*S** stereochem-

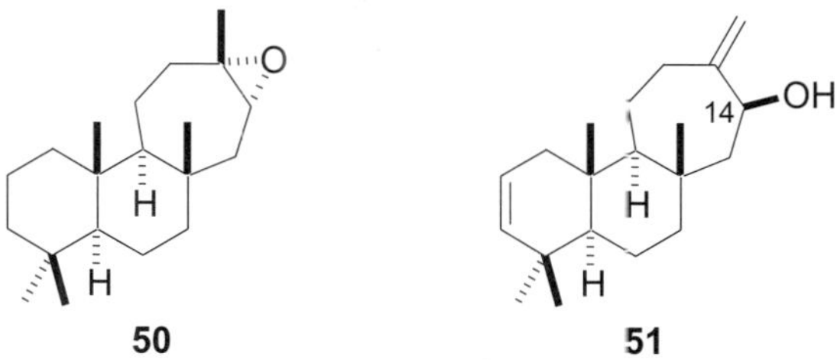

Scheme 34 Structures proposed for barekoxide (**50**) and laukarlaol (**51**) [149–152]

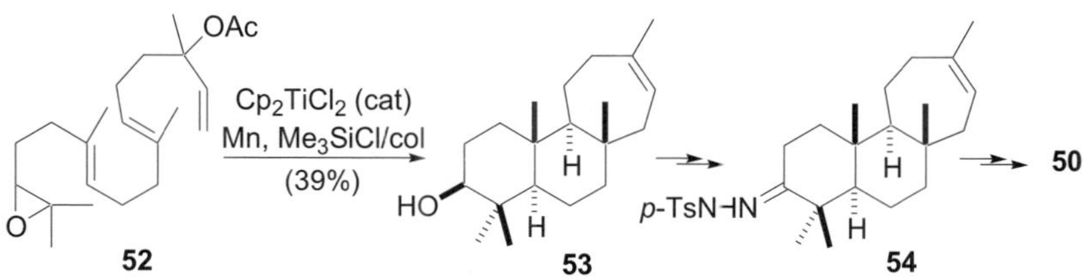

Scheme 35 Titanocene-catalyzed synthesis of barekoxide (**50**) [69]

istry (**51**) originally proposed for this natural product should be revised to the 14*R** one depicted in **57**.

The intrinsic tendency shown by epoxylinalyl, epoxynerolidyl, and epoxygeranyllinalyl systems to undergo 7-*endo* radical cyclizations led us to discuss the possible implications of this phenomenon in the biosynthesis of some terpenoids containing seven-membered rings [69]. In 1990 Urones et al. discovered a new type of diterpenoid with a novel *trans/anti/trans*-fused five/six/seven-membered tricyclic skeleton called valparane, which had eluded chemical synthesis for 15 years [153]. The success obtained in the preparation of **48** (see Scheme 33) prompted us to choose valparane **58** as a synthetic target to check the usefulness of our method to build this type of terpenoid. Thus, we treated tricyclic alcohol **53** with PCl_5 and obtained an excellent 95% yield of the expected product **58** (Scheme 37) [69]. The spectroscopic data for this product matched those of the semisynthetic derivative prepared from natural valparane **59** by Urones' group, thus confirming the structure proposed for the natural product [154].

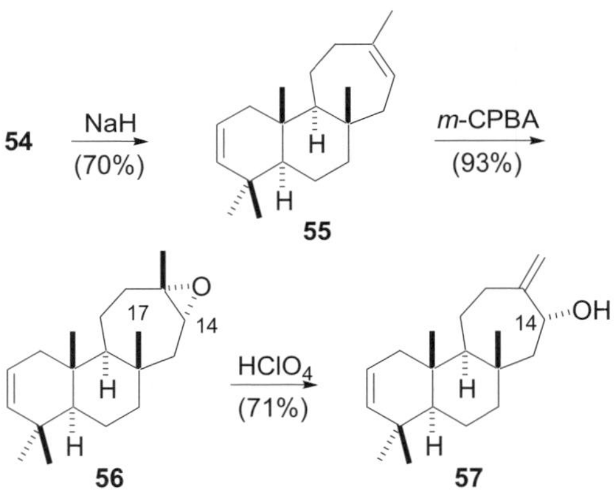

Scheme 36 Titanocene-catalyzed synthesis of authentic laukarlaol (**57**) [69]

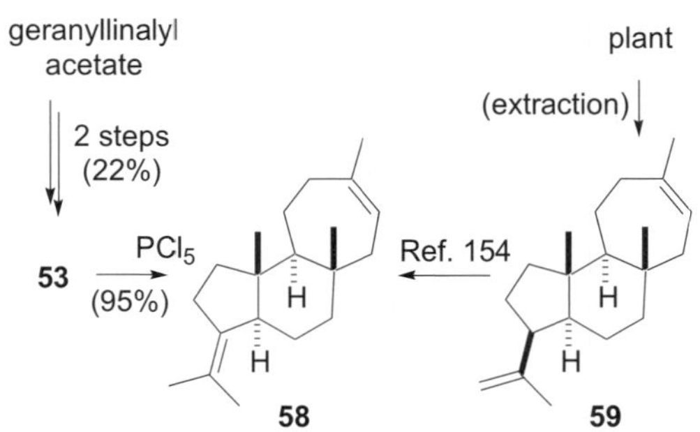

Scheme 37 Titanocene-catalyzed synthesis of valparane **58** and its chemical correlation with natural valparane **59** [69, 154]

4.8 Transannular Cyclizations

We have also taken advantage of the Me_3SiCl/col combination as a titanocene-regenerating agent for developing a catalytic version for the transannular cyclization of 1,10-epoxygermacrolides [68]. Thus, we have been able to synthesize eudesmanolides with an exocyclic double bond such as (+)-9β-hydroxyreynosin or (+)-β-cyclopyrethrosin (**60**) using only substoichiometric proportions of commercial Cp_2TiCl_2 (Scheme 38) [68]. We have also developed an improved synthetic procedure for the biologically active eudesmanolide (+)-tuberiferine [68].

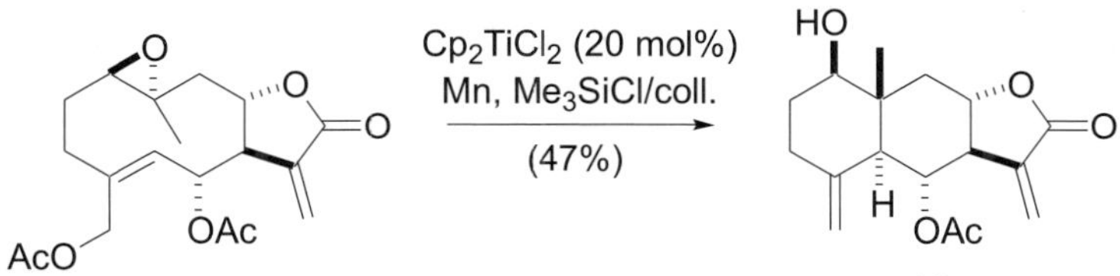

60

Scheme 38 Titanocene-catalyzed synthesis of (+)-β-cyclopyrethrosin (**60**) [68]

5 Concluding Remarks

Among the Cp_2TiCl-promoted and/or -catalyzed chemical transformations described so far, those based on the homolytic epoxide opening have been profusely used in the synthesis of natural products during the past 10 years. At first, owing to its selectivity and the extremely mild experimental conditions employed, the procedure was chosen for deoxygenation and reductive epoxide opening on densely functionalized substrates, but during the last 5 years it has been increasingly employed to achieve radical cyclizations, affording straightforward strategies for the synthesis of complex natural products. In these fields the method has already largely proved its synthetic usefulness. Nevertheless, other Cp_2TiCl-mediated reactions have as yet been scarcely applied. It is presumed, however, that these promising titanocene(III)-based transformations will also be demonstrated to be powerful synthetic tools in the near future.

Acknowledgements We are grateful to the "Junta de Andalucia" for the financial support to our research group (FQM339), including the project from the "Consejería de la Presidencia" (AM43/04). J.J. thanks the University of Granada and J.L.O.-L. the Spanish Ministry of Education and Science for their grants.

References

1. Nicolaou KC, Vourloumis D, Winssinger N, Baran PS (2000) Angew Chem Int Ed 39:45
2. Enemærke RJ, Hjøllund GH, Daasbjerg K, Skrydstrup T (2001) CR Acad Sci 4:435
3. Dunlap MS, Nicholas KM (2001) J Organomet Chem 630:125
4. Enemærke RJ, Larsen J, Skrydstrup T, Daasbjerg K (2004) J Am Chem Soc 126:7853
5. Sekutowski D, Jungst R, Stucky GD (1978) Inorg Chem 17:1848
6. Stephan DW (1992) Organometallics 11:996
7. Nugent WA, RajanBabu TV (1988) J Am Chem Soc 110:8561
8. Gansäuer A, Pierobon M, Bluhm H (1998) Angew Chem Int Ed 37:101
9. Gansäuer A, Bluhm H, Pierobon M (1998) J Am Chem Soc 120:12849
10. Gansäuer A, Bluhm H (2000) Chem Rev 100:2771
11. Gansäuer A, Pierobon M (2001) Rearrangements of cyclopropanes and epoxides. In: Renaud P, Sibi MP (eds) Radicals in organic synthesis, vol 2. Wiley, Weinheim, p 207
12. Gansäuer A, Rinker B (2002) Tetrahedron 58:7017
13. Gansäuer A, Narayan S (2002) Adv Synth Catal 344:465
14. Gansäuer A, Rinker B (2002) Titanocene-catalyzed epoxide opening. In: Marek I (ed) Titanium and zirconium in organic synthesis. Wiley, Weinheim, p 435
15. Gansäuer A, Lauterbach T, Narayan S (2003) Angew Chem Int Ed 42:5556
16. Larock RC (1999) Comprehensive organic transformations, 2nd edn. Wiley, New York
17. Smith MB (2002) Organic synthesis, 2nd edn. McGraw-Hill, New York
18. Rickborn B (1991) Acid-catalyzed rearrangements of epoxides. In: Trost BM, Fleming I, Pattenden G (eds) Comprehensive organic synthesis, vol 3. Pergamon, Oxford, p 733
19. Mitsunobu O (1991) Synthesis of alcohols and ethers. In: Trost BM, Fleming I, Winterfeldt E (eds) Comprehensive organic synthesis, vol 6. Pergamon, Oxford, p 1
20. RajanBabu TV, Nugent WA (1989) J Am Chem Soc 111:4525
21. RajanBabu TV, Nugent WA, Beattie MS (1990) J Am Chem Soc 112:6408
22. RajanBabu TV, Nugent WA (1994) J Am Chem Soc 116:986
23. Murai S, Murai T, Kato S (1991) Reduction of epoxides. In: Trost BM, Fleming I (eds) Comprehensive organic synthesis, vol 8. Pergamon, Oxford, p 871
24. Schobert R (1988) Angew Chem Int Ed 27:855
25. Yadav JS, Shekaram T, Gadgil VR (1990) Chem Commun 843
26. Rao AVR, Bhanu MN, Sherma GVM (1993) Tetrahedron Lett 34:707
27. Yadav S, Srinivas D, Shekaram T (1994) Tetrahedron Lett 35:3625
28. Chakraborty TK, Dutta S (1997) J Chem Soc Perkin Trans 1 1257
29. Weigand S, Brückner R (1997) Synlett 225
30. Chakraborty TK, Dutta S (1998) Tetrahedron Lett 39:101
31. Jørgensen KB, Suenaga T, Nakata T (1999) Tetrahedron Lett 40:8855
32. Hardouin C, Chevallier F, Rousseau B, Doris E (2001) J Org Chem 66:1046
33. Chakraborty TK, Das S, Raju TV (2001) J Org Chem 66:4091
34. Chakraborty TK, Tapadar S (2001) Tetrahedron Lett 42:1375
35. Chakraborty TK, Das S (2002) Tetrahedron Lett 43:2313
36. Chakraborty TK, Tapadar S (2003) Tetrahedron Lett 44:2541
37. Bermejo F, Sandoval C (2004) J Org Chem 69:5275
38. Fernández-Mateos A, Martín de la Nava E, Pascual Coca G, Ramos Silvo A, Rubio González R (1999) Org Lett 1:607

39. Fernández-Mateos A, Mateos Burón I, Rabanedo Clemente R, Ramos Silvo AI, Rubio González R (2004) Synlett 1011
40. Merlic CA, Xu D (1991) J Am Chem Soc 113:9855
41. Merlic CA, Xu D, Nguyen MC, Truong V (1993) Tetrahedron Lett 34:227
42. Dötz KH, Gomes da Silva E (2000) Tetrahedron 56:8291
43. Parrish JD, Little RD (2002) Org Lett 4:1439
44. Gansäuer A, Lauterbach T, Geich-Gimbel D (2004) Chem Eur J 10:4983
45. Fernández-Mateos A, Mateos Burón L, Martín de la Nava EM, Rabanedo Clemente R, Rubio González R, Sanz González F (2004) Synlett 2553
46. Friedrich J, Dolg M, Gansäuer A, Geich-Gimbel D, Lauterbach T (2005) J Am Chem Soc 127:7071
47. Barrero AF, Oltra JE, Cuerva JM, Rosales A (2002) J Org Chem 67:2566
48. Curran DP, Porter NA, Giese B (1996) Stereochemistry of radical reactions. Wiley, Weinheim
49. Renaud P, Sibi MP (eds) (2001) Radicals in organic synthesis. Wiley, Weinheim
50. Spiegel DA, Wiberg KB, Schacherer LN, Medeiros MR, Wood JL (2005) J Am Chem Soc 127:12513
51. Barrero AF, Rosales A, Cuerva JM, Gansäuer A, Oltra JE (2003) Tetrahedron Lett 44:1079
52. Oller-López JL, Campaña AG, Cuerva JM, Oltra JE (2005) Synthesis 2619
53. Justicia J, Rosales A, Buñuel E, Oller-López JL, Valdivia M, Haïdour A, Oltra JE, Barrero AF, Cárdenas DJ, Cuerva JM (2004) Chem Eur J 10:1778
54. Barrero AF, Cuerva JM, Herrador MM, Valdivia MV (2001) J Org Chem 66:4074
55. Leca D, Fensterbank L, Lacôte E, Malacria M (2004) Angew Chem Int Ed 43:4220
56. Leca D, Song K, Albert M, Grangeio Gonçalves M, Fensterbank L, Lacôte E, Malacria M (2005) Synthesis 1405
57. Gansäuer A, Bluhm H (1998) Chem Commun 2143
58. Gansäuer A, Lauterbach T, Bluhm H, Noltemeyer M (1999) Angew Chem Int Ed 38:2909
59. Gansäuer A, Bluhm H, Lauterbach T (2001) Adv Synth Catal 343:785
60. Gansäuer A, Bluhm H, Rinker B, Narayan S, Schick M, Lauterbach T, Pierobon M (2003) Chem Eur J 9:531
61. Gansäuer A, Barchuk A, Fielenbach D (2004) Synthesis 2567
62. Gansäuer A, Rinker B, Barchuk A, Nieger M (2004) Organometallics 23:1168
63. Gansäuer A, Pierobon M (2000) Synlett 1357
64. Gansäuer A, Pierobon M, Bluhm H (2001) Synthesis 2500
65. Gansäuer A, Pierobon M, Bluhm H (2002) Angew Chem Int Ed 41:3206
66. Gansäuer A, Rinker B, Pierobon M, Grimme S, Gerenkamp M, Mück-Lichtenfeld C (2003) Angew Chem Int Ed 42:3687
67. Gansäuer A, Rinker B, Ndene-Schiffer N, Pierobon M, Grimme S, Gerenkamp M, Mück-Lichtenfeld C (2004) Eur J Org Chem 2337
68. Barrero AF, Rosales A, Cuerva JM, Oltra JE (2003) Org Lett 5:1935
69. Justicia J, Oller-López JL, Campaña AG, Oltra JE, Cuerva JM, Buñuel E, Cárdenas DJ (2005) J Am Chem Soc 127:14911
70. Fuse S, Hanochi M, Doi T, Takahashi T (2004) Tetrahedron Lett 45:1961
71. Handa Y, Inanaga J (1987) Tetrahedron Lett 28:5717
72. Enemærke RJ, Larsen J, Hjøllund GH, Skrydstrup T, Daasbjerg K (2005) Organometallics 24:1252
73. Gansäuer A (1997) Chem Commun 457
74. Gansäuer A, Bauer D (1998) J Org Chem 63:2070

75. Hirao T, Hatano B, Asahara M, Muguruma Y, Ogawa A (1998) Tetrahedron Lett 39:5247
76. Barden MC, Schwartz J (1996) J Am Chem Soc 118:5484
77. Moisan L, Hardouin C, Rousseau B, Doris E (2002) Tetrahedron Lett 43:2013
78. Parrish JD, Shelton DR, Little RD (2003) Org Lett 5:3615
79. Jana S, Guin C, Roy SC (2004) Tetrahedron Lett 45:6575
80. Rosales A, Oller-López JL, Justicia J, Gansäuer A, Oltra JE, Cuerva JM (2004) Chem Commun 2628
81. Mandal SK, Jana S, Roy SC (2005) Tetrahedron Lett 46:6115
82. Barrero AF, Herrador MM, Quílez del Moral JF, Arteaga P, Arteaga JF, Piedra M, Sánchez E (2005) Org Lett 7:2301
83. Zhou L, Hirao T (2003) J Org Chem 68:1633
84. Jana S, Guin C, Roy SC (2005) Tetrahedron Lett 46:1155
85. Spencer RP, Schwartz J (2000) Tetrahedron 56:2103
86. Hansen T, Daasbjerg K, Skrydstrup T (2000) 41:8645
87. Samuel E, Vedel J (1989) Organometallics 8:237
88. Martre AM, Mousset G, Danciu V, Cosoveanu V (1998) Electrochim Acta 43:3217
89. Meunier-Prest R, Lamblin G, Mailfert A, Raveau S (2003) J Electroanal Chem 541:175
90. Hersant G, Ferjani MBS, Bennet SM (2004) Tetrahedron Lett 45:8123
91. Golakoti T, Ogino J, Heltzel CE, Husebo TL, Jensen CM, Larsen LK, Patterson GML, Moore RE, Mooberry SL, Corbett TH, Valeriote FA (1995) J Am Chem Soc 117:12030
92. Hardouin C, Doris E, Rousseau B, Mioskowski C (2002) Org Lett 4:1151
93. Hardouin C, Doris E, Rousseau B, Mioskowski C (2002) J Org Chem 67:6571
94. Bhaskar KV, Mander LN (1996) Tetrahedron Lett 37:719
95. Clive DLJ, Magnuson SR (1995) Tetrahedron Lett 36:15
96. Clive DLJ, Magnuson SR, Manning HW, Mayhew DL (1996) J Org Chem 61:2095
97. Maiti G, Roy SC (1996) J Chem Soc Perkin Trans 1 403
98. Mandal PK, Maiti G, Roy SC (1998) J Org Chem 63:2829
99. Ziegler FE, Sarpong MA (2003) Tetrahedron 59:9013
100. Mandal PK, Roy SC (1999) Tetrahedron 55:11395
101. Rana KK, Guin C, Roy SC (2000) Tetrahedron Lett 41:9337
102. Rana KK, Guin C, Roy SC (2001) Synlett 1249
103. Roy SC, Rana KK, Guin C (2002) J Org Chem 67:3242
104. Banerjee B, Roy SC (2005) Synthesis 2913
105. Nakai K, Kamoshita M, Doi T, Yamada H, Takahashi T (2001) Tetrahedron Lett 42:7855
106. Barrero AF, Cuerva JM, Alvarez-Manzaneda EJ, Oltra JE, Chahboun R (2002) Tetrahedron Lett 43:2793
107. Barrero AF, Alvarez-Manzaneda EJ, Alvarez-Manzaneda R (1989) Tetrahedron Lett 30:3351
108. Hoshino T, Shimizu K, Sato T (2004) Angew Chem Int Ed 43:6700
109. Trost BM, Shen HC, Surivet JP (2003) Angew Chem Int Ed 42:3943
110. Ruano G, Grande M, Anaya J (2002) J Org Chem 67:8243
111. Anaya J, Fernández-Mateos A, Grande M, Martiáñez J, Ruano G, Rubio-Gonzalez MR (2003) Tetrahedron 59:241
112. Ruano G, Martiáñez J, Grande M, Anaya J (2003) J Org Chem 68:2024
113. Haruo Y, Hasegawa T, Tanaka H, Takahashi T (2001) Synlett 1935
114. Rosales A, Estévez R, Cuerva JM, Oltra JE (2005) Angew Chem Int Ed 44:319
115. Barrero AF, Oltra JE, Álvarez M (1998) Tetrahedron Lett 39:1401

116. Connolly JD, Hill RA (1991) Dictionary of terpenoids, vol 1. Chapman and Hall, London
117. Fraga BM (2003) Nat Prod Rep 20:392
118. Justicia J, Oltra JE, Barrero AF, Guadaño A, González-Coloma A, Cuerva JM (2005) Eur J Org Chem 712
119. Justicia J, Oltra JE, Cuerva JM (2004) Tetrahedron Lett 45:4293
120. Justicia J, Oltra JE, Cuerva JM (2005) J Org Chem 70:8265
121. Simpson T (1998) Top Curr Chem 195:1
122. Justicia J, Oltra JE, Cuerva JM (2004) J Org Chem 69:5803
123. González AG, Martín JD, Rodríguez ML (1973) Tetrahedron Lett 14:3657
124. González AG, Martín JD, Rodríguez ML (1976) An Quim 72:1004
125. Gansäuer A, Justicia J, Rosales A, Rinker B (2005) Synlett 1954
126. Xing X, Demuth M (1999) Synlett 987
127. Xing X, Demuth M (2001) Eur J Org Chem 537
128. Trost BM (1991) Science 254:1471
129. Trost BM (1995) Angew Chem Int Ed 34:259
130. Abe I, Rohmer M, Prestwich GD (1993) Chem Rev 93:2189
131. Wendt KU, Schulz GE, Corey EJ, Liu DR (2000) Angew Chem Int Ed 39:2812
132. Hoshino T, Sato T (2002) Chem Commun 291
133. Goldsmith DJ (1962) J Am Chem Soc 84:3913
134. van Tamelen EE (1975) Acc Chem Res 8:152
135. Taylor SK (1992) Org Prep Proc Int 24:247
136. Dennison ST, Harrowven DC (1996) J Chem Ed 73:697
137. Huang AX, Xiong Z, Corey EJ (1999) J Am Chem Soc 121:9999
138. Marson CM (2000) Tetrahedron 56:8779
139. Zhang J, Corey EJ (2001) Org Lett 3:3215
140. Yuan M, Schreiber JV, Corey EJ (2002) J Am Chem Soc 124:11290
141. van Tamelen EE, Willet J, Schwartz M, Nadeau R (1966) J Am Chem Soc 88:5937
142. Sharpless KB, van Tamelen EE (1969) J Am Chem Soc 91:1848
143. Eliel EL, Wilen SH, Doyle MP (2001) Basic organic stereochemistry. Wiley-Interscience, New York
144. Naya Y, Kotake M (1968) Tetrahedron Lett 9:1645
145. Warmers U, König WA (1999) Phytochemistry 52:99
146. Cool LG (2001) Phytochemistry 58:969
147. Salomon CE, Magarvey NA, Sherman DV (2004) Nat Prod Rep 21:105
148. Blunt JW, Copp BR, Munro MHG, Northcote PT, Prinsep MR (2004) Nat Prod Rep 21:1
149. Rudi A, Kashman Y (1992) J Nat Prod 55:1408
150. Kuniyoshi M, Marma MS, Higa T, Bernardinelli G, Jefford CW (2000) Chem Commun 1155
151. Kuniyoshi M, Marma MS, Higa T, Bernardinelli G, Jefford CW (2001) J Nat Prod 64:696
152. Su JY, Zhong YL, Zeng LM, Wu HM, Ma K (1995) Phytochemistry 40:195
153. Urones JG, Marcos IS, Basabe P, Alonso CA, Diez D, Garrido NM, Oliva IM, Rodilla JS, Slawin AMZ, Williams DJ (1990) Tetrahedron Lett 31:4501
154. Urones JG, Marcos IS, Basabe P, Alonso C, Oliva IM, Garrido NM, Martín DD, Lithgow AM (1993) Tetrahedron 49:4051

Top Curr Chem (2006) 264: 93–134
DOI 10.1007/128_023

Published online: 5 January 2006

Radical Chemistry on Solid Support

Andrea M. McGhee · David J. Procter (✉)

The School of Chemistry, University of Manchester,
Oxford Road, Manchester M13 9PL, UK
david.j.procter@manchester.ac.uk

Abstract This review covers recent advances in the field of radical chemistry on solid phase. Intermolecular processes using both immobilized radicals with solution-phase acceptors and immobilized acceptors with radicals in solution are discussed, as are radical cyclization reactions on polymer supports. Progress in the development of solid-phase asymmetric radical processes and the design of linkers cleaved by radical processes are also discussed.

Keywords Radicals · Solid phase · Linkers · Supported reagents

Abbreviations

Ac	Acetyl
ACCN	1,1′-Azobis(cyclohexanecarbonitrile)
AIBN	2,2′-Azobis(2-methylpropionitrile)
Bn	Benzyl
Boc	*t*-Butoxycarbonyl
Bz	Benzoyl
CAN	Ceric ammonium nitrate
Cbz	Benzyloxycarbonyl

Cy	Cyclohexyl
DAST	Diethylaminosulfur trifluoride
DCC	1,3-Dicyclohexyl carbodiimide
DCE	Dichloroethane
DEAD	Diethyl azodicarboxylate
DIAD	Diisopropyl azodicarboxylate
DIC	1,3-Diisopropyl carbodiimide
DMAP	4-Dimethylaminopyridine
DMF	*N*,*N*-Dimethylformamide
DMPU	1,3-Dimethyl-3,4,5,6-tetrahydro-2(1*H*)-pyrimidone
DMSO	Dimethylsulfoxide
Fmoc	9-Fluorenylmethoxycarbonyl
HASC	α Heteroatom-substituted carbonyl
HBTU	*N*,*N*,*N*′,*N*′-Tetramethyl-*O*-(1*H*-benzotriazol-1-yl)uronium hexafluorophosphate
HMPA	Hexamethylphosphoramide
HOBt	1-Hydroxybenzotriazole
*m*CPBA	3-Chloroperbenzoic acid
MOM	Methoxymethyl
Ms	Methanesulfonyl
MW	microwave
NEM	*N*-Ethylmaleimide
NMM	*N*-Methylmorpholine
NMP	1-Methyl-2-pyrrolidone
PEG	Poly(ethylene glycol)
Phth	Phthalate
PMB	4-Methoxybenzyl
PTSA	*p*-Toluene sulfonic acid
ROMP	Ring opening metathesis polymerization
TBDMS	*t*-Butyldimethylsilyl
TBDPS	*t*-Butyldiphenylsilyl
TBS	*t*-Butyldimethylsilyl
TFA	Trifluoroacetic acid
THF	Tetrahydrofuran
THP	Tetrahydropyranyl
TMS	Trimethylsilyl
Ts	4-Toluenesulfonyl

1 Introduction

Radical reactions provide a powerful means for manipulating the structure of organic compounds. A vast range of radical transformations is available to the synthetic organic chemist, ranging from functional group interconversions to asymmetric reactions and complex, sequential cyclization reactions where molecular complexity can be increased spectacularly in a single step. Many of these processes are complementary to more traditional polar processes and are characterized by attractive features such as the mild, neutral reaction conditions involved and the compatibility of the processes to functional groups present in the substrates.

With the advent of solid-phase synthesis, and its application in the high-throughput synthesis of "small-molecule" libraries, came the challenge of adapting the plethora of useful solution-phase radical processes to the manipulation of substrates attached to polymer supports. Many factors influence the efficiency of radical processes on solid phase. For instance, the degree of polymer swelling and the resultant kinetics of the process, the loading of the resin and the nature of the linker employed can all have a dramatic effect on the outcome of radical reactions. In recent years, significant progress has been made, and both inter- and intramolecular radical reactions are nearing routine status as tools for solid-phase synthesis.

This chapter builds on previous reviews of this area by Ganesan and Sibi [1] and Ganesan [2] and covers advances made since 2000.

1.1 Rates of Radical Reactions on Solid Phase

Understanding how the rates of radical reactions change as processes are adapted from solution to solid phase is crucial to the development of efficient radical transformations involving resin-bound substrates. Curran [3] has carried out the first rate studies on inter- and intramolecular radical reactions on solid phase. In particular, he has determined the rates of inter- and intramolecular hydrogen atom capture for two-polymer-supported radicals. The rate constant for the 6-exo-trig cyclization of alkyl bromide **1** was determined to be approximately $1.9 \times 10^5\ s^{-1}$ by treating **1** with varying concentrations of tributyltin hydride and determining the ratio of cyclized and reduced products, **6** and **8** respectively (Scheme 1). The rate constant for competing hydrogen abstraction in solution was shown to be approximately $6.4 \times 10^6\ M^{-1}s^{-1}$. Curran then investigated the same hydrogen atom capture process on solid phase [3]. Alkyl bromide **1** was immobilized using THP resin **2** to give cyclization substrate **3** (Scheme 1). Exposure of **3** to several different concentrations of tributyltin hydride and cleavage gave mixtures of **6** and **8**. Assuming that the rate of cyclization of radical **4** is the same as that measured in solution and that the concentration of tributyltin hydride in the resin is the same as that outside, the rate constant for hydrogen atom transfer to the resin-bound radical was determined to be approximately $2.6 \times 10^6\ M^{-1}s^{-1}$ at 80 °C (cf. $6.4 \times 10^6\ M^{-1}s^{-1}$ in solution). Curran therefore concluded that the relative rate of hydrogen transfer on solid phase is similar to that in solution, therefore suggesting that solution-phase rate constants can be used to predict the outcome of radical reactions on solid phase.

Curran [3] has also studied the rate of intramolecular hydrogen atom abstraction from the polymer backbone using immobilized iodide **9** (Scheme 2). The rate constant for 1,5-hydrogen atom abstraction in solution to give **11**

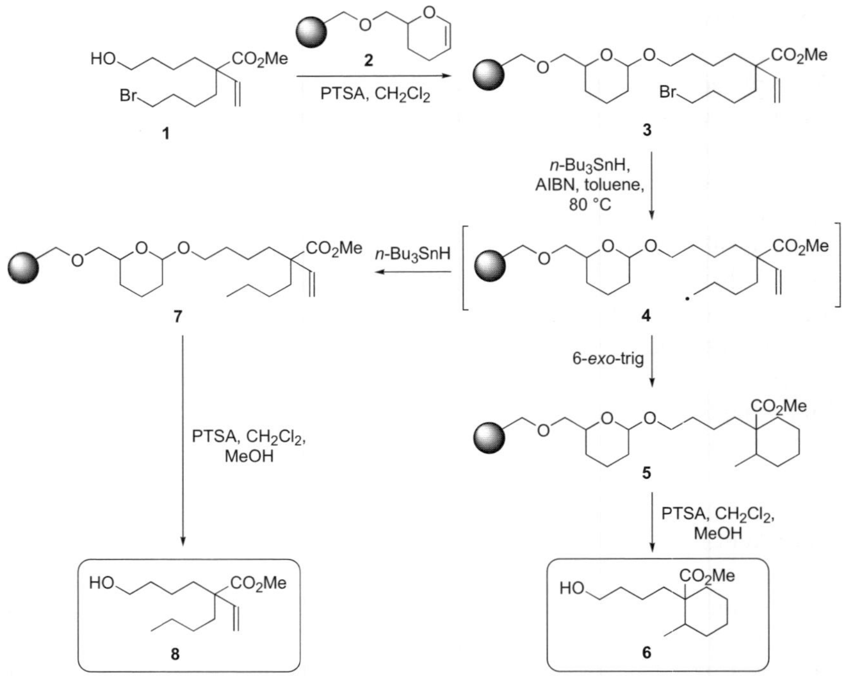

Scheme 1 Determination of rates of inter- and intramolecular radical reactions on solid phase by Curran [3]

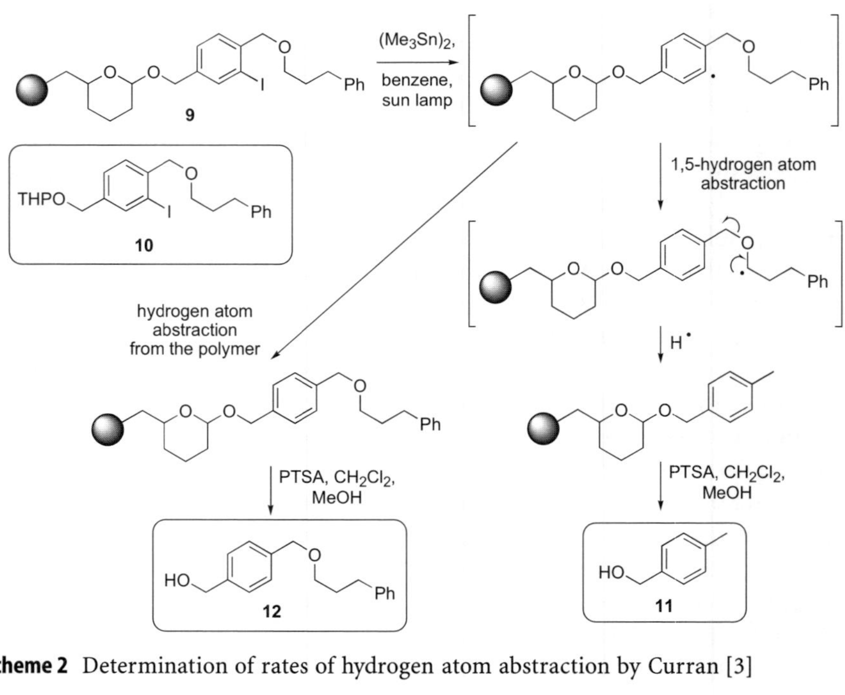

Scheme 2 Determination of rates of hydrogen atom abstraction by Curran [3]

was first measured using analogous iodide **10**, and a value of $4 \times 10^6\ s^{-1}$ was obtained.

Solid-phase studies were then carried out using iodide **9** and hexamethylditin at various concentrations and a ratio of **11** and **12** determined after cleavage from the solid support (Scheme 2) [3]. A rate constant of $3.3 \times 10^6\ s^{-1}$ was thereby obtained for hydrogen atom abstraction from the solid support by the aryl radical. Preparatively, this means approximately half of the resin-bound aryl radicals abstract a hydrogen from the polymer support rather than by 1,5-atom abstraction. The polymer can therefore be thought of as a solvent surrounding the radical. The implications from Curran's study are that only fast radical cyclizations involving aryl radicals will be possible on solid phase, while slower cyclizations and intermolecular reactions using supported aryl radicals will be difficult.

2
Supported Reagents for Radical Chemistry

The use of polymer-supported tin hydrides is now a well-established method for minimizing tin contamination in products resulting from radical reactions. A new version of the Barton–McCombie deoxygenation reaction has been developed by Dumartin [4] which combines the use of a polymer-supported tin hydride with a stoichiometric silane reductant, giving an attractive catalytic process.

In this study, trimethoxylsilane was found to be an effective reagent for the conversion of the polymer-bound phenoxytin intermediate **14** back to tin hydride **13**. Using this reagent system, secondary thionocarbonates were reduced in good yield using 10 mol % of the tin hydride polymer **13** (Scheme 3).

Giacomelli [5] has developed a polymer-supported reagent for the photochemical generation of radicals. *N*-Hydroxy thiazole 2(3)-thione **15**, prepared in four steps and 66% overall yield from 4-hydroxy acetophenone, was attached to *N*-Fmoc-Gly-Wang resin to give the polymer-supported reagent **16** (Scheme 4). Acylation of resin **16** with protected glutamic acid **17** gave **18** which, upon irradiation in the presence of $BrCCl_3$, provided 4-bromo-2-*tert*-butoxycarbonylaminobutanoic acid **19** in 75% yield accompanied by release of CO_2 (Scheme 4).

Supported reagent **16** was also used for the generation and cyclization of alkoxyl radicals (Scheme 5) [5]. Treatment of **16** with a solution of KOH in ethylene glycol gave potassium salt **20** that was alkylated using tosylates **21** and **22**. Irradiation of hydroxamates **23** and **24** in the presence of *t*-BuSH as a radical trap gave tetrahydrofurans **25** and **26** respectively, via *5-exo-trig* cyclization. No purification was required, although the yields of the tetrahydrofuran products were not given.

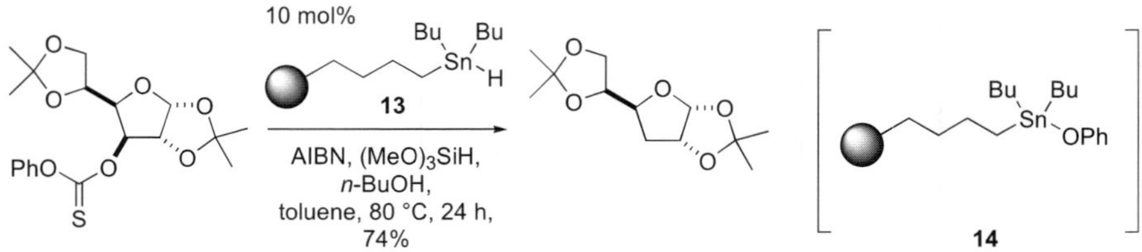

Scheme 3 Barton–McCombie deoxygenation by Dumartin [4]

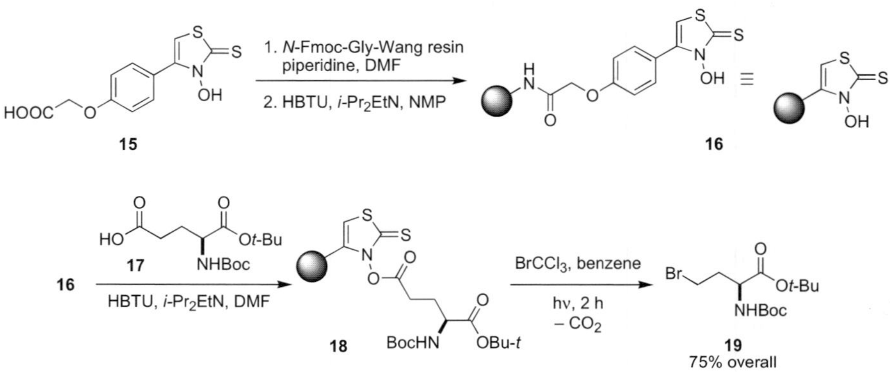

Scheme 4 A polymer-supported reagent for radical generation by Giacomelli [5]

Bowman [6] has reported the synthesis of new, solid-phase triorganogermanium hydrides **28** and **29** by addition of triorganogermanium hydride unit **27** to Merrifield and Quadragel resins (Scheme 6).

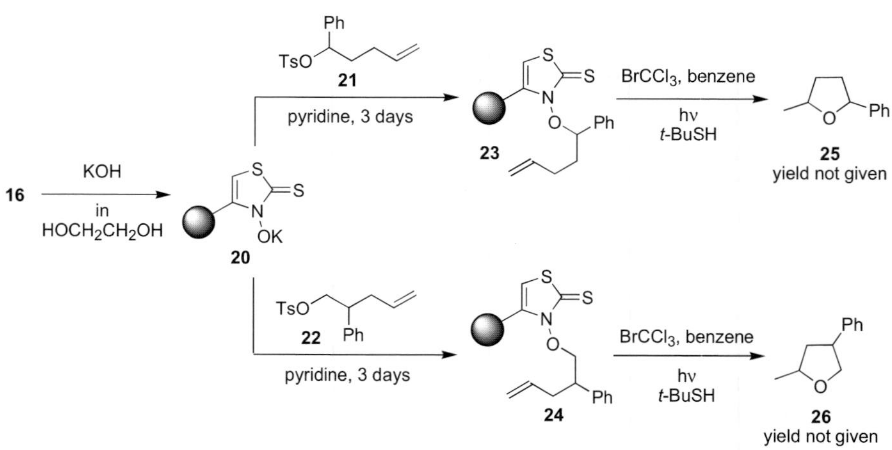

Scheme 5 Generation and cyclization of alkoxyl radicals by Giacomelli [5]

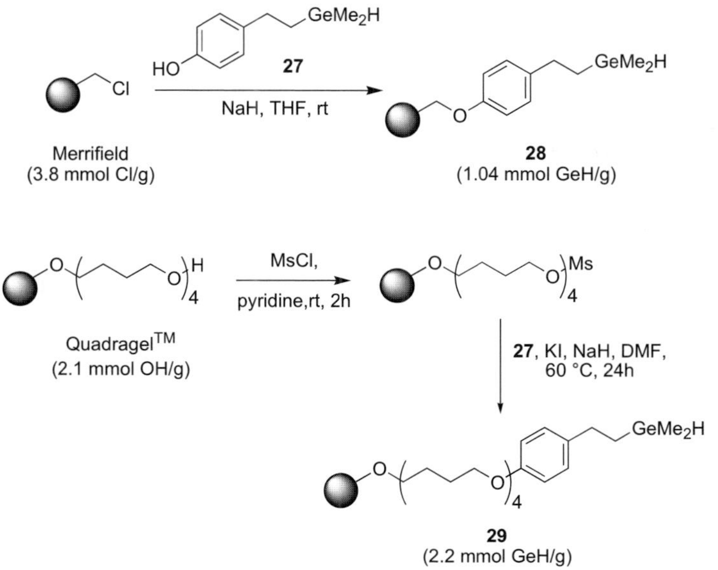

Scheme 6 Generation of resin-bound triorganogermanium hydrides by Bowman [6]

Triorganogermanium hydride resins **28** and **29** were then evaluated in a range of radical transformations [6]. For example, treatment of thioimidazolide **30** with **28** gave the deoxygenated product **31** in moderate yield (Scheme 7), while quadragel-derived hydride resin **29** was used to mediate the radical cyclizations of amide **32** and ether **34** to give lactam **33** and tetrahydrobenzofuran **35** respectively in good yield (Scheme 7).

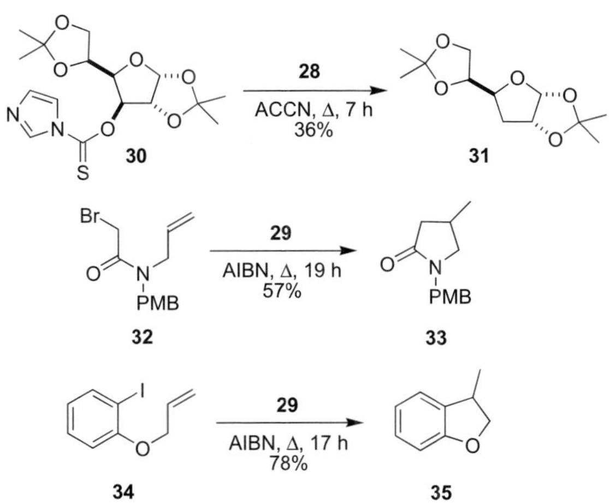

Scheme 7 Utilization of resin-bound triorganogermanium hydrides by Bowman [6]

3
Intermolecular Radical Reactions on Solid Phase

3.1
Intermolecular Radical Reactions Employing Immobilized Radicals

Taddei [7] has studied the use of the Barton decarboxylation reaction in solid-phase chemistry. Feasibility studies began with the preparation of simple immobilized Barton ester **37** from Wang resin-derived acid **36** (Scheme 8). Irradiation of **37** followed by acid cleavage gave **38** in good overall yield. Attempts to prepare bromide **39** by interception of the intermediate radical

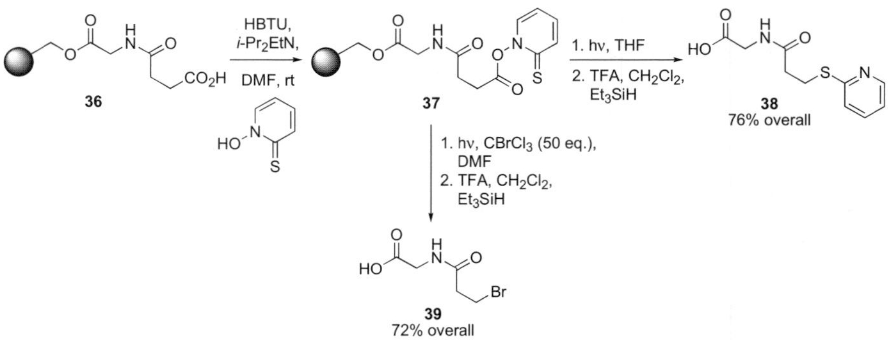

Scheme 8 Barton decarboxylation by Taddei [7]

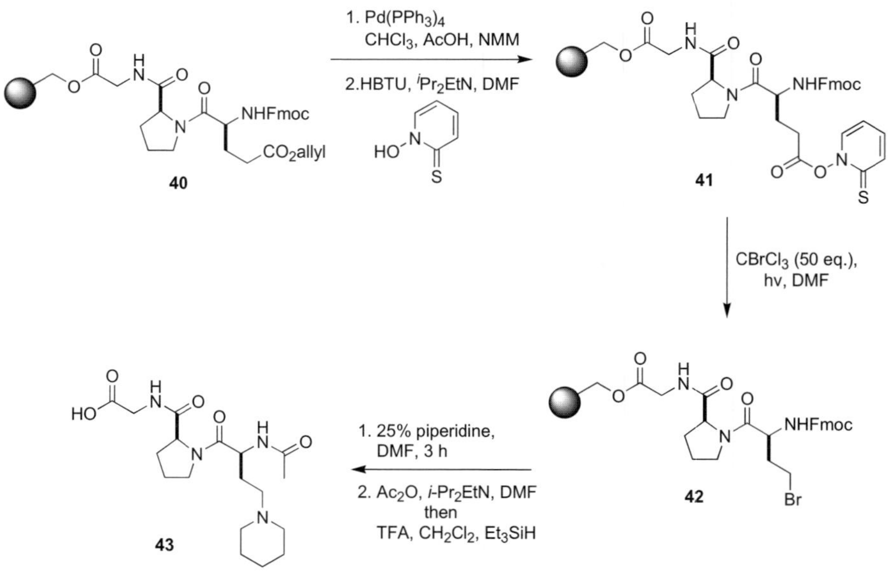

Scheme 9 Barton decarboxylation and radical bromination in peptide formation by Taddei [7]

with $CBrCl_3$ were also successful provided a large excess of the trap was used (Scheme 8). Attempts to employ electron-deficient alkenes as radical traps were less successful.

Taddei [7] employed the radical, bromination step in a solid-phase approach to peptides containing modified amino acids. Immobilized tripeptide **40** was prepared on Wang resin using standard methods. Coupling with 1-hydroxy-2-pyridinethione in the presence of HBTU gave Barton ester **41** (Scheme 9). Irradiation of **41** in DMF with $CBrCl_3$ gave bromide **42** which, when reacted with piperidine, quenched with Ac_2O, and cleaved from resin under acid conditions, gave **43** in 75% yield. A more versatile approach to the synthesis of modified amino acids, wherein photochemical radical decarboxylation was carried out after deprotection of nitrogen and coupling with an *N*-protected amino acid, was also investigated [7].

Huang [8] has utilized selenosulfonate resin **44** in a solid-phase, regio- and stereocontrolled synthesis of vinyl sulfones (Scheme 10). The approach involves intermolecular, radical additions of polymer-bound selenium radicals to alkenes. Treatment of styrene with resin **44**, in the presence of boron trifluoride, gave Markovnikov addition product **45**. Under radical conditions, anti-Markovnikov additions occurred to give immobilized sulfone **46**. Oxidation to the corresponding selenoxides and eliminative cleavage gave vinyl sulfones **47** and **48** in high yields.

Selenosulfonate resin **44** has also been used in radical additions to alkynes in an approach to alkynyl sulfones (Scheme 11) [9]. Radical selenosulfonation

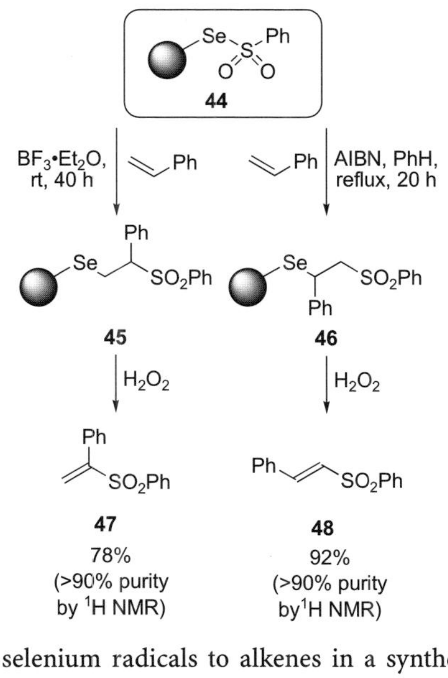

Scheme 10 Addition of selenium radicals to alkenes in a synthesis of vinyl sulfones by Huang [8]

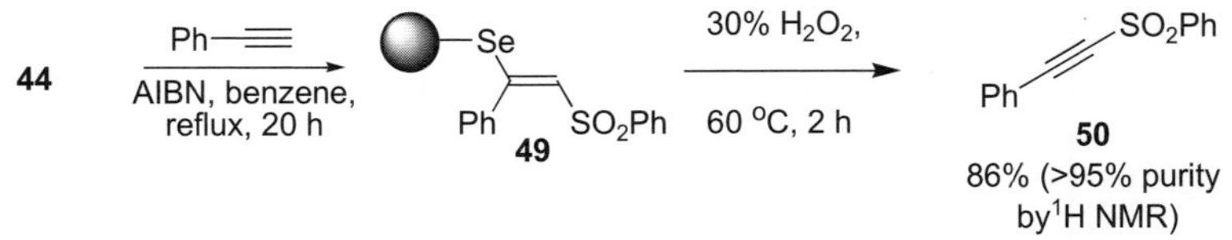

Scheme 11 Addition of selenium radicals to alkynes by Huang [9]

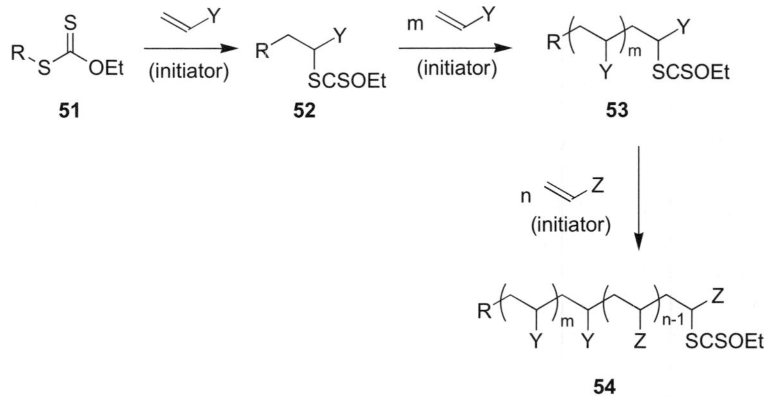

Scheme 12 Radical polymerization by xanthate transfer by Zard [10]

of phenylacetylene proceeded well to provide immobilized vinyl selenides such as **49**. Alkynyl sulfone **50** was obtained in excellent yield and purity upon oxidation and eliminative cleavage.

Zard [10] has developed a general method for the attachment of complex structures to a polystyrene oligomer using xanthate group transfer technology. The general sequence involves heating a mixture of xanthate **51** and an olefin in the presence of a small amount of an initiator to form adduct **52**, by addition of the elements of the xanthate across the alkene double bond (Scheme 12). Since the adduct is itself a xanthate, further addition to the alkene is possible, resulting in polymerization of the alkene. Polymer **53** is also a xanthate and can therefore act as the starting point for another polymerization process providing block polymer **54**. The choice of starting materials and reaction conditions allows the radical process to be used to prepare either monoadduct **52** or macromolecular structures such as **53** or **54**.

Polymer-bound protected arabinose **58** was prepared using this approach (Scheme 13) [10]. Protected arabinose **55** was converted to the chloropheny-lacetate ester **56** before displacement of the halide with potassium *O*-ethyl xanthate gave xanthate **57**. Heating this compound with 15 equivalents of styrene in toluene in the presence of 12 mol % lauroyl peroxide gave **58**, obtained as a white powder by precipitation with methanol. This methodology was extended to various substrates including more complex carbohydrates,

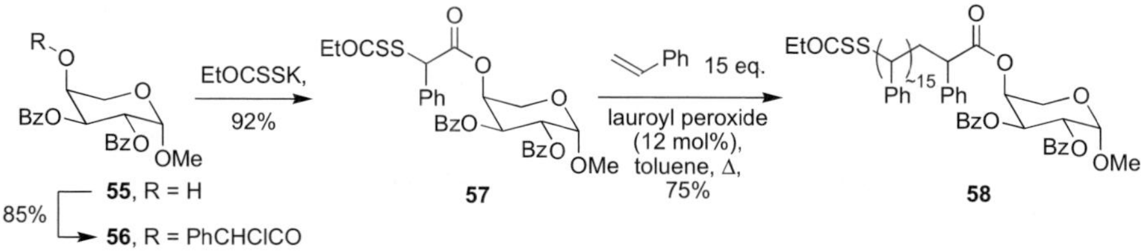

Scheme 13 Synthesis of a polymer-bound arabinose derivative by Zard [10]

cyclitols and nucleosides. In all cases the link to the polymer could be readily cleaved by hydrolysis.

A further example of the utility of this approach involved the polymerization and manipulation of cholanic-acid-derived xanthate **59** (Scheme 14) [10]. Treatment of xanthate **59** with 15 equivalents of styrene provided polymer-bound steroid **60** in 85% yield. In this case a spacer unit was incorporated between the polymer and the substrate. The substrate was then modified at both the xanthate and carboxylic acid termini to provide *N*-cyclopropylamide **61**. Cleavage from the polymer support was achieved using KOH in methanol and THF, and provided amide **62** in 62% yield.

In a related study, Lusinchi [11] reported the synthesis of a new soluble polymer containing a *p*-alkoxybenzyl alcohol linker, analogous to Wang resin. The polymer was utilized in intermolecular radical addition reactions. The polymer was synthesized by subjecting xanthate **65**, formed in three steps from alcohol **63**, to styrene polymerization in the presence of lauroyl peroxide

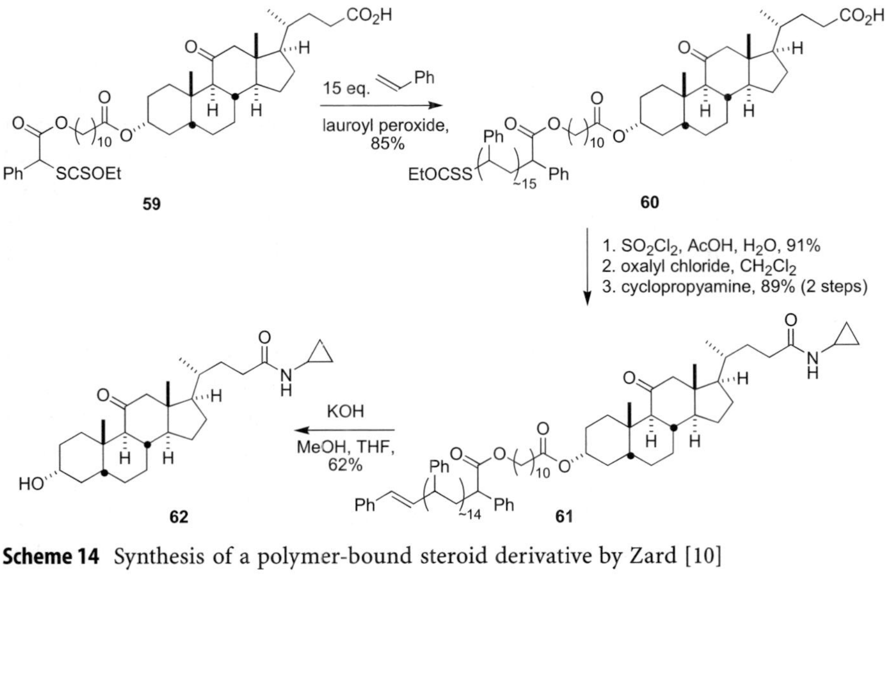

Scheme 14 Synthesis of a polymer-bound steroid derivative by Zard [10]

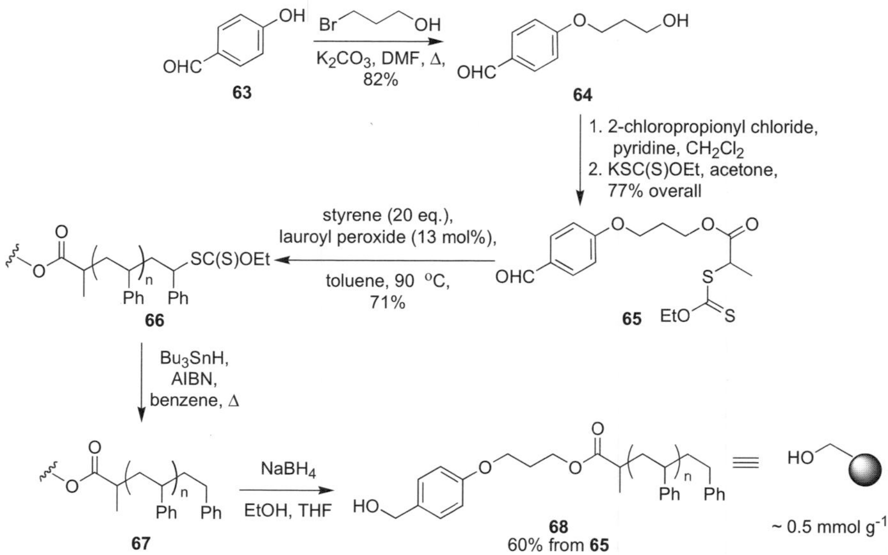

Scheme 15 Synthesis of a soluble "Wang polymer" by xanthate transfer by Lusinchi [11]

(Scheme 15). Removal of the terminal xanthate moiety from polymer **66** was achieved using tributyltin hydride, after which the tin residues could be easily removed by precipitation of the polymer in methanol. The aldehyde group of the resulting polymer **67** was then reduced with sodium borohydride to give the soluble Wang resin analogue **68**.

Lusinchi [11] illustrated the utility of the soluble polymer by examining both the reaction of immobilized xanthates with acceptors in solution and the reaction of immobilized alkenes with xanthates in solution. Firstly, immobilized xanthate **69** was prepared by chloroacetylation of **68**, followed by displacement of chloride with *O*-ethyl xanthate (Scheme 16). Treatment of **69** with excess protected allylamine **70** in the presence of 40 mol % lauroyl peroxide gave, after acid cleavage and column chromatography, adduct **71** in good overall yield. Although ^{1}H NMR indicated that the immobilized xanthate **69** was entirely consumed in the reaction, by-products were obtained due to the product xanthate formed on radical addition reacting further.

In the second case, the soluble polymer **68** was treated with 10-undecenoyl chloride **72** to form the resin-bound alkene **73** (Scheme 16) [11]. Radical addition of xanthate **74**, mediated by lauroyl peroxide (50 mol %), gave the product xanthate **75** and tetralone **76** as a 9 : 1 mixture, in an overall yield of 70%. By-product **76** is formed by intramolecular addition of the intermediate alkyl radical to the aromatic ring competing with intermolecular chain transfer. It was not possible to achieve complete consumption of **73** in this case possibly due to the decrease in the rate of the intermolecular radical addition step as the alkene was consumed. These experiments show, however, that xan-

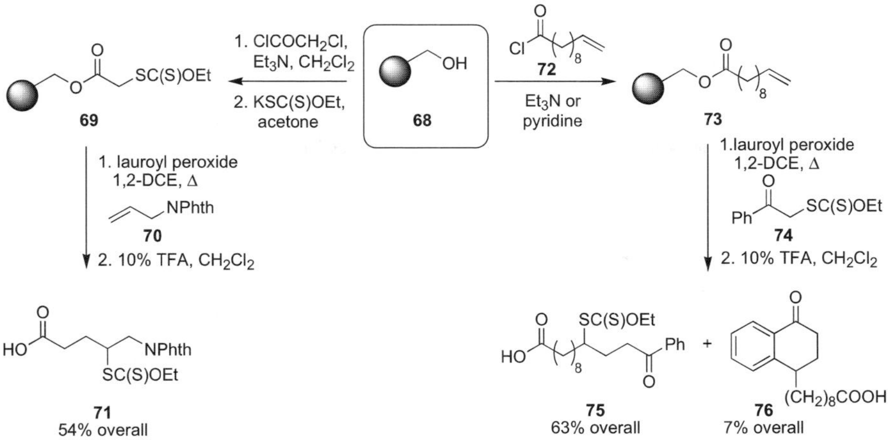

Scheme 16 Solid-phase reactions of xanthates with olefins by Lusinchi [11]

thate radical additions to immobilized acceptors are more efficient than the opposite mode of addition.

The use of soluble polymers in intermolecular radical additions with xanthates gave far better results than the analogous reactions using Wang resin [11]. In addition, the radical reactions could be monitored using ^{1}H NMR.

Enholm [12] has carried out free radical allyl transfer reactions using a non-cross-linked polystyrene, soluble polymer **77**. Model reactions of bromoester **78** with allyltin reagents gave products **79** and **80** in excellent yield (Scheme 17).

Enholm [12] has also prepared an enantiomerically pure soluble polymer support **82** by coupling xylose-derived chiral auxiliary **81** with **77** (Scheme 18). The chiral support was then treated with bromopropionic acid **83** to give substrate **84**. Free radical allyl transfer from allyltributyltin under thermal conditions provided **85** in 93% yield, and basic cleavage from the resin gave (*R*)-(-)-2-methylpent-4-enoic acid **86** in 80% yield and 97% ee, with a 92% yield of recovered **82**. Previous studies of the same process in solution had found the addition of Lewis acids to be crucial for high selectivities to be obtained. Interestingly, the addition of Lewis acids to the reaction on polymer support led to cleavage of the carbohydrate from the polymer backbone. En-

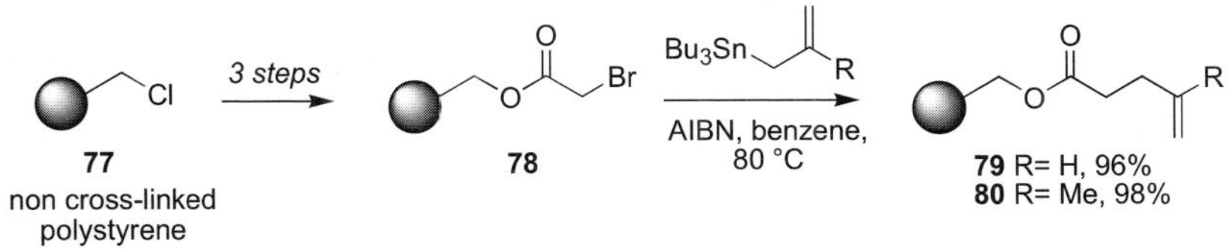

Scheme 17 Allyl transfer on a soluble polymer support by Enholm [12]

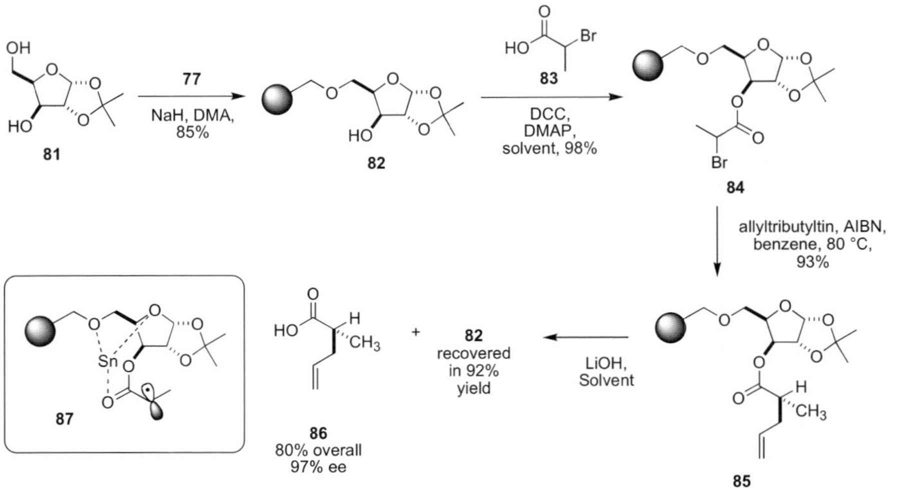

Scheme 18 Asymmetric allyl transfer by Enholm [12]

holm attributed the high selectivities observed in the absence of a Lewis acid additive to n-Bu_3SnBr, formed during the allylation process, acting as a Lewis acid in chelated radical intermediate **87** [12].

Enholm [13] has also described the synthesis of soluble "designer supports" by the ring-opening metathesis polymerization (ROMP) of norbornyl derivatives. Reduction of norbornene-1-carboxaldehyde **88** to the corresponding alcohol **89**, followed by treatment with either 2-bromopropionic acid or 2-bromo-2-phenylacetic acid in the presence of DCC, provided the esters **90** or **91** respectively (Scheme 19). Polymerization of **90** and **91** was carried out with Grubbs' catalyst and halted after 25 s by capping with excess ethyl vinyl ether to give polymers **92** and **93** respectively.

These polymers underwent radical allylation [13] using allyltributyltin to give adducts **94** and **95**, and reduction with tributyltin hydride to give the products **96** and **97** (Scheme 20). All products were obtained in high yield as white crystalline materials which could be easily separated from tin byproducts. The products were released from the polymer support by hydrolysis with lithium hydroxide.

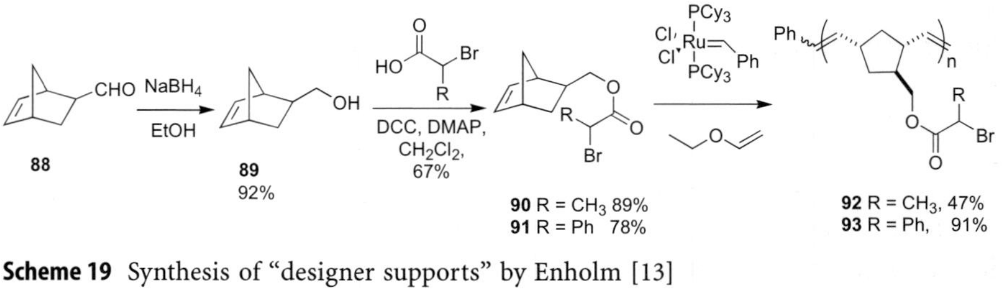

Scheme 19 Synthesis of "designer supports" by Enholm [13]

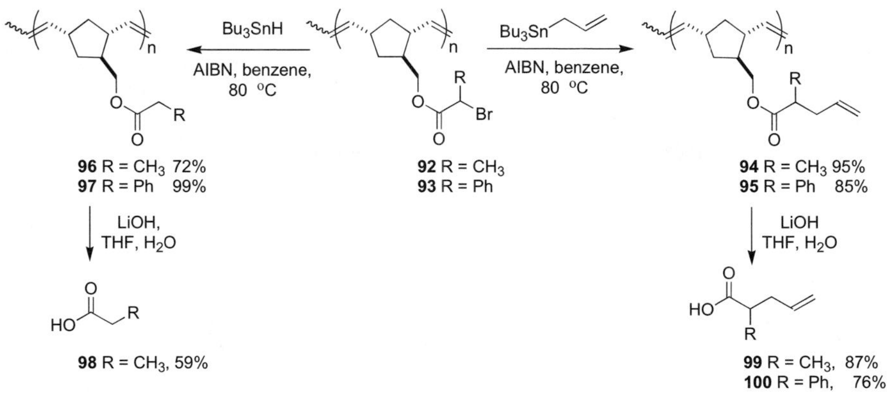

Scheme 20 Allylation and reduction of substrates bound to "designer supports" by Enholm [13]

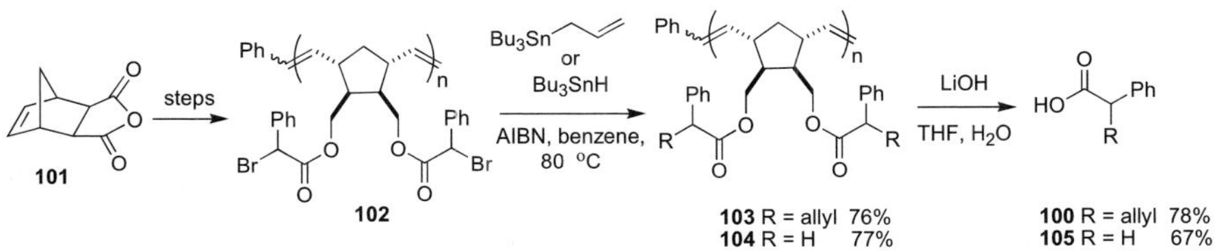

Scheme 21 Allylation and reduction of substrates bound to "designer supports" by Enholm [13]

Similar polymers were formed from norbornene-2,3-dicarboxylic anhydride **101** [13], which allowed for two reactive sites to be incorporated into each monomer unit, effectively providing a loading capacity of 200% (Scheme 21). Treatment of polymer **102** with allyltributyltin or tributyltin hydride resulted in reaction of both bromides in every monomer unit giving products **103** or **104** in 76% and 77% yield respectively. Hydrolysis with LiOH released the expected products **100** and **105** from the polymer support. In principle, it should be possible to recover and reuse the polymers prepared in Enholm's studies.

3.2 Intermolecular Radical Reactions Employing Immobilized Acceptors

Kim and Kim [14] have reported the addition of alkyl radicals derived from alkyl iodides to phenylsulfonyl oxime ether **107** attached to Wang resin (Scheme 22). Treatment of **107** with a range of alkyl iodides and hexamethylditin with irradiation at 300 nm gave the expected adducts such as **109**, in moderate to good yield after cleavage from the solid support.

The authors also investigated the feasibility of a radical cyclization-capture sequence [14] using immobilized phenylsulfonyl oxime ether **107** and iodide **110** (Scheme 23). Upon treatment with hexamethylditin, **110** undergoes a sequence of two 5-*exo* cyclizations followed by capture of the resultant radical by resin **107**. Cleavage from the support gave **111** in moderate yield.

Naito [15, 16] has studied the tin-free, intermolecular carbon–carbon bond forming radical reactions of oxime ethers on solid phase. Oxime ethers **112** anchored to either Wang resin or TentaGel resin underwent smooth radical additions upon treatment with triethylborane and isopropyliodide to give the desired alkylated product **114**, after cleavage from support, provided a large excess of alkyl iodide was employed (Scheme 24). The ethyl addition product **115**, formed by competitive addition of the ethyl radical from triethylborane, was observed regardless of the nature of the support [15, 16]. The radical reaction of the glyoxime ether immobilized on TentaGel resin is the first ex-

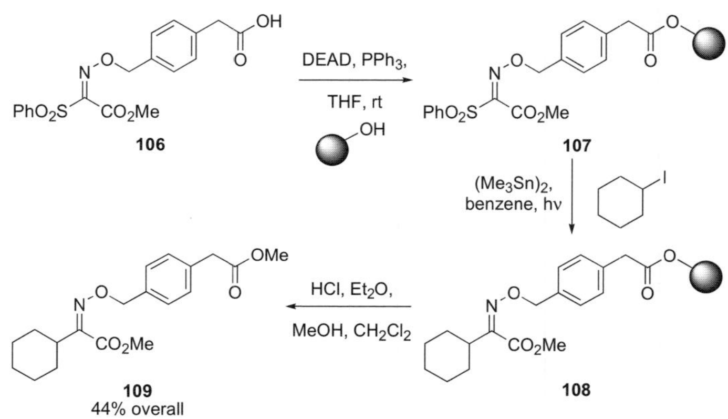

Scheme 22 Alkyl radical additions to phenylsulfonyl oxime ethers by Kim and Kim [14]

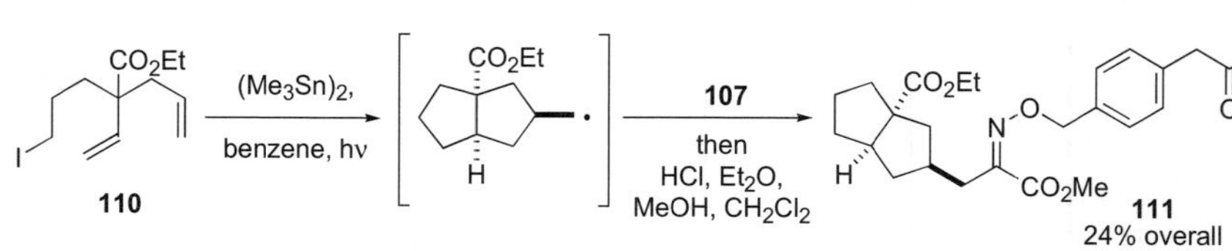

Scheme 23 Cyclization-capture sequence by Kim and Kim [14]

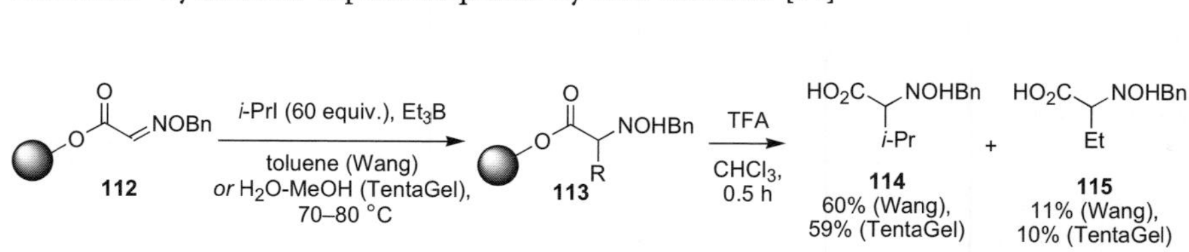

Scheme 24 Alkyl radical addition to oxime ethers by Naito [15, 16]

ample of a solid-phase radical reaction in aqueous media. Radical additions to resin-bound glyoxylic oxime ethers provide a straightforward approach to unnatural α-amino acids.

Plourde [17] has described the radical addition of alkyl, substituted alkyl and benzyl thiols to polymer-supported cyclitol allyl ethers. Treatment of immobilized allyl ether **117** with benzyl thiol in the presence of AIBN provided, after base-induced cleavage from the support and purification, sulfide **119** in high yield and purity (Scheme 25). Similar reactions with hydroxyl- and carboxyl-substituted alkyl thiols also resulted in good yields of products as single regioisomers [17].

To further investigate the scope and utility of the reaction, Plourde [17] carried out thiol addition to supported aminocyclitol **122**, formed by reaction of aminocyclitol **121** with Wang resin carbonate **120** (Scheme 26). Treatment of **122** with benzyl thiol and AIBN followed by cleavage from support gave the product sulfide **123** in good yield and purity.

Kumar [18] has reported the intermolecular radical addition of haloalkanes to polymer-bound alkenes and applied the technique in a solid-phase

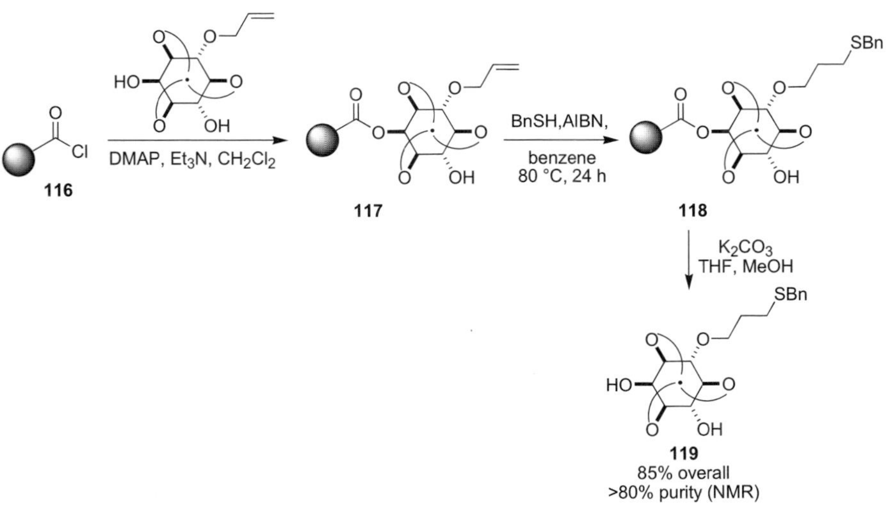

Scheme 25 Thiol addition to cyclitol allyl ethers by Plourde [17]

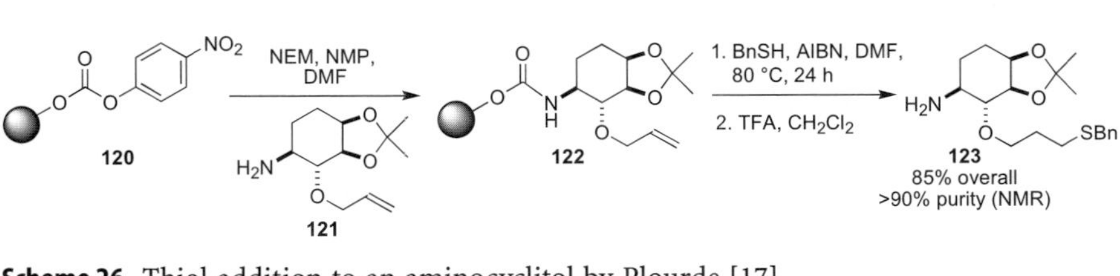

Scheme 26 Thiol addition to an aminocyclitol by Plourde [17]

synthesis of pyrethroid analogues. Unsaturated acids were immobilized using either Merrifield or Wang resin. Treatment of the immobilized acceptor **124** with various haloalkanes and catalytic quantities of benzoyl peroxide at elevated temperatures gave radical 1,2-addition products **125** (Scheme 27). Butyronitrile was employed as a co-solvent except in the addition of CCl_4, where the reaction was carried out neat. Hydrolytic cleavage of the products from the support showed the additions had proceeded in good yield to give products in high purity [18].

The addition products **125** cyclized upon treatment with base to form the corresponding cyclopropane carboxylates which underwent spontaneous dehydrohalogenation at room temperature to form dihaloalkenes **126** as a mixture of *cis/trans* diastereoisomers. Hydrolytic cleavage from the support then gave the product acids **127** [18].

Kumar [18] also carried out analogous additions to immobilized acrylate **128**. 1,2-Radical addition products **130** (Scheme 28) were again obtained in excellent yield and in high purity after cleavage from the support.

Caddick [19] has reported the use of a novel polymer-supported tetrafluorophenol-linked acrylate as an activated acceptor for intermolecular radical reactions. Treatment of immobilized acrylate **132** with a variety of alkyl iodides in the presence of tributyltin hydride and AIBN gave the corresponding esters **133** (Scheme 29). Nucleophilic cleavage using amines gave amides **134** in good overall yield whilst regenerating phenol resin **131**.

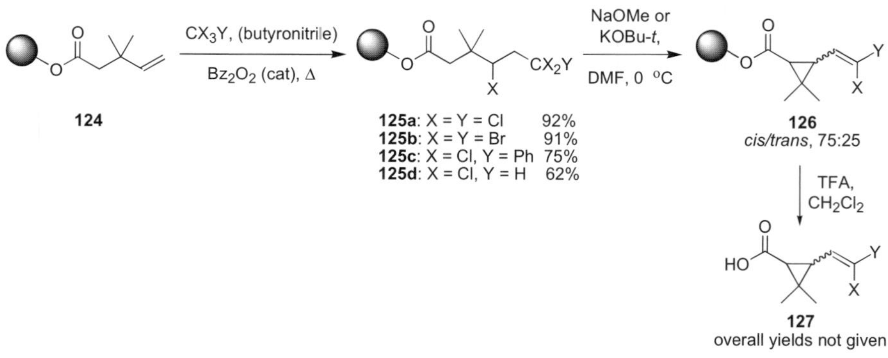

Scheme 27 Radical addition of haloalkanes to alkenes by Kumar [18]

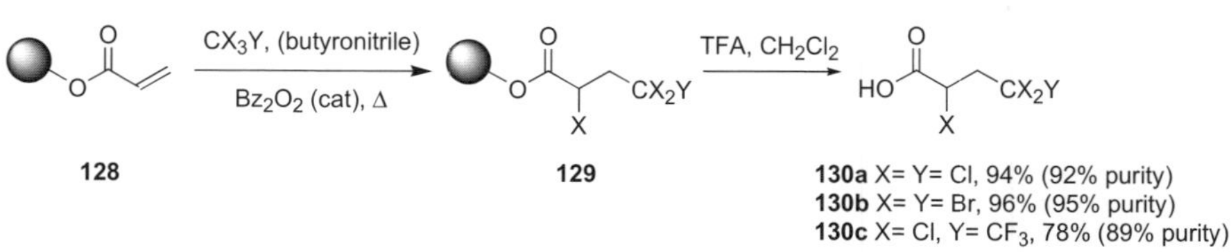

Scheme 28 Radical addition of haloalkanes to an immobilized acrylate by Kumar [18]

Caddick [19] applied this methodology in the first solid-phase radical synthesis of C-glycosides (Scheme 30). Intermolecular addition of the radical derived from iodide **135** to acceptor **132** followed by cleavage with an amino acid derivative such as phenylalanine ethyl ester gave glycopeptide **137** in good overall yield.

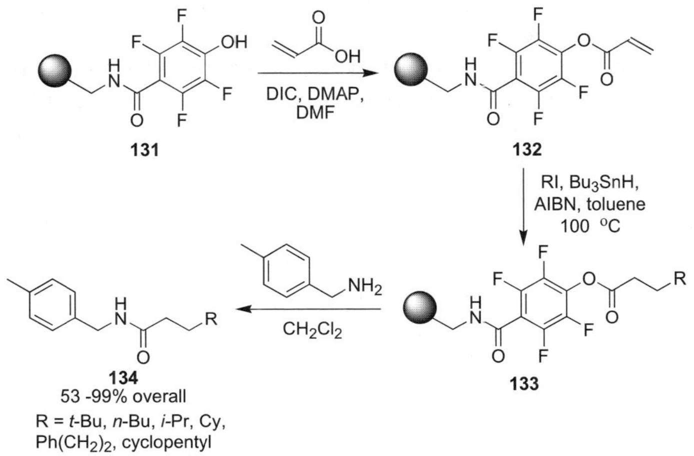

Scheme 29 Alkyl radical additions to tetrafluorophenol-linked acrylate by Caddick [19]

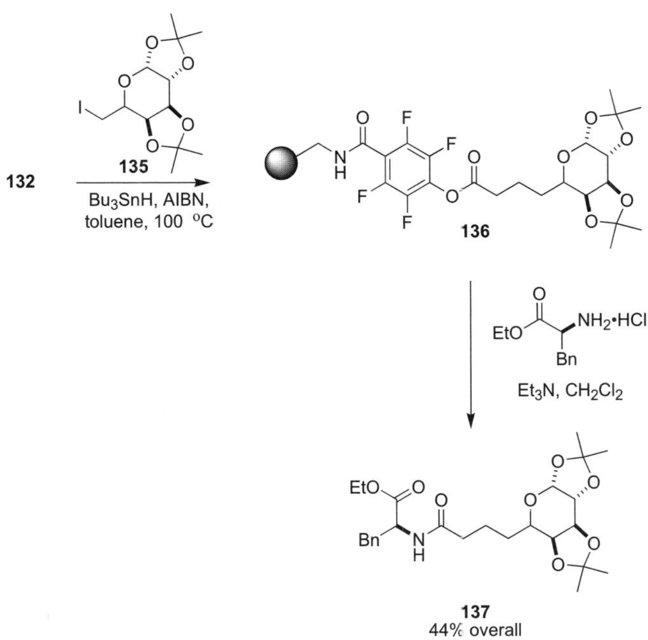

Scheme 30 Solid-phase C-glycoside synthesis by Caddick [19]

In addition, Caddick [19] has extended the methodology to the successful addition of the glycosyl radical derived from glycosyl bromide **138** to acceptor **132** allowing the synthesis of glycopeptide **140** in moderate overall yield (Scheme 31).

Linhardt [20] has employed a related process using samarium(II) iodide for the solid-phase synthesis of C-sialosides. Sialyl donor **141**, immobilized using an amino-functionalized, controlled-pore glass support, was treated with samarium(II) iodide in the presence of ketone and aldehyde electrophiles, e.g. reaction of **141** with cyclopentanone gave adduct **142** (Scheme 32). Cleavage from the support gave C-glycoside **143** in good over-

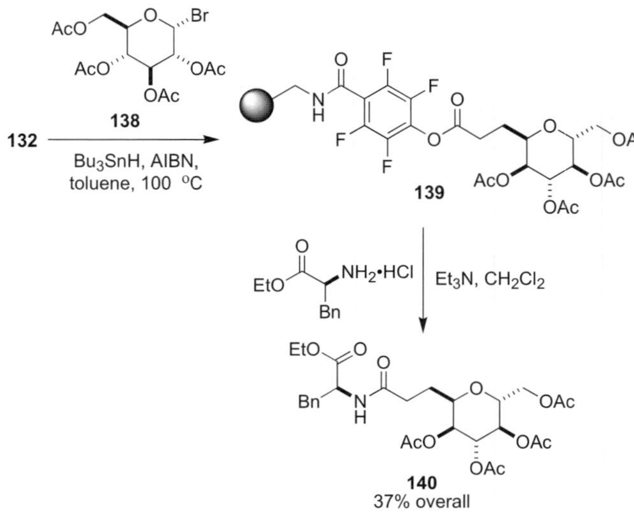

Scheme 31 Glycosyl radical addition to a tetrafluorophenol-linked acrylate by Caddick [19]

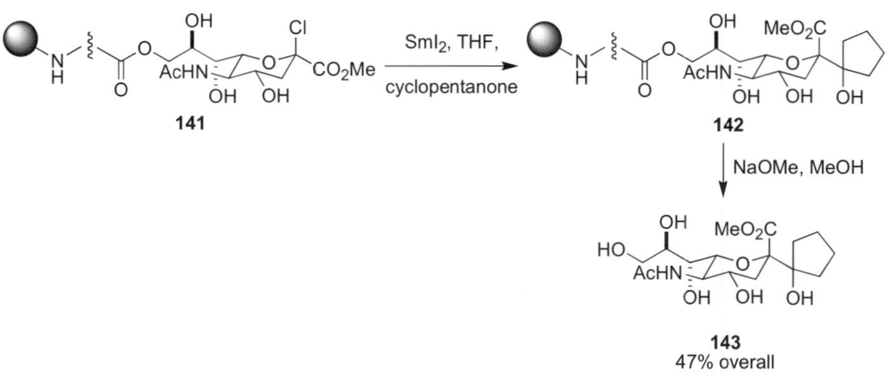

Scheme 32 C-Sialoside synthesis by Linhardt [20]

all yield. The C-glycosylation step proceeds via one-electron reduction of **141** to give a glycosyl radical, which is then reduced further to the corresponding organosamarium.

Building on their earlier studies (see Scheme 24 [16]) Naito [16, 21, 22] has demonstrated that a high degree of stereocontrol is possible in additions of alkyl radicals to immobilized chiral oxime ethers such as **146** (Scheme 33). Substrates were prepared from oxime ether **144** by treatment with (1*R*)-(+)-2,10-camphorsultam followed by deprotection to give **145**. The resulting alcohol **145** was then treated with glutaric anhydride and the resulting carboxylic acid attached to Wang resin to give substrate **146** (0.83 mmol/g). Addition of ethyl radical to **146** was achieved by treatment with triethylborane in CH_2Cl_2 at – 78 °C, while addition of isopropyl radical required the use of triethylborane in *i*-PrI and toluene (4 : 1 v/v) at 0 °C. In both cases the product carboxylic acid was cleaved from resin by treatment with TFA in CH_2Cl_2. Products **147** and **148** were obtained in good yields and in > 95%de and 92%de respectively. Similar diastereoselectivities were observed for radical additions mediated by diethylzinc rather than triethyl borane [16, 21, 22].

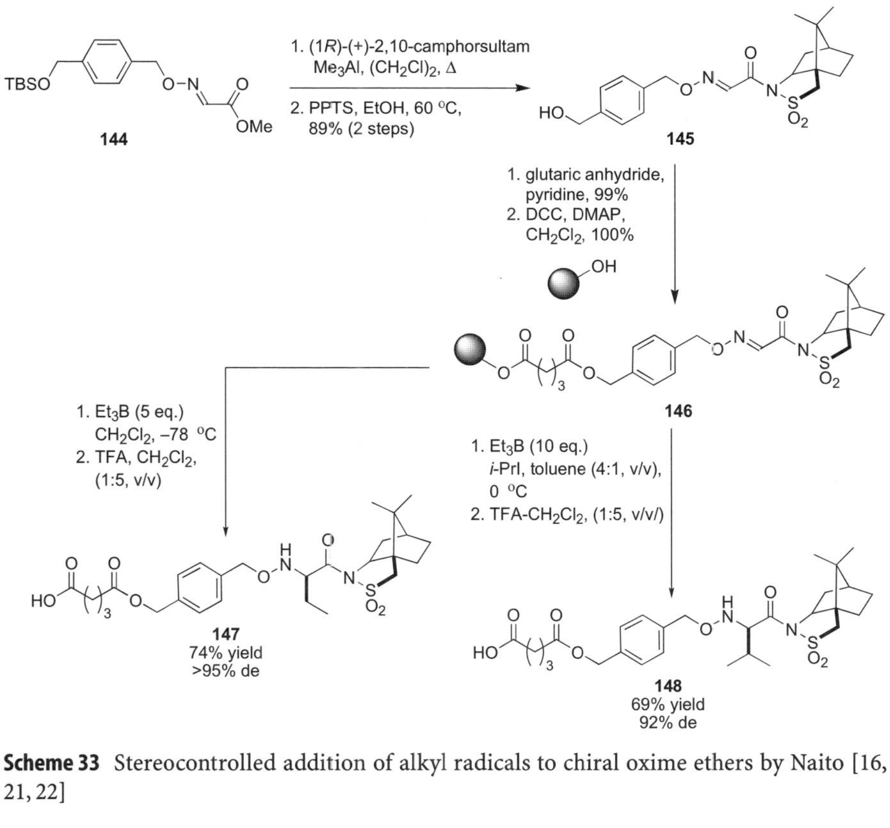

Scheme 33 Stereocontrolled addition of alkyl radicals to chiral oxime ethers by Naito [16, 21, 22]

Procter [23] has reported the intermolecular, reductive coupling of aldehydes and ketones with α,β-unsaturated esters, immobilized using an ephedrine chiral resin, to give enantiomerically enriched γ-butyrolactones.

For example, treatment of acrylate and crotonate ephedrine resins **150** and **151**, with cyclohexanecarboxaldehyde **149**, employing SmI_2 in THF with *t*-butanol as a proton source, gave **152** and **153** respectively, in moderate yield and good to high enantiomeric excess (Scheme 34). The process can be considered an example of an asymmetric catch-release process, where a substrate immobilized using a chiral support captures a reactive intermediate, in this case a ketyl radical anion, from solution [23]. The chiral support controls the asymmetry of the capture step and leads to a diastereomeric, resin-bound intermediate that breaks down to release a non-racemic product.

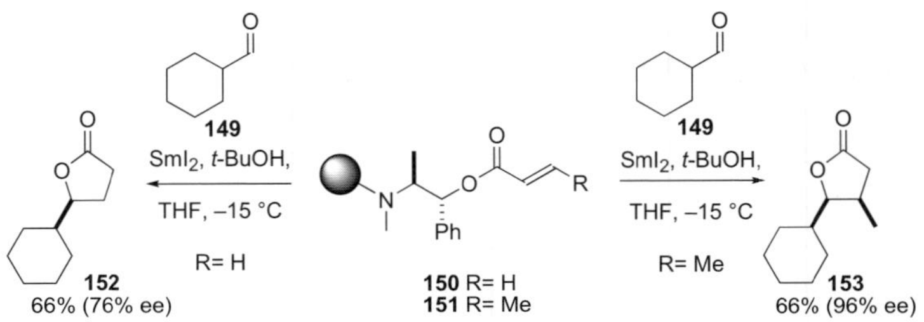

Scheme 34 Asymmetric catch-release for the formation of enantiomerically enriched γ-butyrolactones by Procter [23]

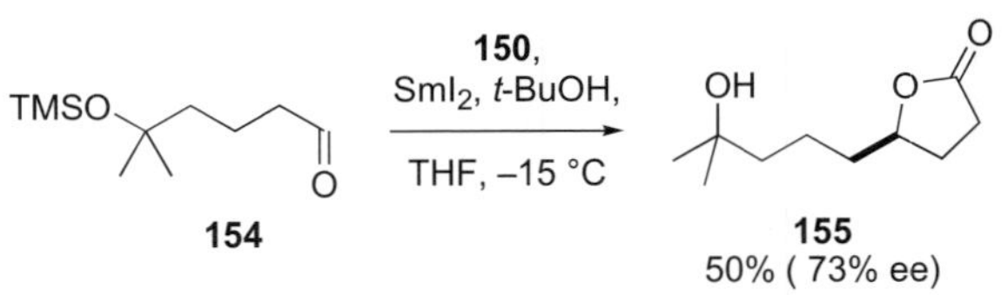

Scheme 35 Natural product synthesis using asymmetric catch-release by Procter [23]

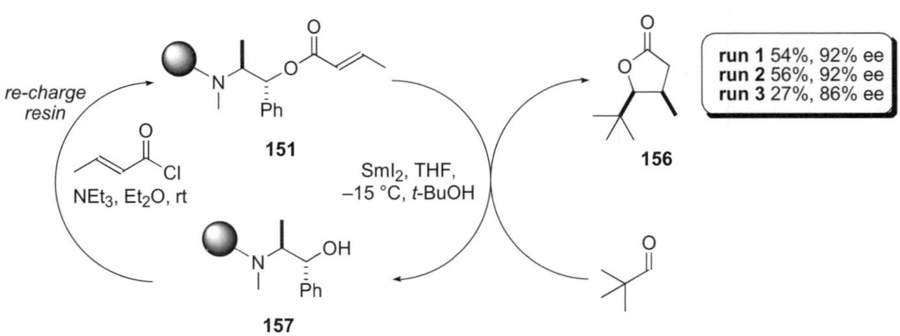

Scheme 36 Resin recycling studies by Procter [24]

The approach has been used in a short, asymmetric synthesis of γ-butyrolactone **155** [23], a moderate DNA-binding metabolite isolated from Streptomyces GT61115, starting from aldehyde **154** (Scheme 35).

In further studies, Procter [24] has investigated the feasibility of recycling the chiral ephedrine resin. For example, employing crotonate resin **151** and 2,2-dimethylpropanal gave lactone **156** in 54% yield and 92% ee (Scheme 36). Recovery of the ephedrine resin **157** and re-esterification with crotonyl chloride gave recycled **151**. Retreatment with 2,2-dimethylpropanal gave **156** in virtually identical yield and enantiomeric excess. Recovery and reuse for a third time, however, led to substantially lower yield, although the enantioselectivity of the process was still high (86%ee) [24].

4 Intramolecular Radical Reactions on Solid Phase

Lown [25] has reported a solid-phase radical cyclization approach to open-chain analogues of the cyclopropylindole class of antitumour antibiotics exemplified by duocarmycin SA (Fig. 1).

Aniline **158** was immobilized using bromo-Wang resin to give cyclization substrate **159** (Scheme 37) [25]. Heating a suspension of **159** in deoxygenated benzene in the presence of tributyltin hydride and AIBN gave **160**. *N*-Acetylation and TFA cleavage with concomitant, partial carbamate deprotection gave **161** in good overall yield.

Enholm [26] has reported the first examples of asymmetric radical cyclizations on soluble polymer supports. The stereocontrol element employed consists of a (+)-isosorbide group attached by a 4-carbon chain to each subunit of a soluble succinimide-derived ROMP backbone. Treatment of the radical cyclization substrate **162** with tributyltin hydride in the presence of zinc chloride followed by hydrolysis of the resulting polymer-supported ester **163** gave the desired product **164** in 80% yield and > 90% ee (Scheme 38). The use of alternative Lewis acids, such as magnesium bromide etherate and ytterbium (III) triflate, resulted in lower enantioselectivities, 84% and 72% respectively. No such decrease in selectivity was observed in analogous reactions carried out off-support [27], suggesting that the polymer backbone is somehow responsible for this phenomenon.

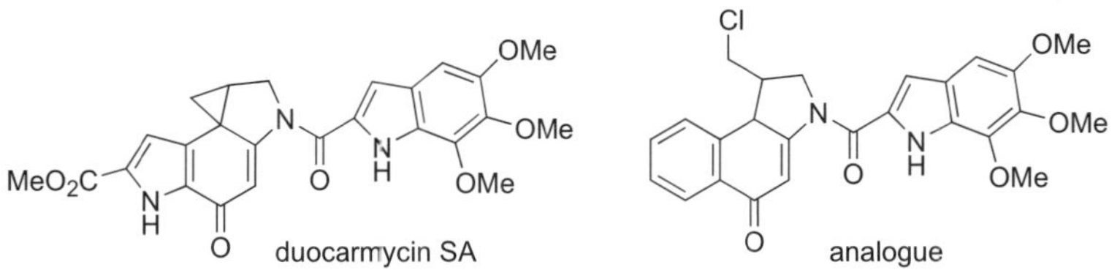

Fig. 1 Duocarmycin and an open-chain analogue

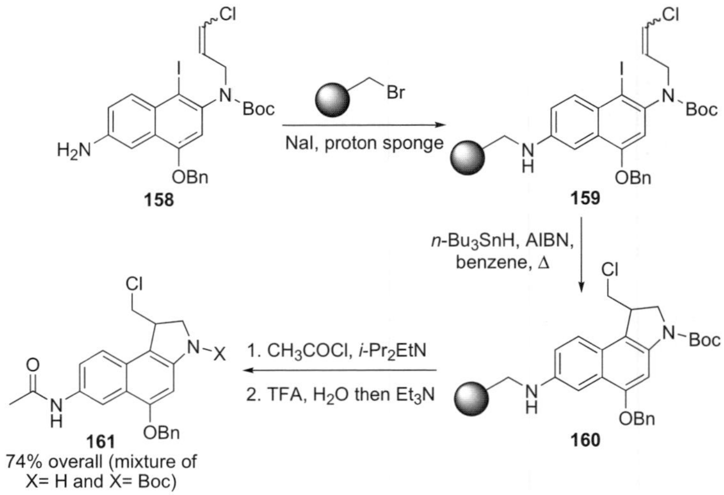

Scheme 37 Solid-phase synthesis of Duocarmycin analogues by Lown [25]

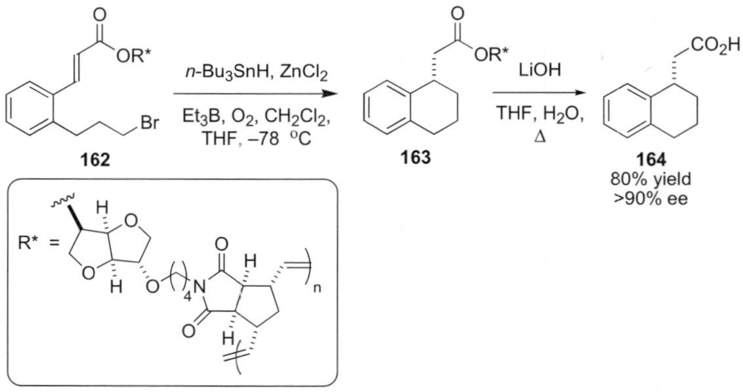

Scheme 38 Auxiliary controlled radical cyclization by Enholm [26]

Enholm's strategy [26] involves incorporation of all the structural features required for the 6-heptenyl radical cyclization reaction into the monomer subunit prior to metathesis polymerization, thus ensuring maximum possible loading. Although several steps are needed to prepare the monomer unit and recycling the support is not possible, this work does show the potential of using chiral linking units to control the stereochemistry of radical cyclizations on polymer supports.

Fukase [28] has described the solid-phase synthesis of 2-oxindoles via the radical cyclization of resin-bound *N*-(2-bromophenyl)acrylamides, such as **167**, using Bu_3SnH and AIBN (Scheme 39). DMF was found to be the best solvent for cyclization on solid phase, though the corresponding solution-phase cyclization in DMF did not proceed at all [28]. This finding is attributed to a reagent concentration effect of the polymer support induced by DMF, in which Bu_3SnH is quite insoluble. The reaction was found to proceed smoothly under microwave conditions, giving the desired oxindoles, such as **168**, after short reaction times (< 1 h) compared to conventional heating methods (approximately 24 h). Several oxindoles were prepared from commercially available 2-bromoanilines and acryloyl chlorides using this approach [28].

Routledge [29] has evaluated hypophosphite salts as alternatives to tributyltin hydride for radical cyclizations on solid phase. The cyclizations of substrates on hydrophobic (polystyrene), hydrophilic (PEG grafted) and macroporous supports using *N*-ethylpiperidine hypophosphite (EPHP) were investigated. Although satisfactory results were obtained with the more hydrophilic supports, optimal results were obtained using the polytetrahydrofuran cross-linked resin, JandaJel. Treatment of JandaJel-derived aryl iodide **169** with

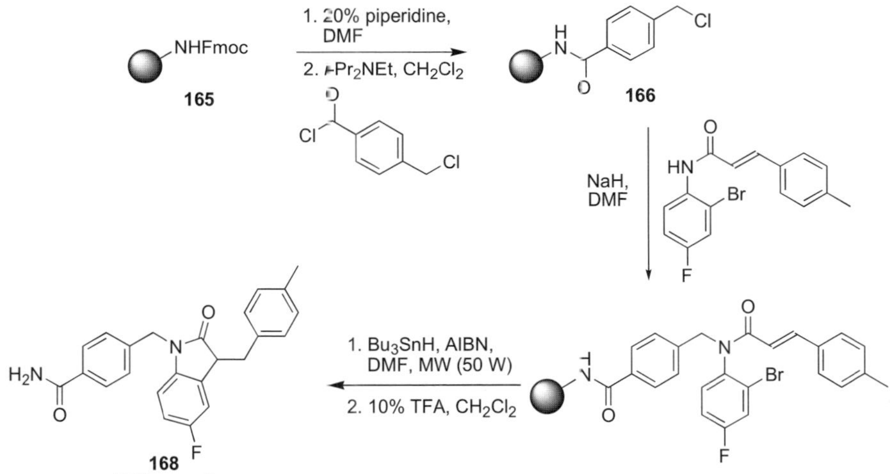

Scheme 39 Oxindole synthesis by Fukase [28]

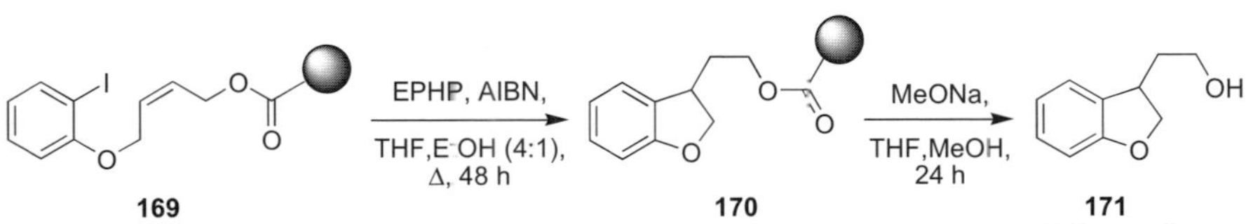

Scheme 40 Radical cyclization using a hypophosphite reagent by Routledge [29]

EPHP gave dihydrobenzofuran **171** in excellent yield after cleavage from the support (Scheme 40).

Alkyl iodide **172** was also found to undergo efficient cyclization under these conditions [29] to give tetrahydrofuran **173** as a mixture of diastereoisomers in high yield after cleavage from the support (Scheme 41).

Bowman [30] has studied the cyclization of aryl radicals onto azoles immobilized on a polymer support. The pyrazole radical precursor was attached to Wang resin and the cyclization of the resulting substrate **175** (Scheme 42) was investigated using both tributylgermanium hydride and tris(trimethylsilyl)silane. Upon treatment with tributylgermanium hydride and cleavage of the product mixture from the polymer support, cyclized product **176** was obtained accompanied by unreacted starting material. Cyclization of **175** using tris(trimethylsilyl)silane gave an improved yield of **176** and acyclic reduction product **177**.

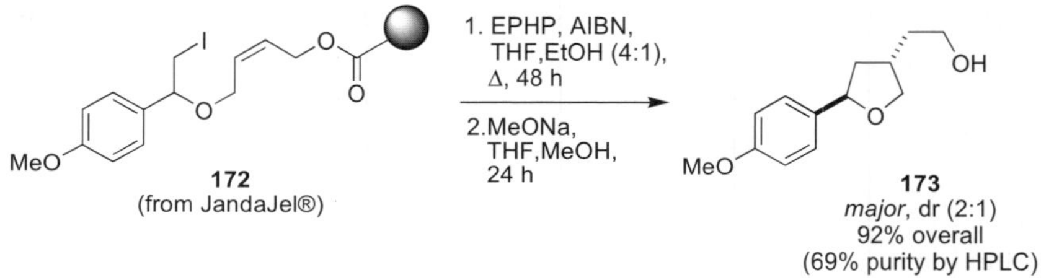

Scheme 41 Furan synthesis using a hypophosphite reagent by Routledge [29]

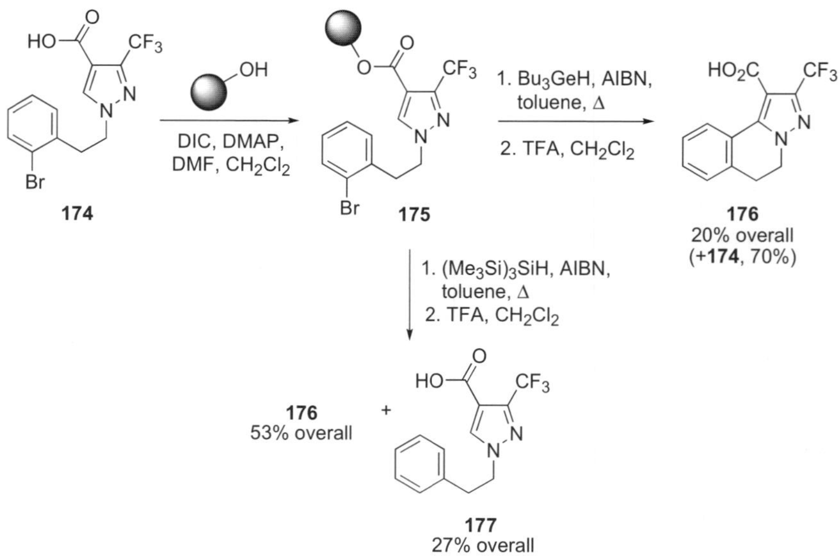

Scheme 42 Radical cyclization onto an azole system by Bowman [30]

5
Sequential Intermolecular and Intramolecular Radical Reactions on Solid Phase

Naito [31] has described the sequential radical addition-cyclization reactions of polymer-supported oxime ethers, triggered by the addition of stannyl and silyl radicals. Oxime ether substrates such as **178** gave substituted lactams, such as **179** and **180** via sequential radical addition and cyclization (Scheme 43). Reaction of oxime ether **178** with stannyl radical using triethylborane as radical initiator [31] proceeded effectively to give the cyclized product **181** in 64% yield after cleavage from resin by treatment with TFA (Scheme 43). Reaction of **178** with tris(trimethylsilyl)silane [31] also proceeded well, providing the cyclized product **182** in 50% yield. Use of the less reactive triethylsilane did not provide the desired product, and this was attributed to competitive intermolecular addition of an ethyl radical, generated from triethylborane, to the oxime ether group. The cyclized products **181** and **182** were obtained as diastereomeric mixtures, but the relative stereochemistry was not discussed.

Stannyl radical addition-cyclization of oxime ether **183** [31] was also examined (Scheme 44). In this case vinyl stannane **185** was obtained in 77% yield.

Reactions of oxime **186**, containing an electron-deficient carbon–carbon double bond, were also investigated (Scheme 45) [31]. In the case of stannyl radical addition-cyclization using triethylborane as the radical initiator, the reaction proceeded as efficiently as in previous cases giving the cyclized product **188** in 64% yield, after base-induced cleavage from the resin and linker.

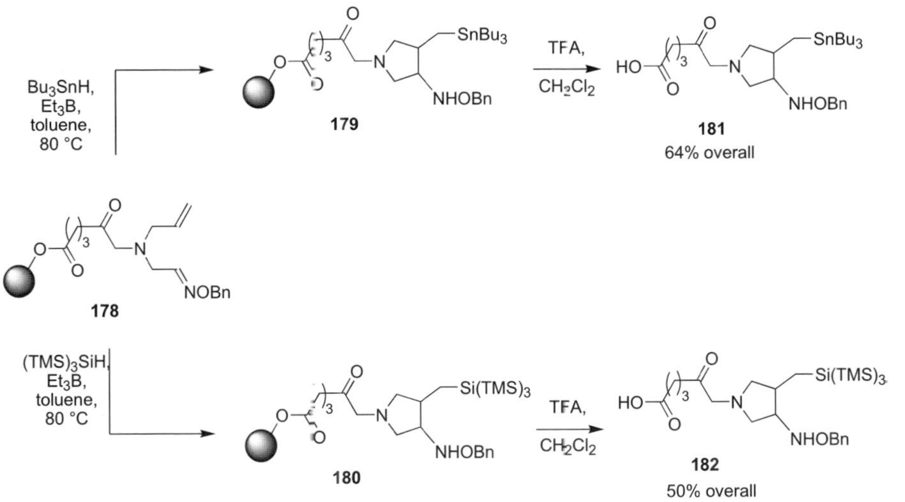

Scheme 43 Stannyl radical addition-cyclization by Naito [31]

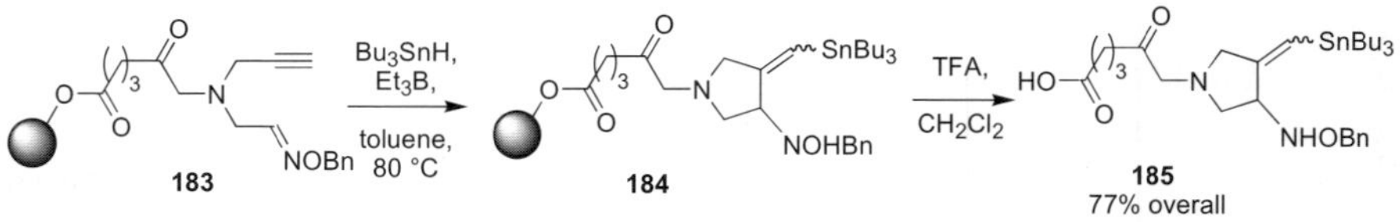

Scheme 44 Stannyl radical addition-cyclization to form a vinyl stannane by Naito [31]

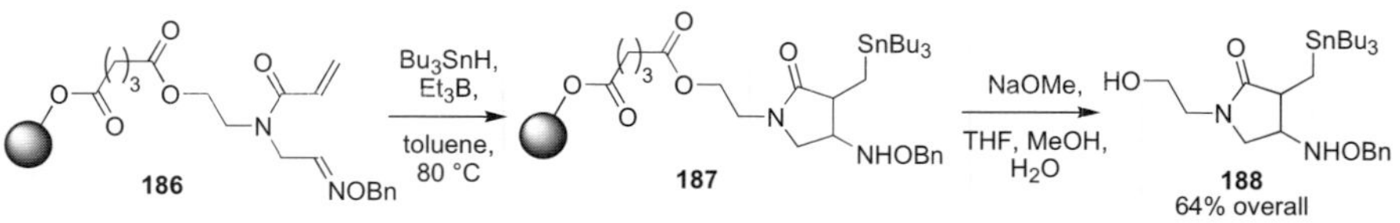

Scheme 45 Stannyl radical addition-cyclization with an electron-deficient acceptor by Naito [31]

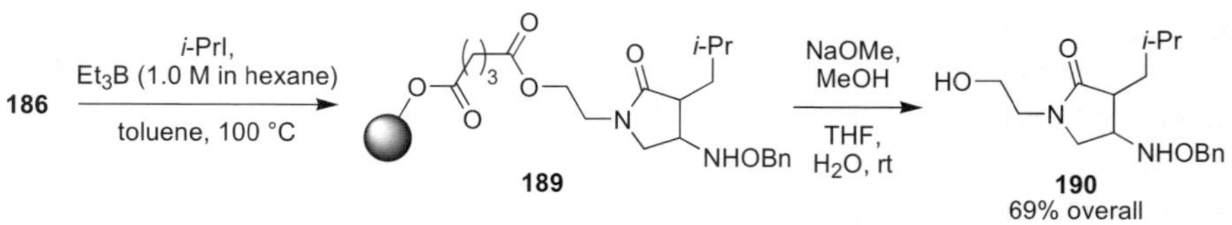

Scheme 46 Alkyl radical addition-cyclization by Naito [16, 32]

Naito has also described analogous tandem radical addition-cyclization processes under iodine atom-transfer reaction conditions [16, 32]. Treatment of **186** with *i*-PrI (30 eq.) and triethylborane (3×3 eq.) in toluene at 100 °C gave, after cleavage from the resin, the desired lactam product **190** in 69% yield (Scheme 46). Similar reactions involving cyclohexyl iodide, cyclopentyl iodide, and butyl iodide were also reported as well as the reaction with ethyl radical from triethylborane [16, 32]. The relative stereochemistry of the products was not discussed.

Interestingly, treatment of enantiomerically pure oxime ether **191** under the same iodine atom-transfer conditions gave mostly lactam **193** after cleavage from the support [32]. Clearly the addition of ethyl radical was competing with iodine atom transfer from isopropyl iodide. It was proposed that Lewis acidic triethylborane co-ordinates to the substrate on the polymer support and hence is concentrated on the surface of the resin. Naito has observed similar phenomena in previous studies [21]. By using a large excess of iospropyl iodide, the desired product **192** was obtained in good yield and with good diastereoselectivity (Scheme 47).

Harrowven [33] has reported two methods for carrying out the radical cyclization reactions of dienes attached to a Wang polystyrene support. Treatment of solid-supported diene **195** with either thiophenol or *p*-tolylbenzeneselenosulfonate, in the presence of AIBN, triggers 5-*exo*-trig

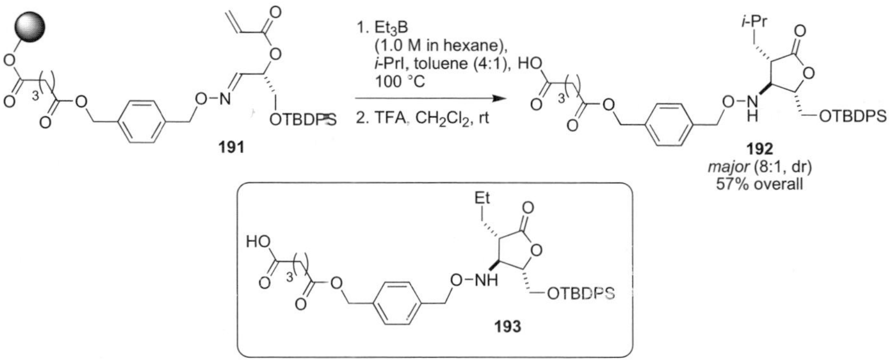

Scheme 47 Diastereoselective addition-cyclization by Naito [32]

radical cyclization to give, after cleavage from the support with sodium methoxide, the highly functionalized cyclopentane products **196** and **197** respectively (Scheme 48). Using thiophenol, the product **196** was obtained in 78% yield and with good selectivity for the *cis* product (*cis* : *trans*, 5 : 1). When *p*-tolylbenzeneselenosulfonate was employed to mediate the radical cyclization, **197** was obtained after cleavage from the support in 74% yield with enhanced selectivity for the *cis* diastereoisomer (*cis* : *trans*, 8 : 1).

This methodology [33] was extended to the cyclization of more complex immobilized di- and tri-substituted dienes such as **198** (Scheme 49). Products were obtained in good yields but diastereoselectivities were significantly reduced. The method employing *p*-tolylbenzeneselenosulfonate is particu-

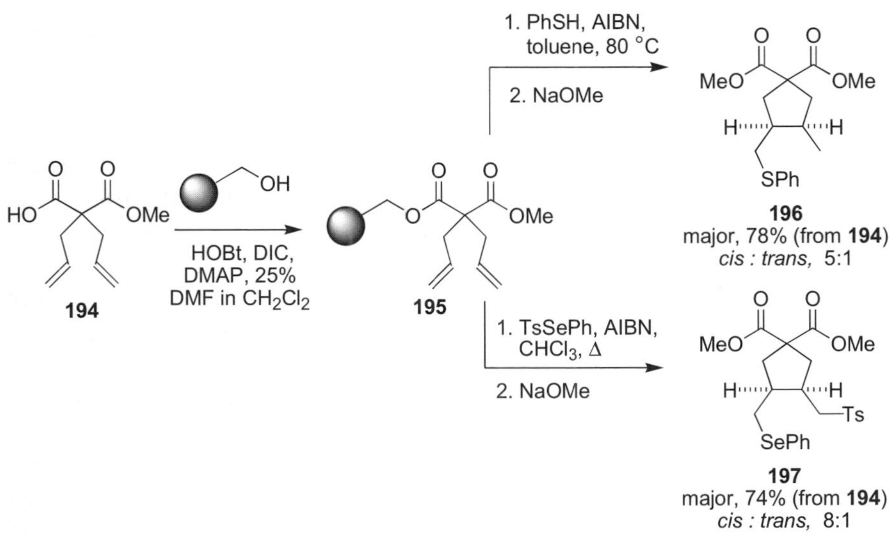

Scheme 48 Diene cyclization by Harrowven [33]

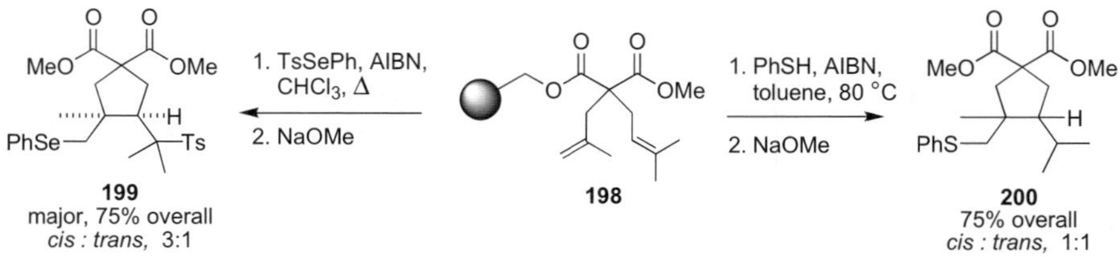

Scheme 49 Cyclization of di- and tri-substituted dienes by Harrowven and Bradley [33]

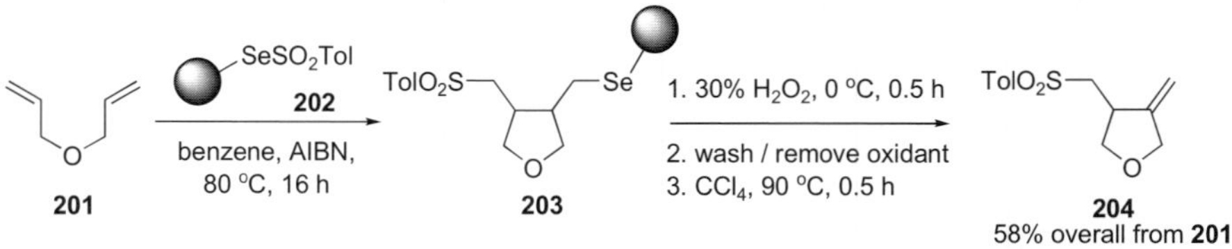

Scheme 50 Radical cyclative capture of 1,6-dienes by Huang [34]

larly suitable for solid-phase library generation as both sulfone and selenide functionalities are simultaneously introduced in the cyclization, providing opportunity for further diversification of the cyclic products.

In a related study, Huang [34] has employed resin-bound selenosulfonate **202** in the radical cyclative-capture of 1,6-dienes such as **201** (Scheme 50). Oxidation and eliminative cleavage provided methylenetetrahydrofurans **204** in moderate yields.

6 Radical Cleavage of Solid-Phase Linkers

6.1 Cleavage Under Electron-Transfer Conditions

The single-electron-transfer reagent samarium(II) iodide (SmI_2) has been used extensively to intitiate radical reactions in solution and has now begun to find application in solid-phase synthesis.

Abell [35] has developed a linker strategy for the synthesis of amides and ureas based on the reduction of N – O bonds by SmI_2. Wang-based hydroxylamine resin **205** was used to prepare a range of immobilized amides and ureas, such as **206**. Traceless cleavage of the linker gave amide **207** in good overall yield and in high purity (Scheme 51).

A similar linker strategy has been used by Taddei [36] in a solid-phase approach to β-lactams. Treatment of immobilized substrates **209** and **212** with SmI_2 resulted in smooth reduction and release of **211** and **213** respectively

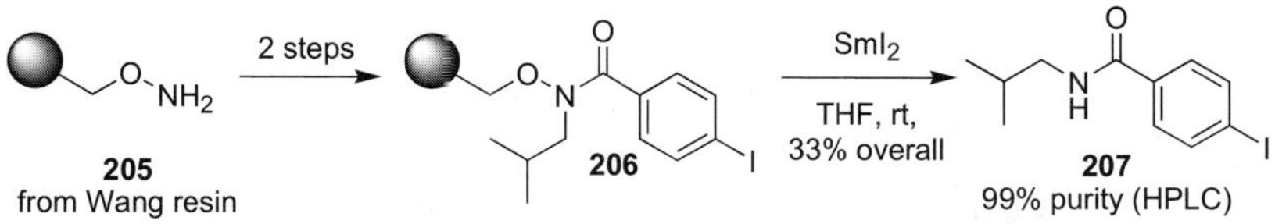

Scheme 51 N – O bond reduction by Abell [35]

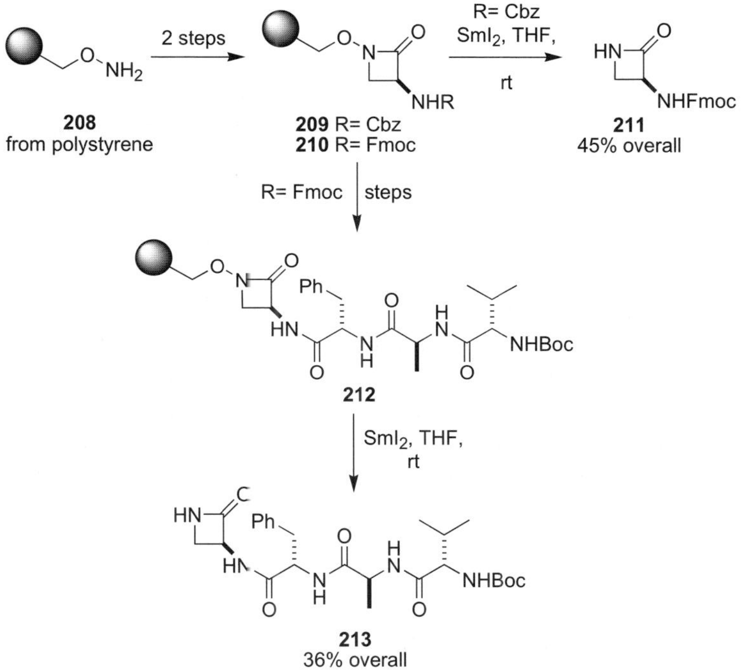

Scheme 52 N – O bond reduction in β-lactam synthesis by Taddei [36]

(Scheme 52), further illustrating the suitability of the electron-transfer cleavage approach for the synthesis of highly functionalized targets.

In addition, Andersson [37] has utilized this type of linker in a solid-phase approach to tertiary amines (Scheme 53). Treatment of alkoxyammonium salt **215** with SmI_2 released amine **216** from the support.

De Clercq [38] has utilized a sulfide linker, cleaved by a radical process initiated by electron transfer, in a solid-phase Julia-type olefination process. Alkylation of an aryl thiol resin followed by *m*CPBA oxidation gave supported sulfone **217** (Scheme 54). Successive treatment of the resin with *n*-butyllithium and an aldehyde followed by trapping of the resultant alkoxide with benzoyl chloride gave resin-bound α-benzoyloxy sulfone **218**. Olefins **219** and **220** were released from the solid support upon reduction with a single-electron-transfer reagent and elimination of the sulfone link-

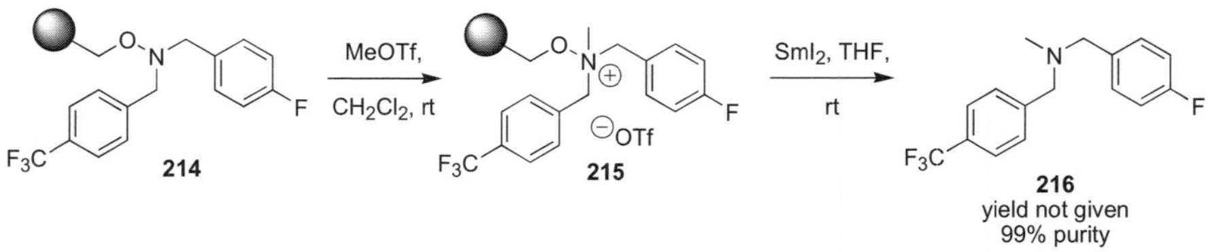

Scheme 53 N – O bond reduction in formation of tertiary amines by Andersson [37]

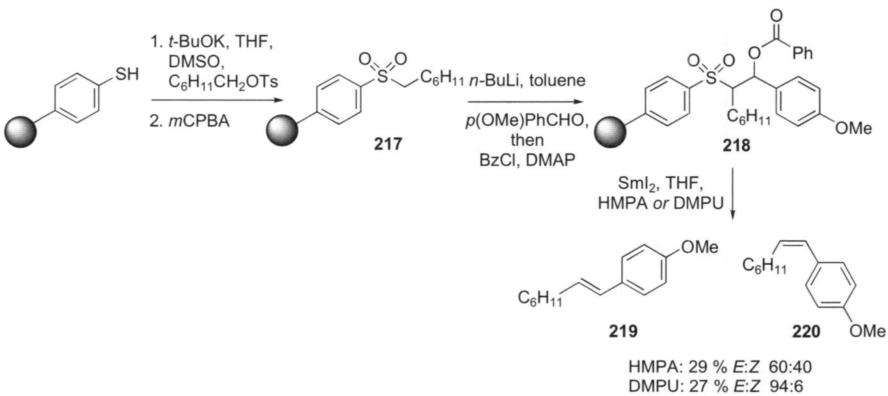

Scheme 54 Reductive cleavage of a sulfone linker by De Clerq [38]

age (Scheme 54). SmI_2 proved to be the most suitable reagent for the process and was used with the promoters HMPA and DMPU [38]. The stereoselectivity of the olefination was found to be strongly dependent upon the additive DMPU, giving rise to higher *E* selectivity than HMPA in the example shown (Scheme 54).

Procter [39, 40] has developed a new class of linker cleaved under mild, neutral electron-transfer conditions using SmI_2. The cleavage of α heteroatom-substituted carbonyl (HASC) linkers is based on the well-established reduction of α heteroatom-substituted carbonyl compounds to the parent ketone, ester or amide, using SmI_2. Lactone **221**, immobilized via an oxygen HASC linker, was converted to a range of ketones and amides, including **222** and **223** (Scheme 55). Treatment with SmI_2 released cyclopropyl ketone **224** and morpholine amide **225** respectively from the support in good overall yield [39, 40].

Cyclopropyl ketone **222** (Scheme 55) was prepared to probe the mechanism of the cleavage reaction [39, 40]. Isolation of **224** where the cyclopropyl ring is intact (Scheme 56) suggests cleavage proceeds via formation of radical **227** rather than ketyl radical anion **226**, formed by single-electron transfer to the ketone carbonyl, as cyclopropylmethyl radical anions are known to undergo facile fragmentation.

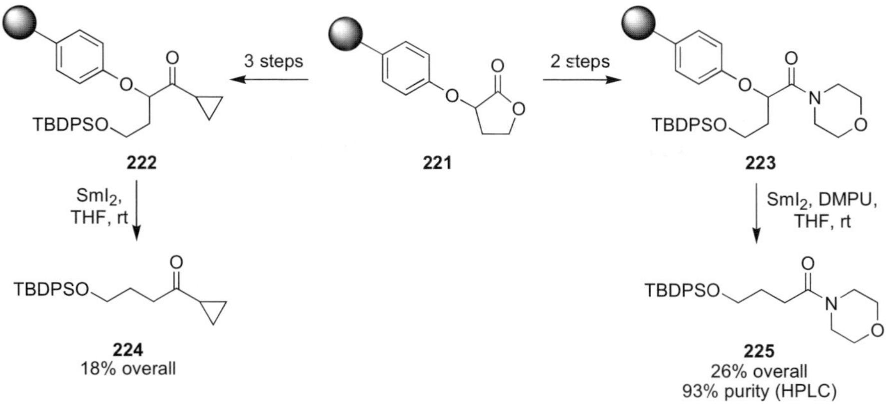

Scheme 55 Reductive cleavage of an oxygen HASC linker by Procter [39, 40]

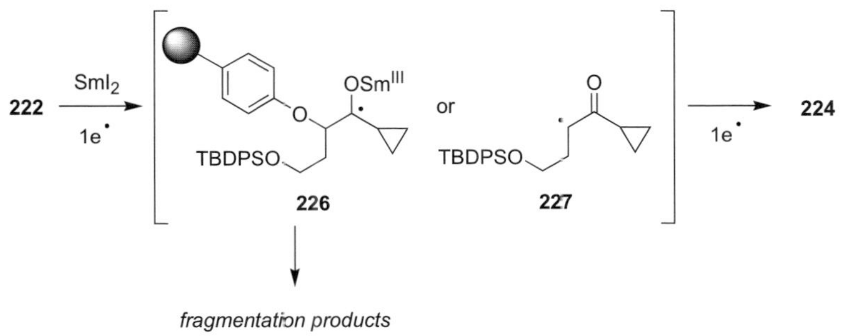

Scheme 56 Cleavage mechanism of HASC linker by Procter [39, 40]

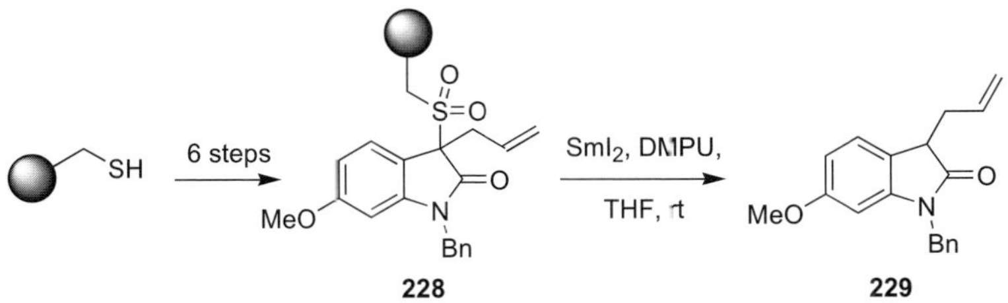

Scheme 57 Reductive cleavage of a sulfur HASC linker by Procter [41]

Procter [41] has also utilized a sulfur HASC linker for the solid-phase synthesis of oxindoles. Treatment of immobilized substrate **228** with SmI_2 released oxindole **229** from the support in good overall yield (Scheme 57).

Floreancig [42] has described a new solid-phase linker system cleaved by oxidative electron transfer. The cleavage process is based on the oxidative fragmentation of homobenzylic ethers. Acetal **230**, immobilized on a soluble oligonorbornene scaffold prepared by ROMP polymerization, was efficiently

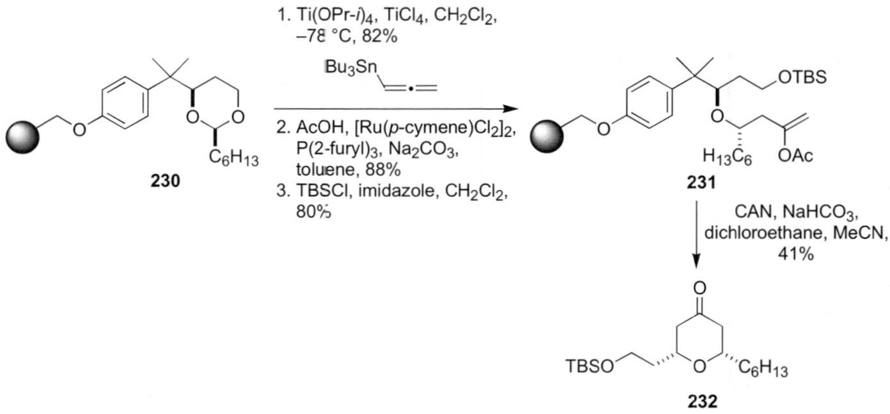

Scheme 58 Linker cleavage and cyclization by oxidative electron transfer by Floreancig [42]

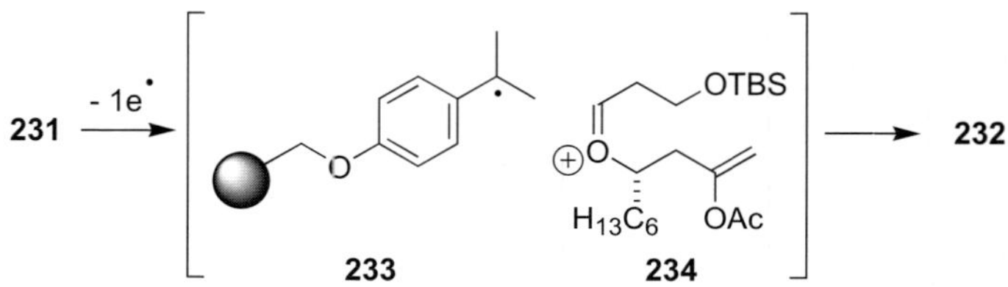

Scheme 59 Mechanism of oxidative cleavage by Floreancig [42]

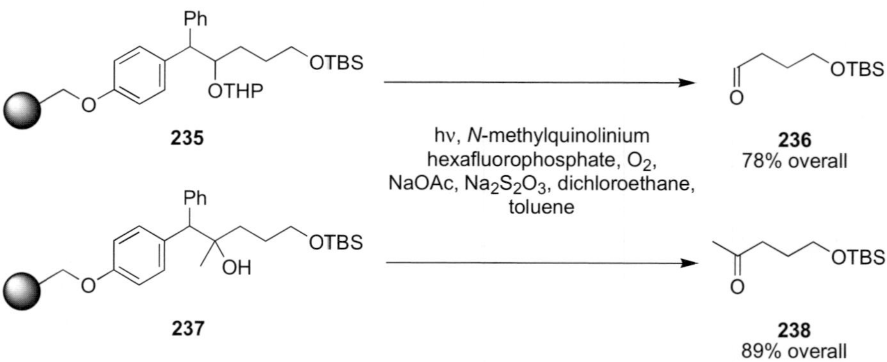

Scheme 60 Oxidative cleavage in aldehyde and ketone formation by Floreancig [42]

converted to vinyl acetate **231** (Scheme 58). On treatment with ceric ammonium nitrate (CAN) a cyclorelease process occurred to give pyranone **232** as a single diastereoisomer.

Cleavage occurs via radical fragmentation and formation of the benzylic radical **233** and oxonium ion **234** (Scheme 59) [42]. The latter then cyclizes to give the observed product **232**.

Floreancig [42] has also shown that the linker strategy is suitable for the synthesis of simpler ketones and aldehydes. Treatment of immobilized substrates **235** and **237** with an alternative reagent system for single-electron oxidation gave aldehyde **236** and ketone **238** respectively in excellent yield (Scheme 60). In the absence of an appended nucleophile, the intermediate oxonium ions, analogous to **234** (Scheme 59), break down to give the corresponding carbonyl compound.

6.2
Linkers Cleaved by Photolysis

The use of linkers designed to undergo cleavage upon photolysis is a common strategy in solid-phase synthesis. Bochet [43] has recently reviewed this area, and it will not be covered in this review.

6.3
Selenium-Based Linkers

Nicolaou [44–47] has pioneered the use of selenium linkers in solid-phase synthesis. A selenium linker [44, 45], cleaved under radical conditions, has been utilized in a cyclative-capture strategy for the solid-phase synthesis of indolines (Scheme 61). Treatment of *o*-allylanilines with selenenyl bromide resin in the presence of a Lewis acid results in cyclative-capture to give immobilized indoline **240**. Further functionalization was then carried out prior

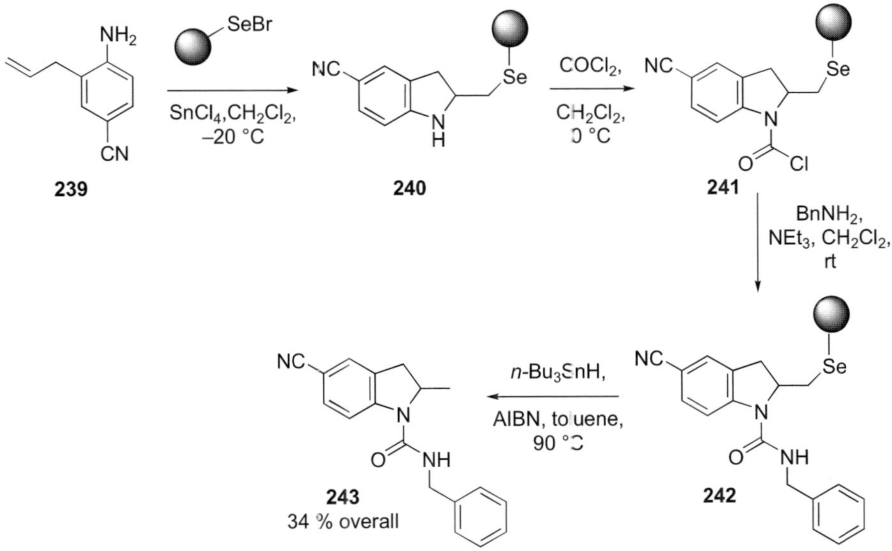

Scheme 61 Radical cleavage of a selenium linker by Nicolaou [44, 45]

to cleavage: acylation of nitrogen with phosgene gave **241** and further reaction with amines gave **242**. Radical cleavage with tributyltin hydride gave functionalised 1-methylindoline **243** in good yield (Scheme 61). A range of 1-methylindolines were synthesized using this approach in overall yields between 14 and 34% [44, 45].

Further complexity can be introduced in the cleavage step [44, 45] when a suitably positioned radical acceptor is present (Scheme 62). For example, resin-bound indoline **244** was coupled with 1-cyclohexenecarboxylic acid to form **245**. Cleavage of the selenium linker with tributyltin hydride gave a carbon-centred radical that underwent 6-*endo*-trig cyclization onto the tethered double bond to give **246**, thus illustrating the feasibility of accessing polycycles using this methodology (Scheme 62) [44, 45].

In an attempt to introduce more diversity upon cleavage, exposure to a radical initiator in the absence of a radical quench [45] led to cleavage and radical rearrangement, allowing access to 2-methyl indoles. For example, treatment of **247** with AIBN gave primary radical **248** initially upon cleavage, which rearranged to give 2-methyl indole **249** in good overall yield (Scheme 63).

Nicolaou [46] has also exploited a selenium linker cleaved using radical conditions in the solid-phase synthesis of 2-deoxyglycosides, 2-deoxyorthoesters and 2,3-allyl orthoesters. Resin-bound tributyltin selenolate **251** was conveniently synthesized in two steps from a selenenyl bromide resin. Treatment of **251** with a trichloroacetimidate donor gave resin-bound se-

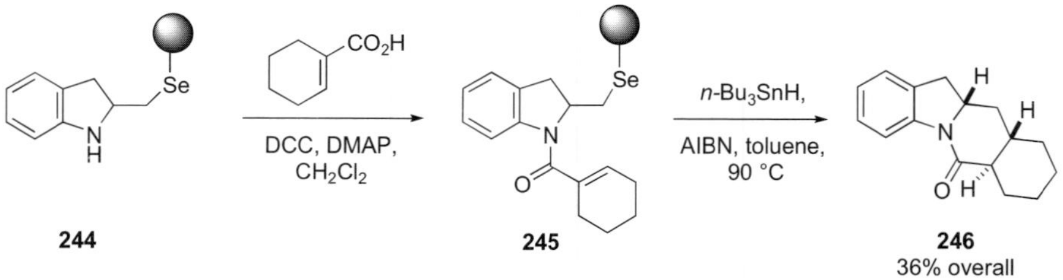

Scheme 62 Selenium linker cleavage-cyclization by Nicolaou [44, 45]

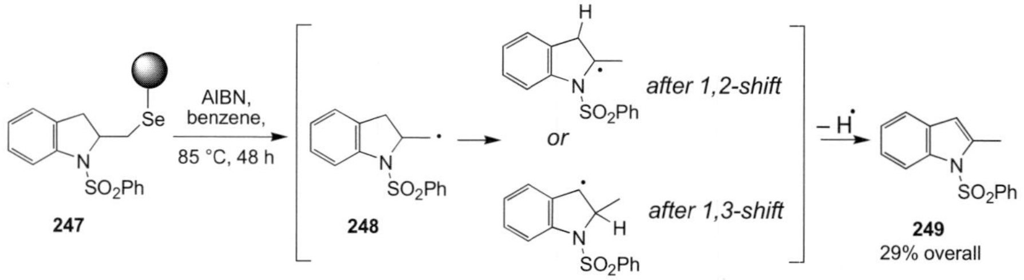

Scheme 63 Selenium linker cleavage in the absence of a radical quench by Nicolaou [45]

lenoglycoside **252** (Scheme 64). Removal of the C2 and C3 acetate protecting groups and reprotection of the C3 hydroxyl gave **253** ready for a 1,2-selenomigration reaction. Treatment of **253** with DAST promoted 1,2-migration of the selenium link to resin in a stereospecific manner to give **254**, now linked to resin at the C2 position of the sugar, whilst simultaneously generating an anomeric fluoride leaving group. Glycosyl fluoride donor **254** then underwent glycosylation with alcohols or sugars. For example, exposure to Lewis acid and monoprotected diol **255** selectively gave the α-glycoside **256** [46]. Reductive cleavage of the selenium linker with tributyltin hydride gave the desired 2-deoxyglycoside **257** in excellent yield. A 3×3 library of 2-deoxyglycosides [46] was synthesized using this approach in overall yields between 11 and 32% (Scheme 64).

Nicolaou [47] has applied this methodology in synthetic studies towards the antibiotic everinomicin 13,384-1. Glycosylation of immobilized fluoride donors such as **258** (Scheme 65) with sugar acceptors allowed the synthesis of disaccharides such as **259**, attached to solid support through a selenium link-

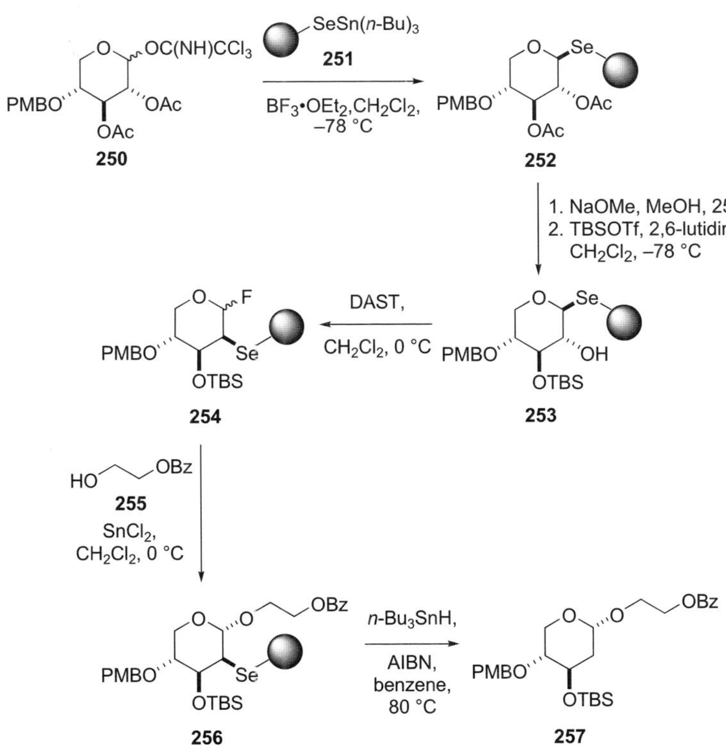

Scheme 64 Selenium linker cleavage in the synthesis of 2-deoxyglycosides by Nicolaou [46]

ing atom. As illustrated in previous model studies, radical cleavage allowed access to 2-deoxy glycosides such as **260**.

Building on Nicolaou's precedent, Engman [48] has reported the first example of the homolytic cleavage of a selenide linker followed by carbonylation of the resultant radical and cyclization (Scheme 66). The loading of presursor **261** was determined by radical cleavage with tri(trimethylsilyl)-

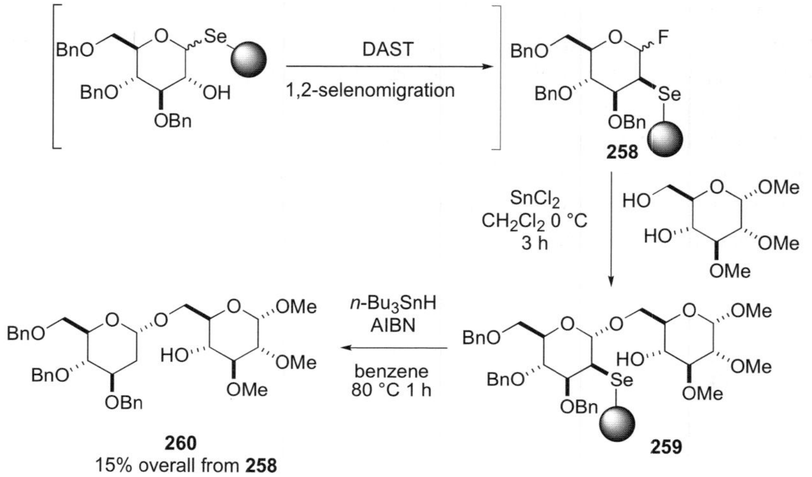

Scheme 65 Selenium linker cleavage in an approach to everinomicin 13,384-1 by Nicolaou [47]

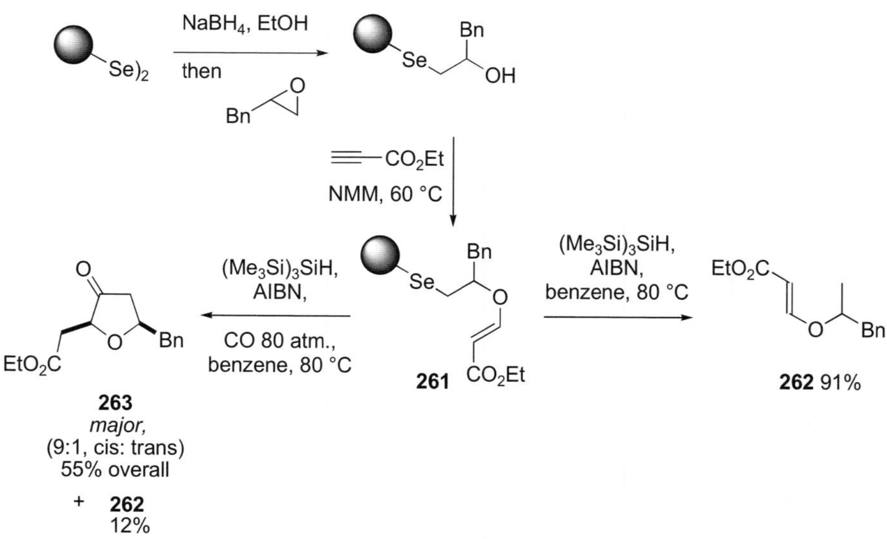

Scheme 66 Selenium linker cleavage, carbonylation and cyclization by Engman [48]

silane and AIBN to give **262** in excellent yield. Treatment of substrate **261** with tri(trimethylsilyl)silane and AIBN under an atmosphere of CO gave lactone **263** in good overall yield and with good diastereoselectivity.

Ruhland [49] recently reported the first use of a tellurium-based linker in solid-phase synthesis. As the homolysis of carbon–tellurium bonds is known to occur at lower temperatures than the homolysis of the corresponding selenides, it was proposed that tellurium linkers might prove valuable for the solid-phase synthesis of temperature-sensitive targets. A library of simple aryl alkyl ethers was prepared using polystyrene-bound telluride complex **264** (Scheme 67), prepared from bromopolystyrene in five steps [49]. Alkylation of **264** with two alkyl halides and subsequent Mitsunobu coupling with three phenols gave ethers **267** and **268** ready for radical cleavage from the polymer support. The cleavage reaction was studied at 60 °C and 90 °C. Rather surprisingly, the yields of products **269** and **270** obtained using the tellurium linker system were found to be consistently lower than those obtained using an analogous selenium linker system [49]. The authors speculate that the presence of

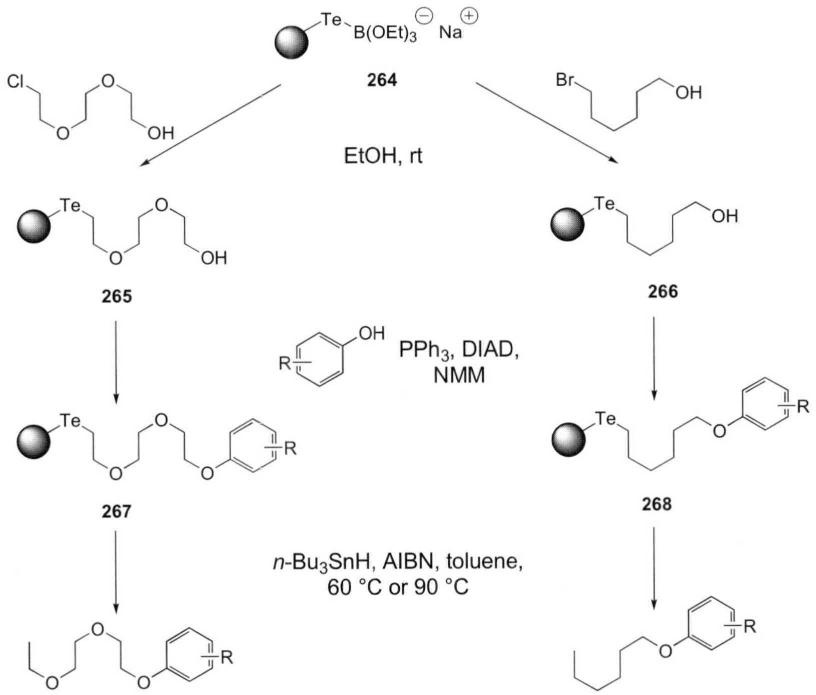

269a R= 4-Ph, 73% (90 °C), 55% (60 °C)
269b R= 2-F, 83% (90 °C), 60% (60 °C)
269c R= 3-OMe, 74% (90 °C), 56% (60 °C)

270a R= 4-Ph, 86% (90 °C), 57% (60 °C)
270b R= 2-F, 73% (90 °C), 48% (60 °C)
270c R= 3-OMe, 84% (90 °C), 62% (60 °C)

Scheme 67 Cleavage of a tellurium-based linker by Ruhland [49]

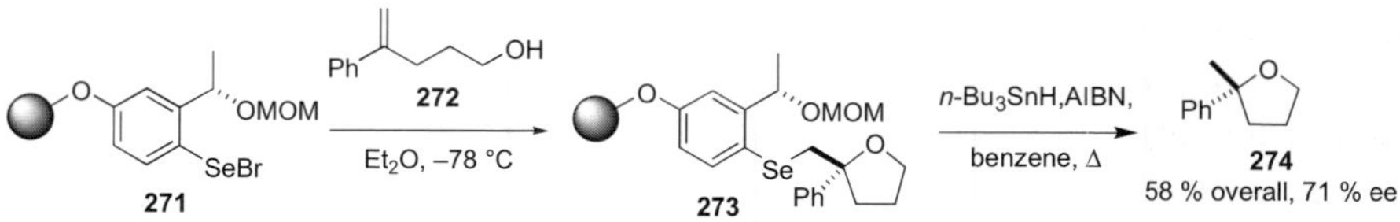

Scheme 68 Cleavage of a chiral selenium linker by Wirth [50]

traces of elemental tellurium in the resin may be interfering with the radical chain process.

Wirth [50, 51] has reported the first polymer-bound enantiomerically pure, electrophilic selenium reagents for asymmetric addition of nucleophiles to alkenes and has utilized radical cleavage to release enantiomerically enriched products from the support (Scheme 68). Chiral selenenyl bromides were prepared and immobilized on polystyrene, TentaGel and mesoporous silica supports. Polystyrene-based, selenenyl bromide resin **271** was found to be most effective in asymmetric selenenylation reactions. Treatment of unsaturated alcohol **272** with chiral resin **271** led to diastereoselective seleniranium ion formation and intramolecular, nucleophilic addition of the tethered alcohol to give tetrahydrofuran **273** immobilized via a chiral selenium linkage. Radical cleavage gave tetrahydrofuran **274** in 58% yield and 71% ee (Scheme 68) [50].

7 Summary

Many examples of inter- and intramolecular radical reactions in solid-phase synthesis have now been reported. Intermolecular processes using both immobilized radicals and solution-phase acceptors, and immobilized acceptors and radicals in solution, have been used with both modes giving acceptable results. Radical cyclization reactions have been widely utilized and incorporated in powerful cyclative-loading and cleavage strategies. The further development of asymmetric radical processes suggests the asymmetric synthesis of compound libraries using solid-phase radical chemistry will soon become a practical option. In addition, the development of new linker strategies where cleavage depends upon radical processes has led to the solid-phase preparation of new compound classes and improved access to more established targets. Interest in radical reactions in solid-phase synthesis has also led to advances in traditional areas of radical chemistry, for example the introduction of polymer-supported reagents to mediate radical reactions in solution is a direct product of the marriage of radical chemistry and solid-phase synthesis.

In summary, the utility of radical processes in solid-phase synthesis is now well established, and the use of such reactions should be considered as a viable strategy for library synthesis. As the frontiers of solid-phase synthesis continue to advance, radical processes promise to be a major impetus behind new developments in the field.

References

1. Ganesan A, Sibi MP (2002) In: Nicolaou KC, Hanko R, Hartwig W (eds) Handbook of combinatorial chemistry, vol 1. Wiley, Weinheim, p 227
2. Ganesan A (2001) In: Renaud P, Sibi MP (eds) Radicals in organic synthesis, vol 2. Wiley, Weinheim, p 81
3. Curran DP, Yang F, Cheong JH (2002) J Am Chem Soc 124:14993
4. Boussaguet P, Delmond B, Dumartin G, Pereyre M (2000) Tetrahedron Lett 41:3377
5. De Luca L, Giacomelli G, Porcu G, Taddei M (2001) Org Lett 3:855
6. Bowman WR, Krintel SL, Schilling MB (2004) Synlett 1215
7. Attardi ME, Taddei M (2001) Tetrahedron Lett 42:3519
8. Qian H, Huang X (2001) Synlett 1913
9. Qian H, Huang X (2002) Tetrahedron Lett 43:1059
10. Quiclet-Sire B, Wilczewska A, Zard SZ (2000) Tetrahedron Lett 41:5673
11. Dublanchet AC, Lusinchi M, Zard SZ (2002) Tetrahedron 58:5715
12. Enholm EJ, Gallagher ME, Jiang S, Batson WA (2000) Org Lett 2:3355
13. Enholm EJ, Gallagher ME (2001) Org Lett 3:3397
14. Jeon GH, Yoon JY, Kim S, Kim SS (2000) Synlett 128
15. Miyabe H, Nishimura A, Fujishima Y, Naito T (2003) Tetrahedron 59:1901
16. Miyabe H, Ueda M, Naito T (2004) Synlett 1140
17. Plourde R, Johnson LL, Longo RK (2001) Synlett 439
18. Kumar HMS, Chakravarthy PP, Rao MS, Reddy PSR, Yadav JS (2002) Tetrahedron Lett 43:7817
19. Caddick S, Hamza D, Wadman SN, Wilden JD (2002) Org Lett 4:1775
20. Baytas SN, Wang Q, Karst NA, Dordick JS, Linhardt RJ (2004) J Org Chem 69:6900
21. Miyabe H, Konishi C, Naito T (2000) Org Lett 2:1443
22. Miyabe H, Konishi C, Naito T (2003) Chem Pharm Bull 51:540
23. Kerrigan NJ, Hutchison PC, Heightman TD, Procter DJ (2003) Chem Commun 1402
24. Kerrigan NJ, Hutchison PC, Heightman TD, Procter DJ (2004) Org Biom Chem 2:2476
25. Jia G, Iida H, Lown JW (2000) Synlett 603
26. Enholm EJ, Cottone JS (2001) Org Lett 3:3959
27. Enholm EJ, Cottone JS, Allais F (2001) Org Lett 3:145
28. Akamatsu H, Fukase K, Kusumoto S (2004) Synlett 1049
29. Chevet C, Jackson T, Santry B, Routledge A (2005) Synlett 477
30. Allin SM, Bowman WR, Elsegood MRJ, McKee V, Karim R, Rahman SS (2005) Tetrahedron 61:2689
31. Miyabe H, Tanaka H, Naito T (2004) Chem Pharm Bull 52:842
32. Miyabe H, Fujii K, Tanaka H, Naito T (2001) Chem Commun 831
33. Harrowven DC, May PJ, Bradley M (2003) Tetrahedron Lett 44:503
34. Qian H, Huang X (2003) J Comb Chem 5:569
35. Myers RM, Langston SP, Conway SP, Abell C (2000) Org Lett 2:1349
36. Meloni MM, Taddei M (2001) Org Lett 3:337
37. Gustafsson M, Olsson R, Andersson C-M (2001) Tetrahedron Lett 42:133
38. D'herde JNP, De Clercq PJ (2003) Tetrahedron Lett 44:6657

39. McKerlie F, Procter DJ, Wynne G (2002) Chem Commun 584
40. McKerlie F, Rudkin IM, Wynne G, Procter DJ (2005) Org Biom Chem 3:2805
41. McAllister LA, Brand S, de Gentile R, Procter DJ (2003) Chem Commun 2380
42. Liu H, Wan S, Floreancig PE (2005) J Org Chem 70:3814
43. Bochet CG (2002) J Chem Soc Perkin Trans 1:125
44. Nicolaou KC, Roecker AJ, Pfefferkorn JA, Cao G (2000) J Am Chem Soc 122:2966
45. Nicolaou KC, Roecker AJ, Hughes R, van Summeren R, Pfefferkorn JA, Winssinger N (2003) Biorg Med Chem 11:465
46. Nicolaou KC, Mitchell HJ, Fylaktakidou KC, Suzuki H, Rodriguez RM (2000) Angew Chem Int Ed 39:1089
47. Nicolaou KC, Fylaktakidou KC, Mitchell HJ, van Delft FL, Rodríguez RM, Conley SR, Jin Z (2000) Chem Eur J 6:3166
48. Berlin S, Ericsson C, Engman L (2003) J Org Chem 68:8386
49. Ruhland T, Torang J, Pedersen H, Madsen JC, Bang KS (2004) Synthesis 2323
50. Uehlin L, Wirth T (2001) Org Lett 3:2931
51. Uehlin L, Wirth T (2001) Chimia 55:65

Top Curr Chem (2006) 264: 135–162
DOI 10.1007/128_022

Published online: 20 December 2005

Modification of Amino Acids, Peptides, and Carbohydrates through Radical Chemistry

Signe Grann Hansen · Troels Skrydstrup (✉)

Department of Chemistry, University of Aarhus, Langelandsgade 140, 8000 Aarhus C, Denmark
ts@chem.au.dk

Abstract This review provides an overview of some of the more recent work directed to exploit radical-based chemistry for the modification of some of Natures most important biomolecules, such as amino acids, peptides, and carbohydrates. Radical reactions are particularly advantageous for carrying out a variety of structural modifications on biomolecules as the reaction conditions are typically compatible with a wide variety of functional groups and solvents. An array of effective synthetic transformations will be discussed including selective side chain and backbone modifications of amino acids and peptides, along with methods for the transformation of carbohydrate substituents, as well as fragmentation and cyclizations reactions for the preparation of either structurally modified carbohydrates or chiral building blocks.

Keywords Amino acids · Carbohydrates · Peptides · Radical reactions

1 Introductory Remarks

The ability to selectively perform functional group transformations on biomolecules such as carbohydrates and peptides is a topic of high interest for the preparation of mimics and analogues thereof, for biomedical research as well as for drug development programs. However, performing modifications at specific sites of the biomolecule is a formidable challenge owing to both the repetitiveness of the functional groups (i.e., hydroxy groups for sugars, amide bonds for peptides, side chains on amino acids etc.) and their close association. For selective transformations one must have a profound understanding of the reactivity differences that can arise for these same groups due to potential topological diversity, and ultimately of reactivity differences which are found in the local environment of the molecule. With such knowledge individual groups can be activated for further transformation or protected to allow activation of less reactive groups, a strategy which has traditionally been applied in carbohydrate chemistry. However, the ability to adapt ionic reactions for the creation of new carbon–carbon or carbon–heteroatom bonds with such groups is not always a simple task, again due to the close vicinity of other similar functional groups which can participate and provide alternative reaction pathways to those that are desired.

Radical reactions provide an interesting option for carrying out chemical modifications of amino acids, peptides, and carbohydrates. In general, because of their early transition states, such reactions are less prone to be influenced by steric factors as compared to ionic reactions. Also, many radical reactions are compatible with a wide variety of functional groups and solvents, which cannot be said for ionic reactions. In this survey, we highlight some of the radical reactions which have been developed and applied to the modification of such biomolecules within the last five years or so. We will focus mainly on new reactions that have been developed or adapted to the transformation of Nature's molecules, rather than focusing on well-established classical radical reactions that have been widely applied in both carbohydrate and peptide chemistry (i.e., radical deoxygenations). Additionally, radical reactions that are applied to the synthesis of amino acids or carbohydrates via de novo approaches will not be covered. Unfortunately, page limitations do oblige us to be selective with respect to the reactions discussed in this account, and therefore we apologize to those whose work may not have been cited. It is our intention to give the reader a flavor of the many possibilities that the application of radical reactions can provide to synthetic chemists.

2
Amino Acids and Peptides

In this chapter, the main focus will be on two types of reactions dealing with both the creation of carbon–carbon bonds on amino acids and peptides and carbon–carbon bond fragmentations, all of which proceed via radical intermediates. Much of the earlier work on the use of free radical reactions in the synthesis of α-amino acids and derivatives has been published in an extensive review in 1997 by Easton [1].

2.1
Carbon–Carbon Bond Forming Reactions

2.1.1
Intramolecular Reactions

A remarkable intramolecular photoinduced radical reaction was published in 1999 by Giese and coworkers, which demonstrated the ability of a functionalized alanine derivative **1** to undergo a Norrish–Yang photocyclization leading to the proline derivatives **3** and **4** in an 85 : 15 ratio (Scheme 1) [2]. Both proline compounds were isolated with high enantioselectivity (**3**, 92% *ee*; **4**, 88% *ee*) when naphthalene, a triplet quencher, was present whereas with benzophenone, a triplet sensitizer, all four cyclization products were obtained with little selectivity. The enantioselective reaction pathway proceeds with retention of configuration, even though the sole sp^3-hybridized chiral center is converted into a prochiral, sp^2-hybridized radical center as illustrated by the diradical intermediate **2**. This reaction provides an interesting example of a memory of chirality [3], whose effect is the result of the inversion/retention ratio at the chiral center. Recent ab initio computations have postulated the existence of a hydrogen bonding network in the diradical intermediate which hinders the rotation required for the production of the enantiomeric product [4].

An alternative photochemically promoted radical cyclization which transforms allylglycine-containing peptides to their corresponding proline derivatives was reported by the Blechert group in 2001 [5]. In this work, one-electron oxidation via a photoinduced electron transfer of α-silylmethylamino derivatives, as exemplified with **5**, provides a corresponding radical cation which readily undergoes fragmentation to a free α-aminomethyl radical **6** [6]. 5-*Exo* cyclization then generates the 4-methylproline-containing

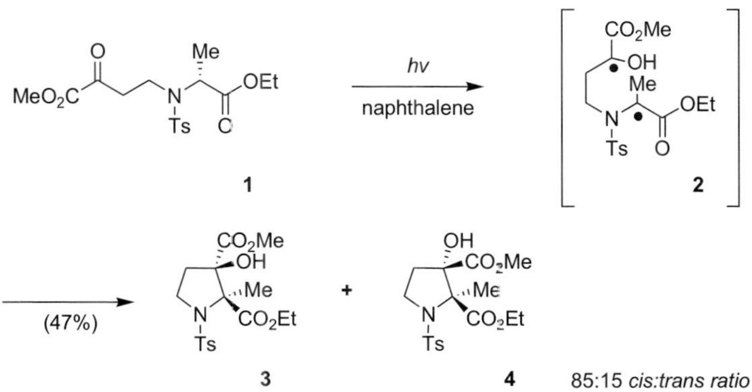

Scheme 1 Asymmetric synthesis of proline derivatives via a Norrish–Yang photocyclization

peptide **7** in yields up to 86%. The presence of proline residues in peptides is well known to induce β turns. Hence, the reaction provides an interesting method for influencing the secondary structure of the peptide through the selective conversion of an acyclic amino acid residue to a cyclic one (Scheme 2).

An elegant macrocyclization approach to the synthesis of cyclic peptide mimics has been achieved through a Korean–American collaboration exploiting a single electron transfer (SET)-promoted photocyclization process [7]. Peptide precursors possessing an N-terminal phthalimido group as a light-absorbing electron donor–acceptor moiety and a C-terminal α-amidosilane or α-amidocarboxylate end were irradiated, generating cyclic peptides in modest to good yields as exemplified in Scheme 3.

As illustrated in Scheme 4, the reactions are proposed to proceed via an initial photoinduced SET event between the phthalimido acceptor group and the

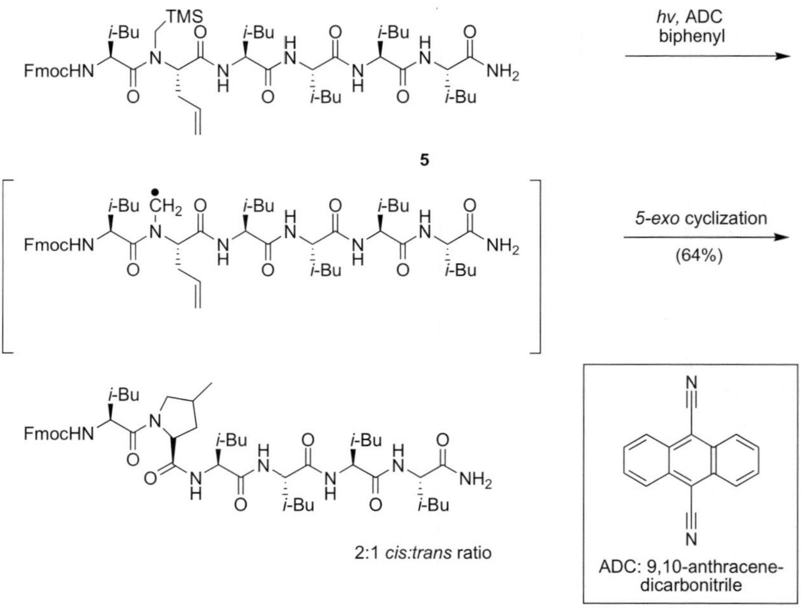

Scheme 2 A photochemically promoted radical cyclization of an allylglycine containing peptide

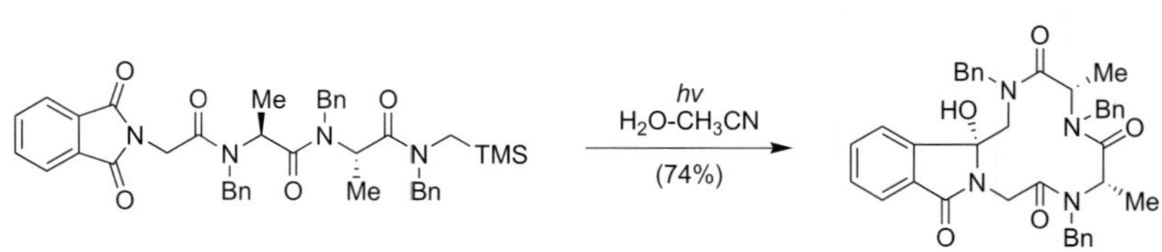

Scheme 3 Cyclic peptide synthesis via a photo-induced single electron transfer and biradical coupling

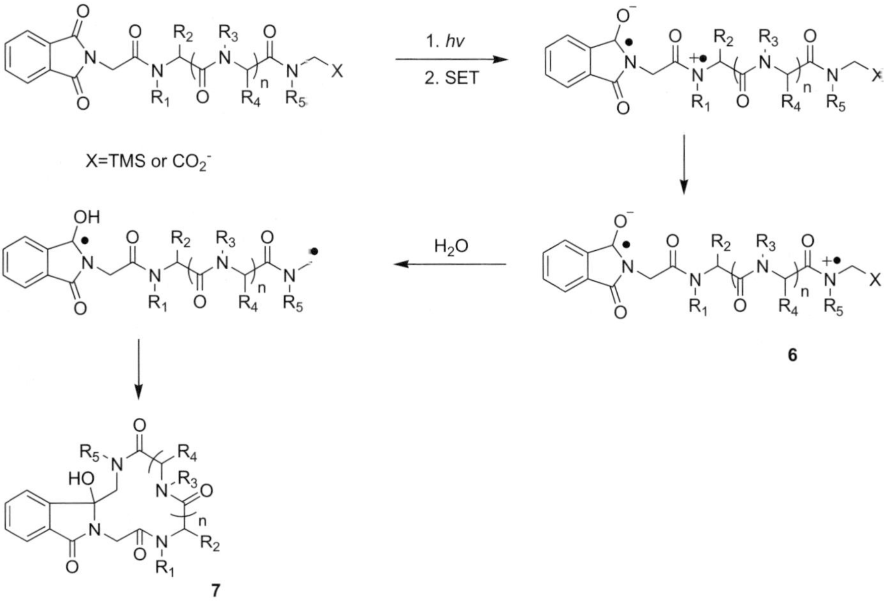

Scheme 4 Mechanistic proposal for the photo-induced peptide cyclization

adjacent amide donor site. Subsequent migration of the radical cation center to the C-terminal end followed by fragmentation to the α-aminomethyl radical and finally biradical cyclization eventually lead to the macrocyclic ring system **7**. Interestingly, the efficiency of the ring-closing process is apparently independent of the chain length. A plausible explanation was given suggesting that the zwitterionic biradical **6** exists in a folded conformation, thus minimizing the distance of the opposite charged centers. For this to be sound it does require that the rates of fragmentation and biradical coupling lie in the vicinity of those for conformational randomization of the peptide chain.

Iqbal and coworkers have reported another approach to cyclic di- and tripeptide mimics constrained with a disubstituted aromatic linker which exploits a free-radical-mediated macrocyclization [8]. As an example, the tripeptide derivative **8** was promoted to undergo 17-*endo* ring closure under tin hydride conditions, providing the cyclic peptide **9** in 54% yield (Scheme 5). The reasonable yields obtained for these macrocyclization processes is proposed by the authors to occur because of preorganization of the acyclic precursors, possibly owing to a reverse turn (γ/β turn).

A radical cascade reaction has been accomplished by Stalinski and coworkers which converts dipeptide derivatives to nitrogen-containing heterocycles (Scheme 6) [9]. In this work, *N*-bromobenzyl-*N*-propargyl-substituted dipeptides such as **10** were subjected to Stork's catalytic procedure with tributyltin hydride [10]. An aryl radical is formed followed by a 1,5-hydrogen shift, generating the α-centered carbon radical **11**. 5-*Exo*-dig radical cyclization

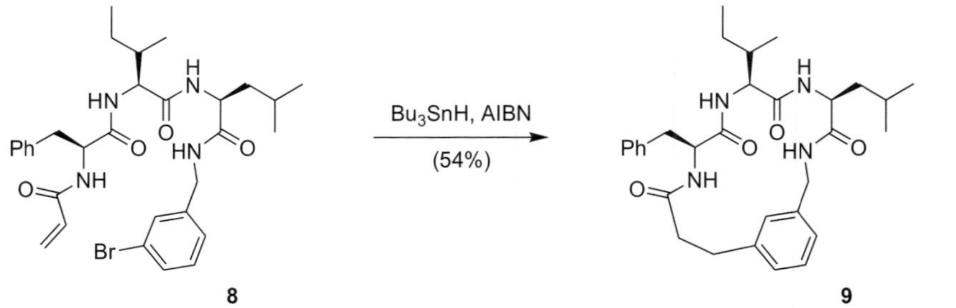

Scheme 5 Synthesis of cyclic peptide mimics via a free-radical mediated macrocyclization

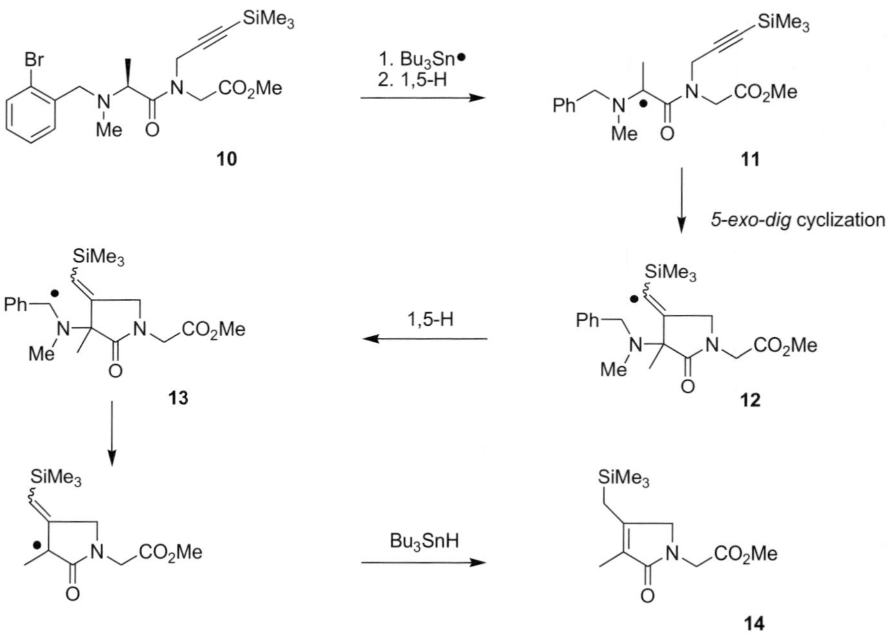

Scheme 6 Mechanism for the transformation of dipeptide derivatives to nitrogen-containing heterocycles via a radical cascade reaction

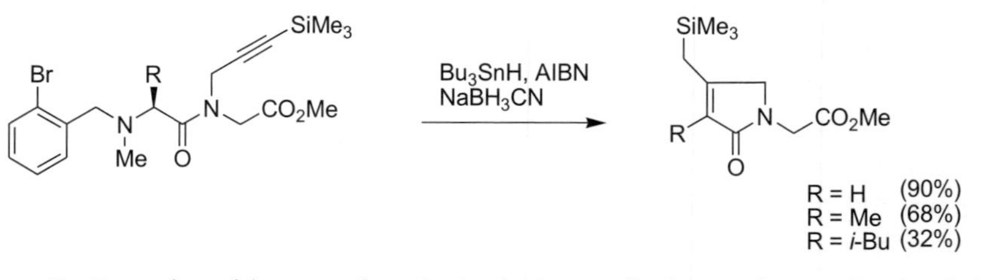

Scheme 7 Examples of heterocycles obtained via a radical cascade reaction involving dipeptide derivatives

produces a vinyl radical **12** which can undergo another 1,5-hydrogen transfer with the benzylic amine to produce an α-amino radical **13**. Finally, a β fragmentation and hydrogen abstraction from tributyltin hydride leads to the unsaturated γ – lactam **14**. The yields of these reactions are good with glycine and alanine at the N-terminal end, but dropped to below 50% when other amino acids were explored (Scheme 7).

2.1.2 Intermolecular Reactions

Skrydstrup and collaborators demonstrated a novel method for the introduction of side chains onto glycine residues of small peptides [11]. In this work, a pyridyl sulfide group was introduced into a series of di-, tri-, and tetrapeptides via a two-step procedure involving an initial radical-induced bromination of glycine residues with *N*-bromosuccinimide followed by nucleophilic substitution with 2-mercaptopyridine (Scheme 8, Route A). The yields were

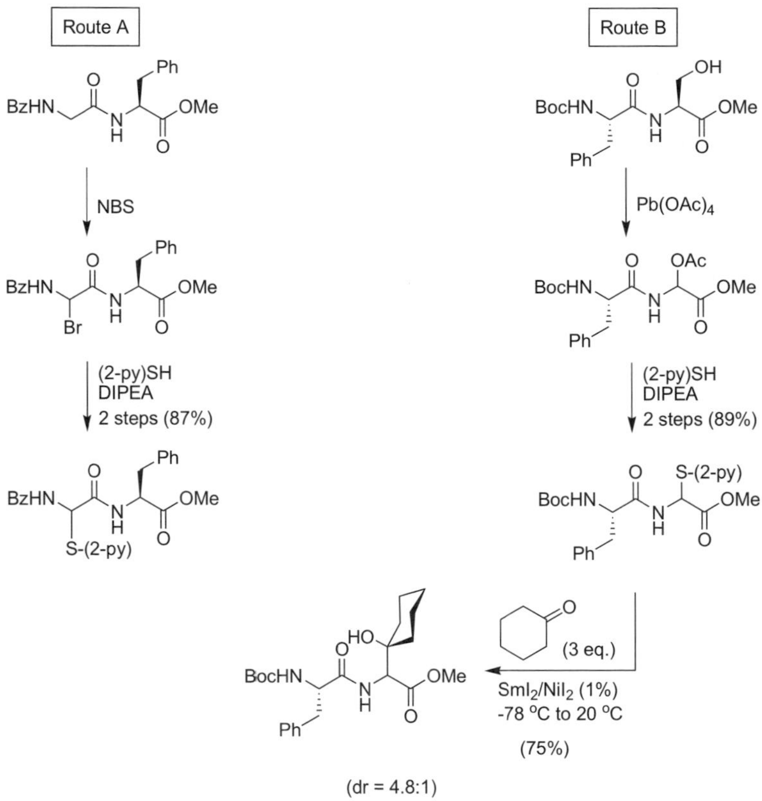

Scheme 8 Approaches to side chain introduction onto glycine or serine/threonine units of small peptides involving an SmI_2-mediated reductions step

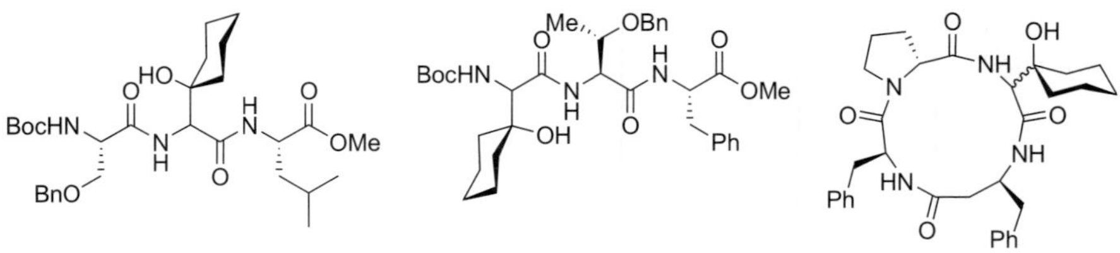

90%, dr = 1:1 82%, dr = 1:1 55%, dr = 2:1

Scheme 9 Examples of non-natural peptides synthesized via the direct introduction of side chains onto small peptides

generally good for the dipeptides but dropped with the longer peptides, which was illustrated by the extended reaction times required in the bromination step.

Alternatively, similar compounds could be prepared by a lead tetraacetate promoted degradation of serine residues to a glycine acetate and subsequent nucleophilic displacement with 2-mercaptopyridine (Scheme 8, Route B) [12]. The alkylation step was then achieved by the treatment of a solution of the peptide and carbonyl compound with samarium diiodide in the presence of 1 mol % NiI_2. Examples of the modified peptides are illustrated in Scheme 9. Yields of the peptides reached 90% and do not seem to be dependent on the chain length of the peptide.

The diastereoselectivities of these coupling reactions are low or essentially nonexistent, which is somewhat surprising considering the high oxophilicity of lanthanide metal ions, and consequently their good complexing abilities with amide functionalities, suggesting the potential for asymmetric induction from the adjacent chiral amino acids. Nevertheless, the method does allow for the preparation of peptides containing both an unnatural D- or L-amino acid unit from a single reaction, signifying that the methodology could be applicable for the generation of peptide-based libraries. The last example also reveals the possibility of performing these coupling reactions with small cyclic peptides of biological interest.

Scheme 10 is representative of the mechanism of these coupling reactions involving a captodatively stablized glycyl radical **15** from the initial reduction of the pyridyl sulfide group by the divalent lanthanide reagent. Further reduction of this carbon radical by a second equivalent of samarium diiodide leads to a Sm(III) enolate intermediate **16** of unknown geometry, which ultimately reacts with the carbonyl compound to give **17**.

Variations on the same theme were explored in the preparation of polycyclic β-lactams from the SmI_2-promoted cyclization of C4-keto-functionalized 1-[(benzoyloxy)(ethoxycarbonyl)methyl]-2-azetidinones (Scheme 11) [13]. Cyclization of the azetidinone **18** afforded stereoselectively the tricyclic [4.5.6] core structure of the potent antibiotic sanfetrinem as the major compound, whereas in many of the other cases tested (e.g., with **19**), cyclization

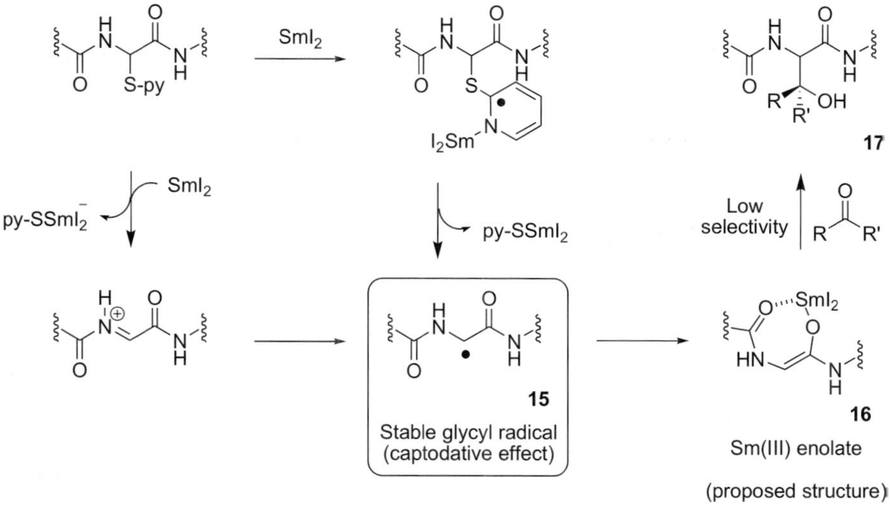

Scheme 10 Mechanism for the SmI_2-promoted reduction of the thiopyridine containing peptides and their coupling to carbonyl compounds

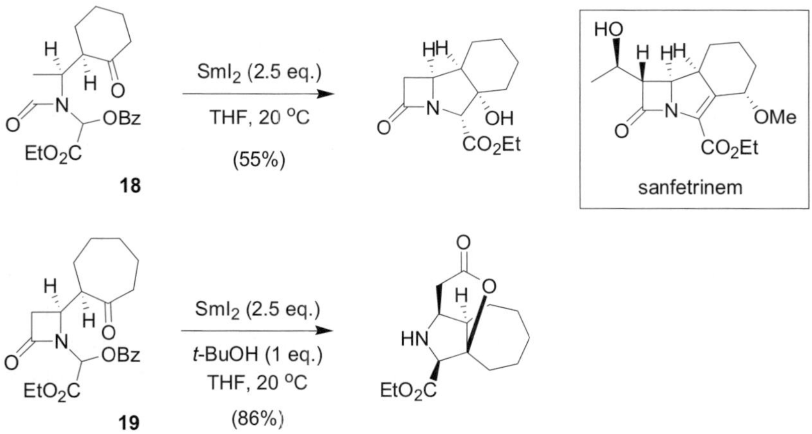

Scheme 11 An SmI_2-mediated reductive cyclization and rearrangement of β-lactams to functionalized proline derivatives

was followed by an N-to-O acyl migration involving cleavage of the β-lactam ring with the formation of functionalized proline derivatives. The cyclization reactions were also proposed to proceed via a glycine radical intermediate.

A different approach for the introduction of side chains onto glycine residues via similar radical intermediates was also reported by the same group in their exploitation of Zard's xanthate chemistry [14] with small peptide substrates (Scheme 12) [15]. The xanthate group was introduced in a similar fashion as for the synthesis of the corresponding pyridyl sulfides

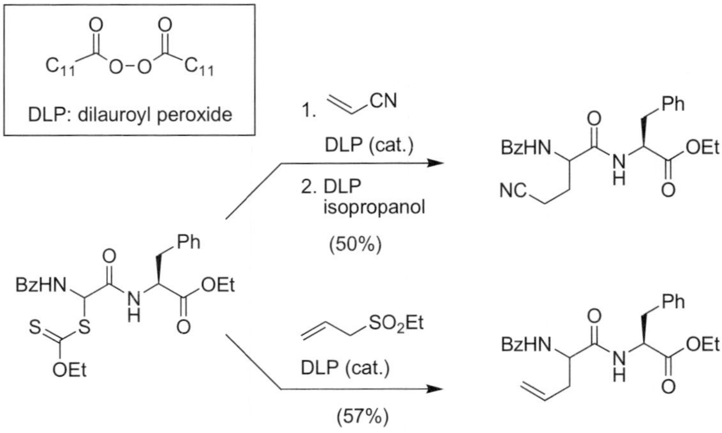

Scheme 12 Side chain introduction via xanthate-containing peptides

(see Scheme 8). Treatment of these xanthates with dilauroyl peroxide (DLP) in the presence of acrylonitrile leads to a C-alkylated peptide possessing a xanthate group in the side chain. Reductive desulfurization was conveniently carried out by refluxing a solution of the xanthate in isopropanol with a stoichiometric amount of DLP [16]. The nice feature of these combined two steps for the alkylation of peptides is that no toxic reagents, such as the tributyltin hydride traditionally used in radical chemistry, are required. In a second approach, glycyl xanthates were allylated with allyl ethyl sulfones without the necessity for a desulfurization step (Scheme 12, bottom), in accord with work published by Zard et al. [17].

The Skrydstrup team has demonstrated the ability of amino acids as their thioesters to undergo efficient acyl-like radical additions with electron-deficient alkenes in the presence of samarium diiodide, providing a facile entry to the common structure of a series of protease inhibitors [18]. It is noteworthy that earlier attempts to couple acyl radicals generated from amino acids to similar alkenes resulted in fast decarbonylation before the C – C bond forming step leading to γ-amino acids [19]. The absence of decarbonylation observed for the 4-pyridyl thioesters such as **20** was explained by the generation of a ketyl radical anion **21** via a SmI_2-mediated carbonyl reduction rather than an acyl radical intermediate. Subsequent addition to the alkene, reduction to the enolate, and protonation upon workup lead to a thiohemiacetal which ultimately generates the ketone product. Three examples are given in Scheme 13.

This coupling procedure with the thioesters proved sensitive to the substitution pattern of both the amino acid and alkene. In contrast, coupling reactions with the *N*-acyl oxazolidinone derivatives such as **22** proved to be much more effective (Scheme 14) [20]. Mechanistic studies suggested that an alternative pathway was operating in these cases, where reduction of the al-

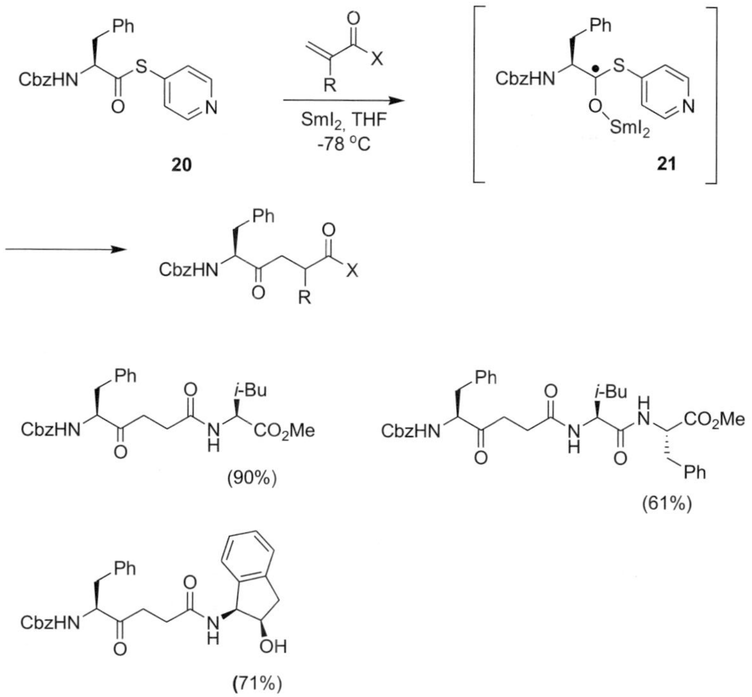

Scheme 13 Synthesis of peptide mimics via an acyl-like radical addition reaction involving 4-pyridylthio esters of amino acids

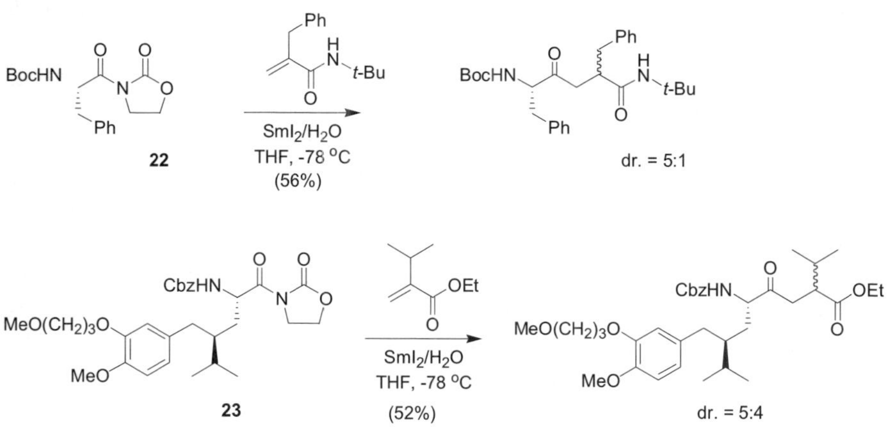

Scheme 14 Alternative acyl-like radical addition reactions with *N*-acyl oxazolidinone derivatives of amino acids

kene to a radical anion was followed by addition of this allylic-type radical to the *N*-acyl carbonyl group activated by a bidentate bound lanthanide ion [21]. Starting from the derivative **23** this reaction has recently been exploited to construct a significant fragment of the potent renin inhibitor, aliskerin (Lindsay and Skrydstrup, unpublished results). Similar attempts with the thioesters only led to traces of the coupling products.

2.2 Fragmentation Reactions

Fragmentation reactions such as decarboxylation of amino acids are well known and can be carried out for example by electrochemical oxidation to carboxyl radical intermediates [22]. The group of Hernández and Suárez have shown quite nicely that the combination of (diacetoxyiodo)benzene (DIB) and iodine provides a convenient alternative when carrying out such reactions with cyclic amino acids (Scheme 15) [23]. The carbon-centered radical **24** generated after decarboxylation undergoes a second oxidation step to give the *N*-acyliminium ion **25**, which can be trapped by nitrogen, oxygen, or carbon nucleophiles.

As an extension, the same laboratory demonstrated the possibility of performing one-pot oxidative decarboxylations with concomitant iodination when excess iodine was used (Scheme 16) [23]. As illustrated with the pipecolinic carbonate **26**, this sequence can be performed followed by trapping of the iminium ion intermediate with 2-hexen-1-ol leading to the functionalized piperidine **27** in 71% yield. Subsequently, a 5-*exo* cyclization can be performed to give the bicyclic *N*,*O*-acetal **28**, thus illustrating the potential of preparing complex systems in few steps.

Another cyclic system was also constructed from the 4-hydroxyproline derivative **29**, which was subjected to the decarboxylation–iodination protocol (Scheme 17). Allylation at C-1 followed by a radical-mediated allylation at C-2 and ring-closing metathesis provided the bicyclic ring system **30** known from many naturally occurring alkaloids.

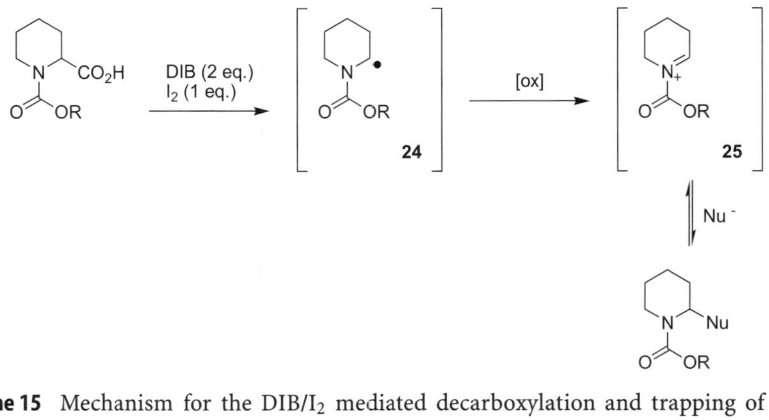

Scheme 15 Mechanism for the DIB/I_2 mediated decarboxylation and trapping of cyclic amino acids

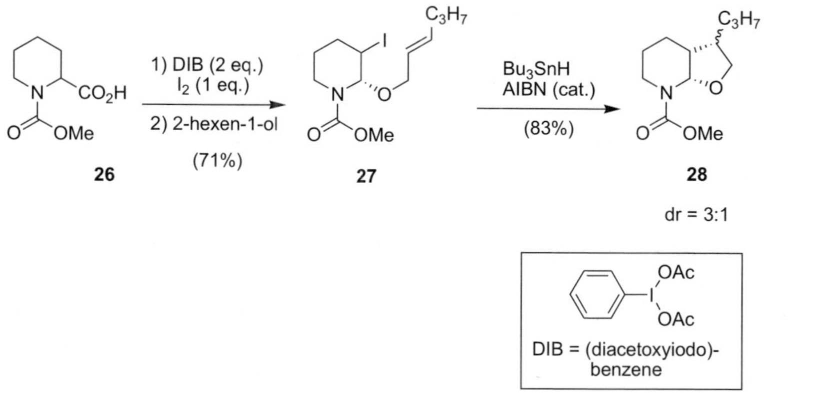

Scheme 16 An example of a one pot DIB/I_2-mediated decarboxylation and iodination sequence followed by a 5-exo radical cyclization step

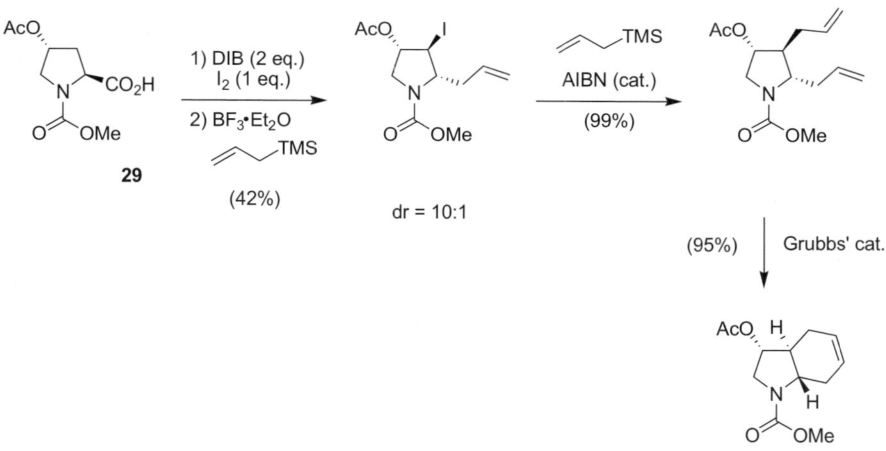

Scheme 17 A further example on the application of the decarboxylation-iodination protocol to the synthesis of bicyclic systems

The Spanish group has in addition exploited the same reagents for the β-scission of serine derivatives (Scheme 18) [24]. Oxidation of the alcohol **31** with DIB/I_2 to the corresponding alkoxy radical is followed by a rapid β fragmentation, liberating formaldehyde and generating the captodatively stabilized glycine radical **32**. A second oxidation step then affords the glycine cation intermediate **33**, which can be trapped with carbon nucleophiles to furnish new amino acids in good overall yields.

This fragmentation step was earlier reported in 1997 by Easton and co-workers in the tributyltin hydride-mediated conversion of nitrate esters of β-hydroxy-α-amino acids to glycyl radicals (Scheme 19a) [25]. Attack of the

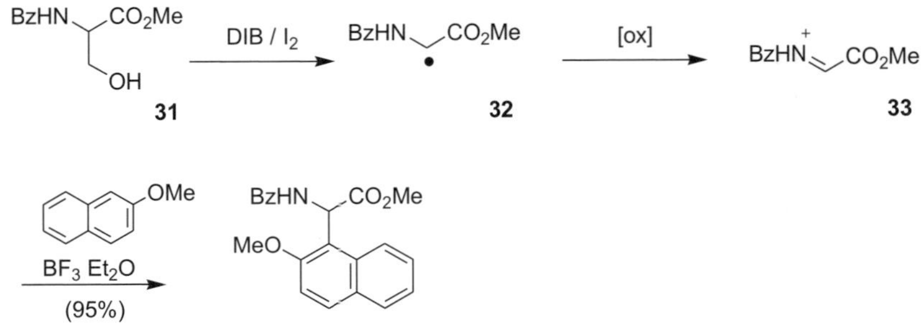

Scheme 18 DIB/I_2 mediated β-scission of serine derivatives

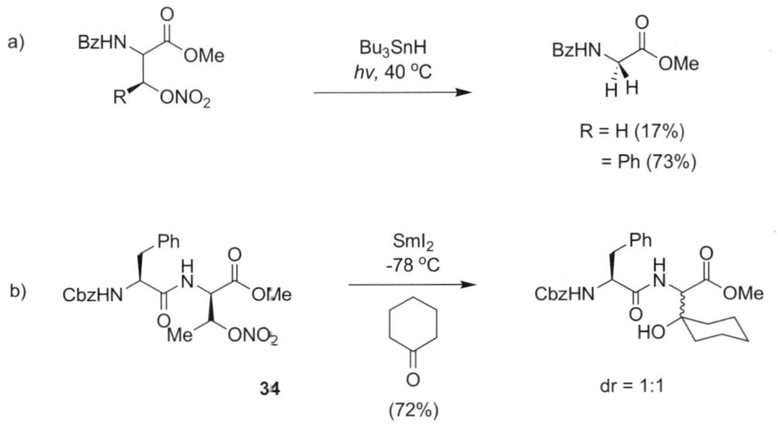

Scheme 19 Examples of the nitrate ester fragmentation of serine and threonine derivatives

tin-centered radical on the nitrate ester liberates an alkoxy radical which readily undergoes β-scission. The fragmentation step of the reactive alkoxy intermediate competes with hydrogen abstraction, and hence better results for β-scission were obtained with threonine compared to serine.

Recently, samarium diiodide was discovered to initiate the same fragmentation process with the threonine-containing peptide **34** (Scheme 19b) as well as to promote C – C bond formation after reduction of the glycyl radical intermediate, as discussed in Sect. 2.1.2 [26]. The method suggests the possibility of functionalizing peptides very simply and in only two steps.

The ability of diacyl peroxides to undergo photochemical decomposition with concomitant C – C bond formation was elegantly exploited by Vederas and coworkers for the preparation of functionalized amino acids (Scheme 20) [27, 28]. Symmetrical and unsymmetrical diacyl peroxides of amino acids were conveniently prepared from suitably protected aspartic and glutamic acids in a few steps. Photolysis of the peroxides with a 254-nm UV lamp at low temperature and in the absence of solvent led to decarboxylation

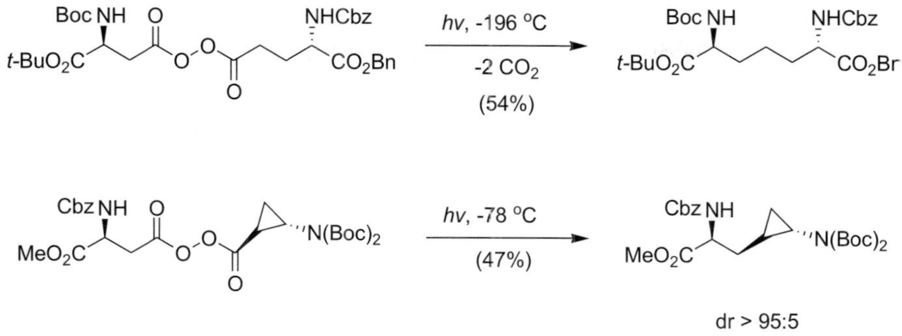

Scheme 20 Modified amino acid synthesis via the photochemical decomposition of diacyl peroxides

and cage recombination of the generated carbon-centered radicals. Under these conditions good yields of the modified amino acids were obtained, and in the case of the unsymmetrical diacyl peroxides no crossover products were detected. A nice application of this method to the synthesis of a fragment of an inhibitor of the vascular cell adhesion molecule 1 (VCAM-1) exhibiting anti-inflammatory properties was also reported.

3 Carbohydrates

Radical chemistry has been widely exploited for the modification of carbohydrates. In particular, tributyltin hydride-promoted deoxygenation provides a convenient method for the effective removal of hydroxyl groups without intervention of other functionalities. Stereoselective carbon–carbon bond formation at the anomeric center has also revealed its applicability for the preparation of carbon analogues of *O*-glycosides. The majority of this work has been reviewed in earlier publications and hence will not be covered in this account. Instead a selection of newer applications is provided here, which have been categorized according to the type of radical reaction carried out.

3.1 Stereoselective Introduction of Sugar Ring Substituents

The direct allylation of radical precursors (e.g., alkyl halides, thioacyl derivatives) with allyltributylstannane in the presence of an initiator represents a well-established protocol for carbon–carbon bond formation [29, 30]. This methodology provides a convenient means for introducing an allyl group to the anomeric carbon of carbohydrates [31]. In their recent work on the preparation of building blocks for *C*-glycoside synthesis, Postema and coworkers extended this radical allylation methodology to the stereoselective introduction of carbon substituents at various positions of the sugar ring in glucose and galactose (Scheme 21) [32]. The stereochemistry of the protected hydroxyl groups in the individual monosaccharides dictates the stereochemical course of these reactions. As an example of this influence, radical allylation of the iodide **35** provided a mixture of equatorial and axial C-4-allylated products in a ratio of 6 : 1.

On the other hand, Mikkelsen and Skrydstrup disclosed that the equatorial product was exclusively formed when allylation of the mannose derivative **36** possessing an axially oriented C-2 acetate was attempted [33]. The absence of the axial product in this latter case was explained by an intervening 1,3-diaxial interaction in the C – C bond forming step (Scheme 22).

An approach to the free radical acylation of carbohydrate derivatives was reported by Kim and coworkers [34]. This group had earlier demonstrated the successful acylation of alkyl iodides with sulfonyl oxime ether derivatives promoted by tributyltin hydride [35–37]. However, application of this chemistry to the synthesis of branched monosaccharides required some modification of the experimental protocol in order to avoid the competing side reactions observed in the C acylation at C-1 of the sugar ring (e.g., simple reduction or 1,2-acyloxy migration of the intermediate anomeric radical). Replacement of the tin hydride reagent with hexamethylditin and photolytic initiation of the reactions provided significantly improved yields of the *C*-glycoside. Other positions on the sugar ring were also successfully functionalized, as depicted in Scheme 23.

Variations on this theme were published by the groups of Zard and Renaud, who disclosed novel techniques for the olefination and azidination of radicals

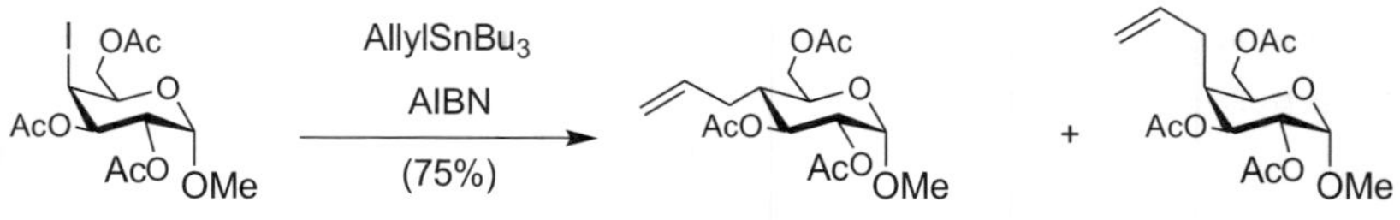

35

gluco:galacto ratio = 6:1

Scheme 21 Stereoselective free radical-mediated allylation of carbohydrates

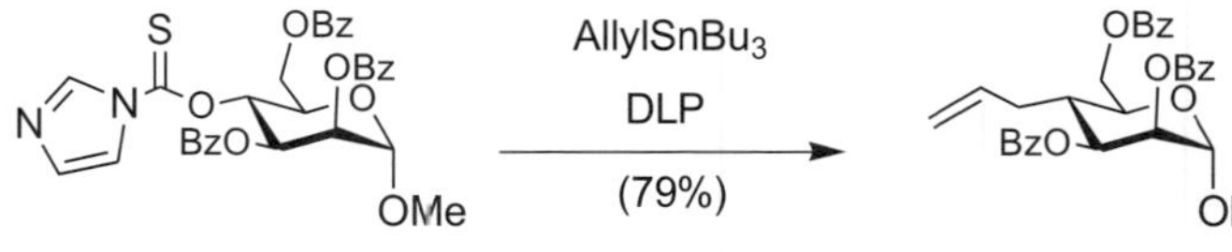

36

Scheme 22 Stereoselective allylation of a mannose derivative

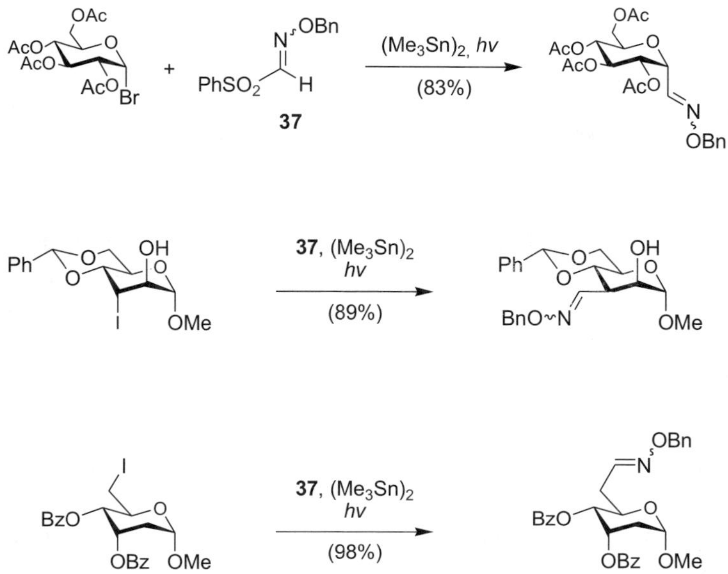

Scheme 23 Free radical-mediated acylation of carbohydrates

from readily available vinyl sulfones and sulfonyl azides, respectively [38, 39]. As exemplified in Scheme 24, the methodology provides an interesting means of preparing stereoselectively α-*C*-vinyl glycosides and α-anomeric azides from the corresponding glycosyl xanthates. As with the above C acylation chemistry, these reactions proceed through similar mechanisms. Alkyl radical addition to the sulfonyl compound leads to the generation of a sulfonyl radical. With the benzenesulfonyl derivatives the sulfonyl radical is sufficiently stable and eventually attacks the ditin reagent, regenerating the trialkyltin radical. In the case of the ethylsulfonyl radical, SO_2 is liberated forming the ethyl radical which

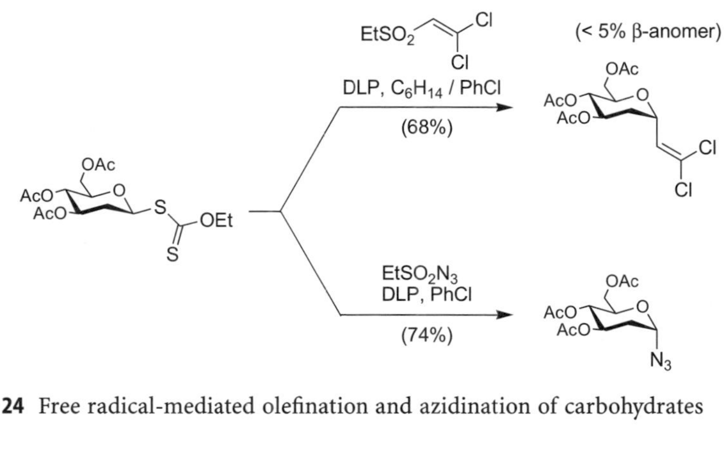

Scheme 24 Free radical-mediated olefination and azidination of carbohydrates

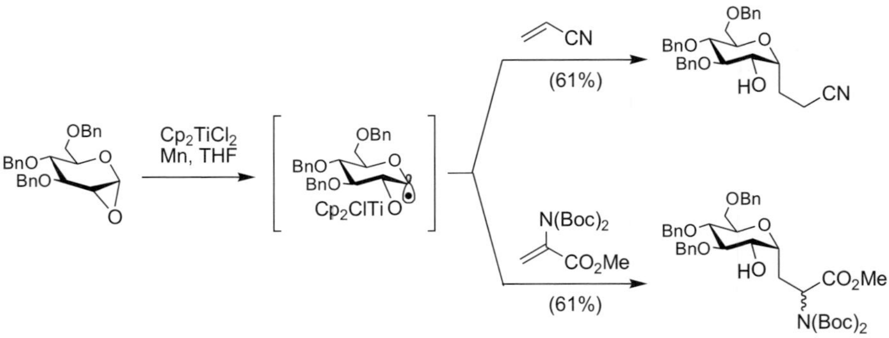

Scheme 25 C-Glycoside synthesis via the reductive opening of 1,2-anhydrosugars

is sufficiently effective for generating carbon radicals from xanthates in the absence of tin hydride or ditin reagents.

Dötz and da Silva as well as Parrish and Little recently provided an alternative procedure for the preparation of *C*-glycosides starting from 1,2-anhydrosugars [40, 41]. Epoxides can selectively undergo a C – O bond cleavage to a corresponding β-alkoxy radical with the low-valent titanium(III) reagent, Cp_2TiCl [42, 43]. The regioselectivity of the opening is guided by the stability of the two possible radicals formed. When 1,2-anhydrosugars, which were easily prepared from the corresponding glycal with dimethyl dioxirane, were treated with this low-valent titanium complex in the presence of a trapping agent, the α-*C*-glycosides were produced in yields attaining 61%, as depicted by the examples in Scheme 25.

3.2
Extension of Sugar Ring Substituents

Two interesting methods appeared in 2004 from the Zard group for the extension of the sugar ring substituents using radical chemistry. In the first report, a mild method for a one-carbon addition of unactivated alkenes was developed. In this work, the xanthate derivative of 1,3-dithiane oxide **38** in the presence of lauroyl peroxide was found to undergo radical addition to alkenes to provide the radical adducts **39**, in most cases in reasonably good yields (Scheme 26) [44]. The corresponding parent 1,3-dithiane was not sufficiently reactive to unactivated alkene addition.

As the product **39** still contains the xanthate group, alternative radical transformations can be performed, which also include its removal with tris(trimethylsilyl)silane. Further manipulations of the dithiane ring comprise desulfurization or hydrolysis, thus confirming the use of such xanthates as the synthetic equivalent of a methyl and formyl radical. Two examples of the use of this chemistry for the extension of alkene-containing sugars were successfully examined, as illustrated in Scheme 27.

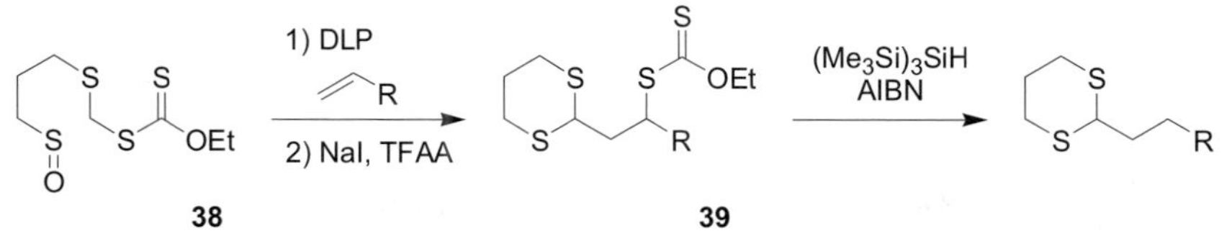

Scheme 26 Method for the free radical-mediated one carbon extension of unactivated alkenes

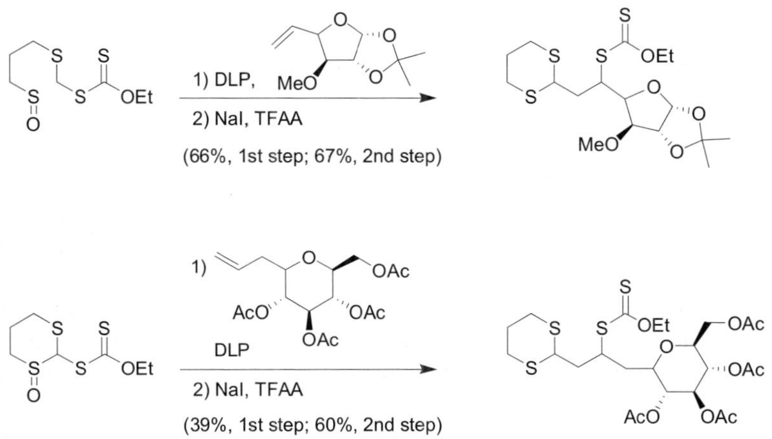

Scheme 27 Application of the one carbon extension approach to carbohydrates

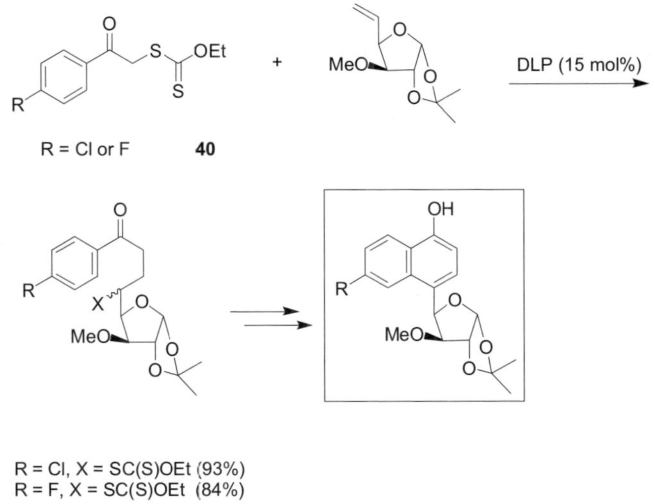

Scheme 28 Application of the xanthate chemistry to the extension of olefin-branched carbohydrates

In the second report, Zard and collaborators examined the addition of the xanthate derivatives **40** to different olefin-branched glycosides as the initial step for the preparation of an important class of *C*-aryl glycosides, namely the gilvocarcins [45]. The addition steps are high yielding and the radical adducts can effectively be desulfurized, ring-closed, and subsequently aromatized (Scheme 28).

3.3 Fragmentation Reactions

A series of reports from the Suárez group over the last 12 years have disclosed an impressive array of fragmentation reactions with carbohydrates initiated by the organohypervalent iodine reagent, diacetoxyiodobenzene (DIB) or iodosylbenzene in combination with iodine. Previous work from the laboratory has demonstrated the ability of hypervalent iodine reagents to promote the oxidative β fragmentation of glycopyranose and glycofuranose derivatives to C-2 radicals [46]. In recent years, several extensions of this chemistry from the same group have been published. As an example, the Spanish group has shown that the C-5 carbon of the furanoside derivative **41** can be transformed into the chloride **42** with DIB and ICl (Scheme 29) [47]. Subsequent treatment with allyltributyltin and azobisisobutyronitrile (AIBN) catalyst provided an interesting method for a formal extension at C-5.

The C-1-protecting group is important for the success of this fragmentation process, as exchange of the benzoyl group to a benzyl group leads to significant amounts of products from intramolecular hydrogen abstraction. This competing pathway was also observed with other substrates such as the protected polyol **43**, which provided both the products of fragmentation (Scheme 30, Route A) and a hydrogen abstraction pathway followed by an oxidation and cyclization step (Route B).

The ability of the oxygen-centered radical to abstract hydrogen from an adjacent carbon allowed the same team to develop a method for the selective removal of methoxy protecting groups in carbohydrates (Scheme 31) [48]. As exemplified by the substrate **44**, treatment with DIB/iodine promotes formation of the alkoxy radical which abstracts a hydrogen from the C-4 methoxy

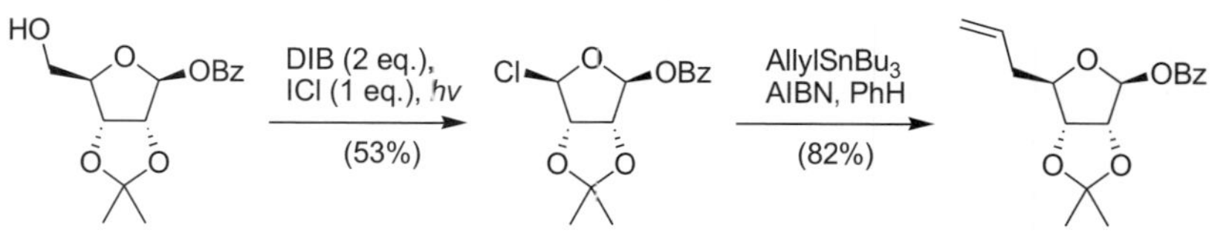

41 **42**

Scheme 29 DIB/I_2-mediated oxidative β-fragmentation of a furanose derivative followed by side chain introduction

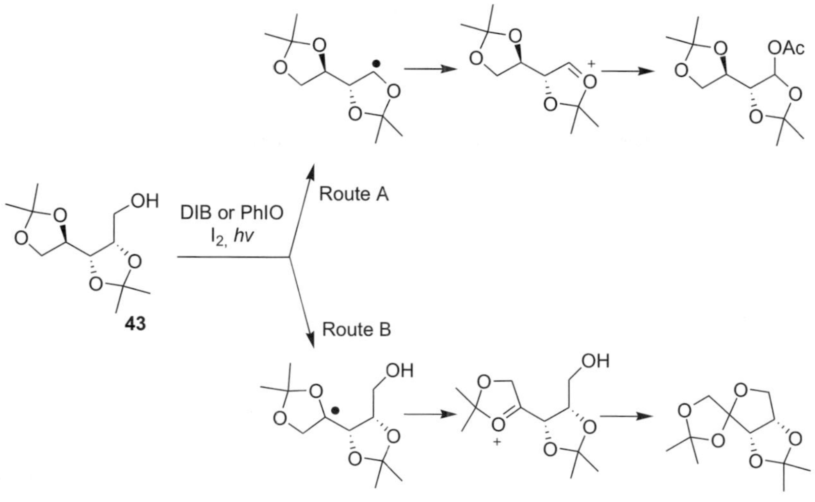

Scheme 30 An example of an oxidative β-fragmentation of a sugar derivative with concomitant hydrogen abstraction

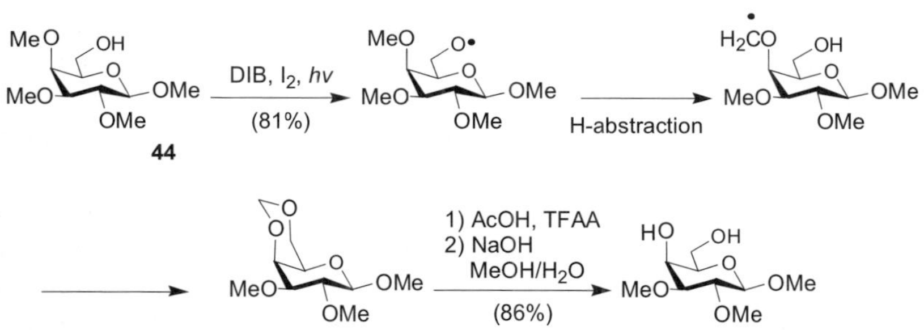

Scheme 31 DIB/I_2 promoted demethylation via an alkoxy radical intermediate

group via a seven-membered cyclic transition state. Oxidation and cyclization lead to the cyclic acetal, which can easily be hydrolyzed to the diol. Several examples of this methodology have been provided.

Sulfone-containing 2-deoxy-glycopyranoses are easily accessible from glycals. When such derivatives were exposed to the hypervalent iodine reagent with iodine, a fragmentation process was carried out leading to the formation of chiral vinyl sulfones after treatment with DBU, as shown for the substrate **45** (Scheme 32) [49]. Initial alkoxy radical fragmentation leads to the primary alkyl radical **46** which was proposed to be trapped by iodine providing the β-iodosulfone **47**. Subsequent base-promoted elimination generates the desired vinyl sulfone **48**.

Other options have also been established for promoting alkoxy radical fragmentations. Two different methods were proposed by Suárez and collab-

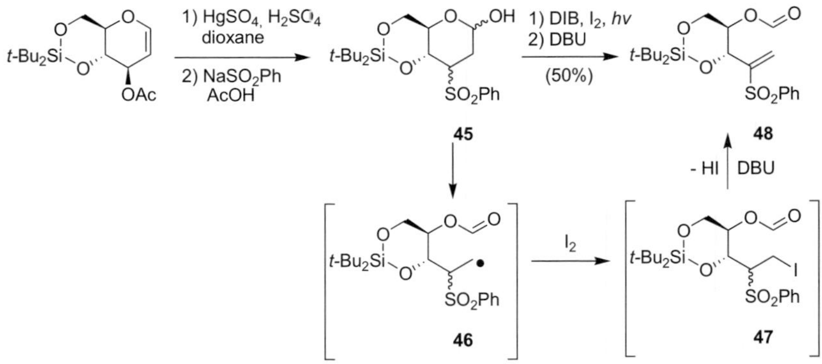

Scheme 32 Synthesis of chiral vinyl sulfones via a DIB/I_2-promoted oxidative fragmentation of carbohydrates

orators. A series of anomeric nitrates of furanoses and pyranoses were found to effectively undergo fragmentation when subjected to tributyltin hydride and AIBN in refluxing benzene (Scheme 33) [50]. The corresponding alditols from this one-carbon scission step were obtained in good yields, thus demonstrating the usefulness of this route to the preparation of chiral synthons from carbohydrates.

The nitrate esters provided satisfactory yields for this fragmentation reaction, whereas the synthesis of these esters posed certain drawbacks depending on the sugar. These included the instability of certain cases examined, as well as the nitrating conditions used (Ac_2O, HNO_3) which proved incompatible with easily oxidized or acid-sensitive substrates. However, a novel means for the generation of nitrate esters from alcohols under nonacidic conditions has recently been disclosed [26]. An alternative and complementary approach to the preparation of alkoxy radicals at the anomeric center was therefore investigated using *N*-phthalimido glycosides [50]. Such compounds can easily be prepared by substitution of glycosyl halides with *N*-hydroxyphthalimide or under Mitsunobu conditions from the corresponding alcohol. Reduction with Bu_3SnH/AIBN also proceeds well and provides products of one-carbon fragmentation. An example of this sequence is shown in Scheme 34 (top), for which the corresponding nitrate ester could not be prepared under the nitration conditions employed. Finally, an extension of this radical chemistry was

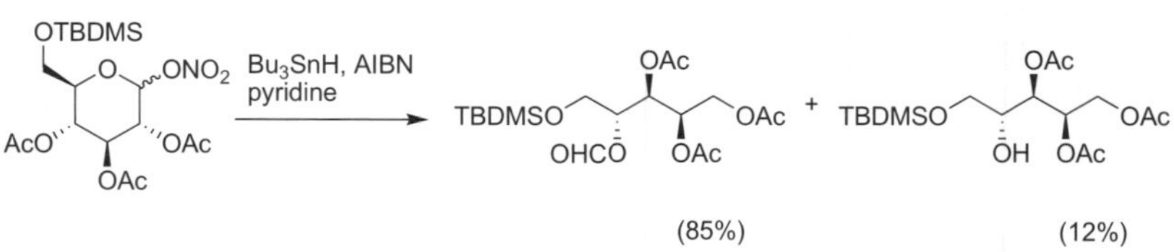

Scheme 33 Fragmentation of anomeric nitrates

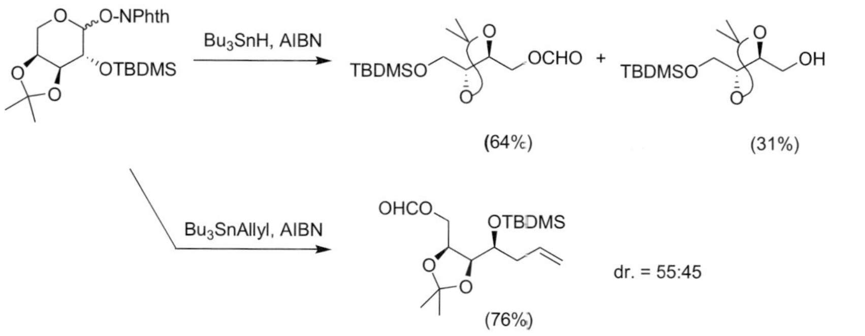

Scheme 34 Fragmentation of *N*-phthalimido glycosides

revealed later where the intermediate radical after fragmentation was trapped with allyltributyltin, providing alternative chiral synthons (Scheme 34, bottom) [51].

An interesting and effective reductive radical fragmentation route to the synthesis of deoxysugars was revealed by Crich and Yao in 2003 [52]. Their work is based on an earlier publication by the Roberts group demonstrating the ability to cleave 4,6-benzylidene-protected glycosides to their corresponding 6-deoxy-4-benzoyl derivative via radical chain reactions promoted by thiols (protic polarity-reversal catalyst) [53]. This reaction provides a more direct route to such compounds, in contrast to the better known two-step sequence involving initial treatment with *N*-bromosuccinimide (NBS) followed by reduction of the intermediate primary bromide [54–56]. However, Crich and Yao discovered that application of the Roberts technology proved incompatible with sugars possessing benzyl ether. In order to remedy this problem an alternative method was devised to generate the key α-benzylidene radical via the modified 4,6-benzylidene protecting group (Scheme 35).

Upon treatment with tributyltin hydride and AIBN, a reductive fragmentation ensued providing the corresponding 6-deoxy compounds in the manno

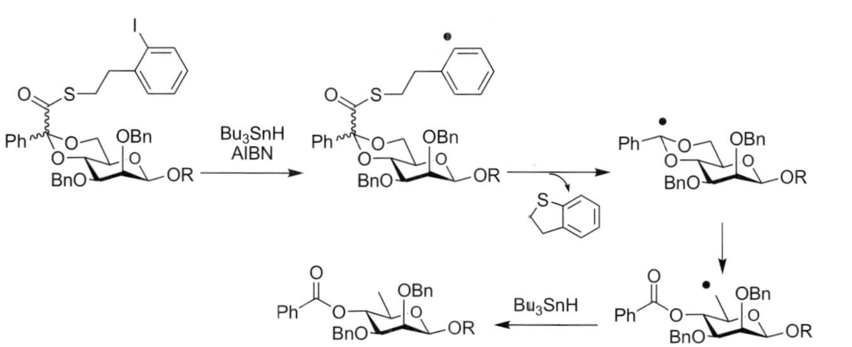

Scheme 35 Mechanism for the reductive fragmentation of a modified 4,6-benzylidene protecting group

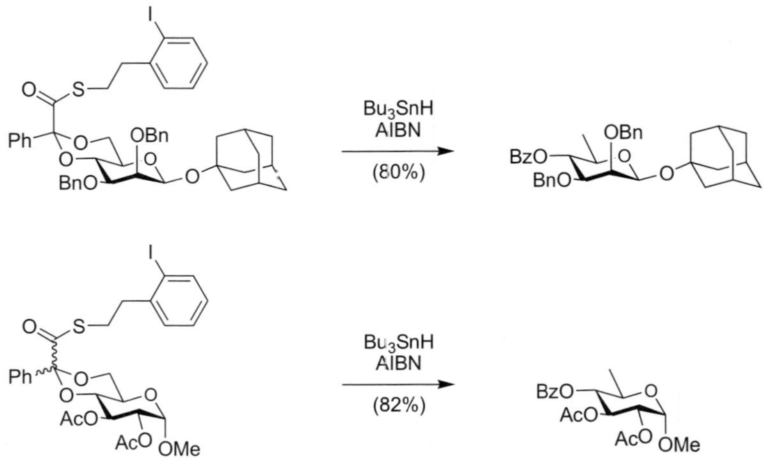

Scheme 36 Applications of the reductive fragmentation sequence for the synthesis of 6-deoxy sugars

and gluco series, whereas the galactosides furnished the 4-deoxy sugar as the major product (Scheme 36). Most rewarding was the compatibility of this diol protecting group with Crich's glycosylation conditions for the preparation of β-mannosides.

In the mid-1990s, Schwartz and collaborators reported in a series of papers the ability of Cp_2TiCl to reduce glycosyl bromides to their corresponding glycosyl radicals, followed by trapping with electron-poor alkenes to give *C*-glycosides or reduction to the anion followed by β elimination to afford glycals [57]. In order to reduce the number of equivalents of the Ti(III)-based reducing agent in the glycal-forming reactions, a catalytic version of this

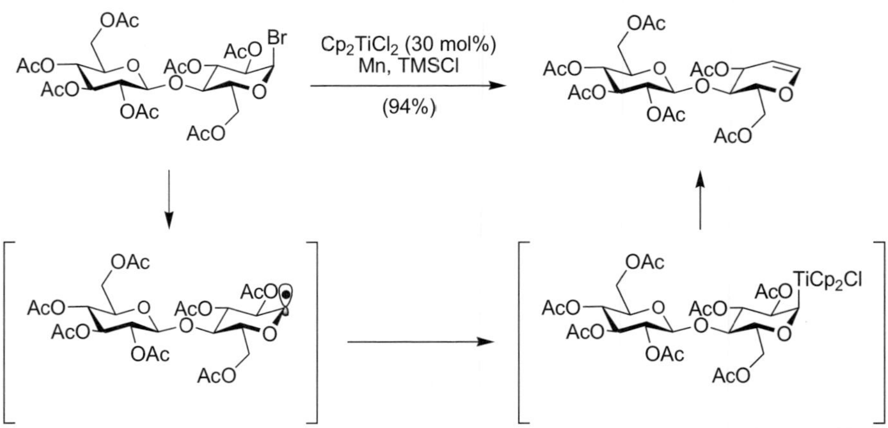

Scheme 37 Synthesis of glycals via a Cp_2TiCl-mediated reduction of glycosyl bromides

reaction was reported by Skrydstrup and Hansen [58]. Gansäuer and collaborators had earlier shown, in both pinacol couplings and reductive epoxide openings, the ability to regenerate the low-valent titanium reagent with the combination of a stoichiometric reductant such as manganese or zinc and a trialkylsilyl chloride or collidinium chloride for the liberation of the Ti(IV) species from the product [59–61]. Adaptation of a similar protocol to glycal synthesis was achieved as exemplified in Scheme 37 by using 30 mol % Cp_2TiCl_2, which is a significant reduction of the catalyst loading considering that 200 mol % is required in the stoichiometric version.

3.4 Cyclization Reactions

Two interesting radical-based cyclizations with carbohydrates have been published by two separate Spanish groups. In the first, Suárez and collaborators applied their alkoxy radical generation protocol to the preparation of functionalized bicyclic systems [62]. As illustrated with the *C*-glycoside **49** in Scheme 38, alkoxy radical generation is initiated by the treatment with DIB/I_2. A 1,5-hydrogen abstraction leads to a carbon-centered radical which is then oxidized to the corresponding oxonium ion intermediate **50**. Finally, ring closure with the primary alcohol affords the dioxabicycloheptane **51** in good yield. The reaction sequence proved to be quite effective for a number of *C*-glycosides tested, and provides a useful methodology for the preparation of chiral synthons.

An alternative ring-closing protocol from the Chiara group was shown in a report from 2005 [63]. Earlier work from this group had demonstrated the

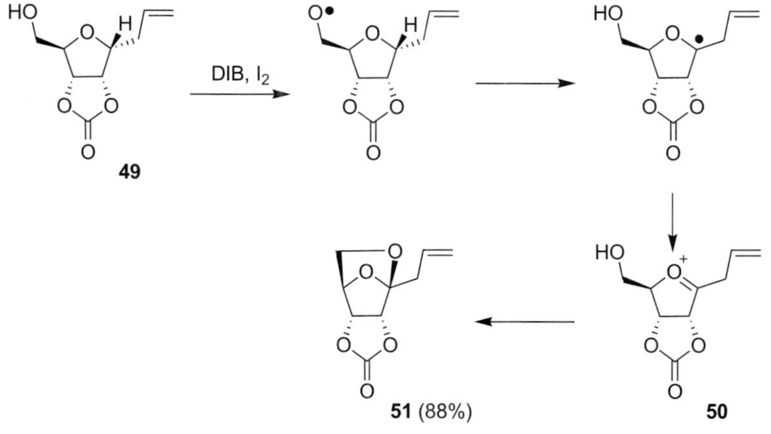

Scheme 38 An example of a DIB/I_2-mediated oxidation-hydrogen abstraction-cyclization sequence

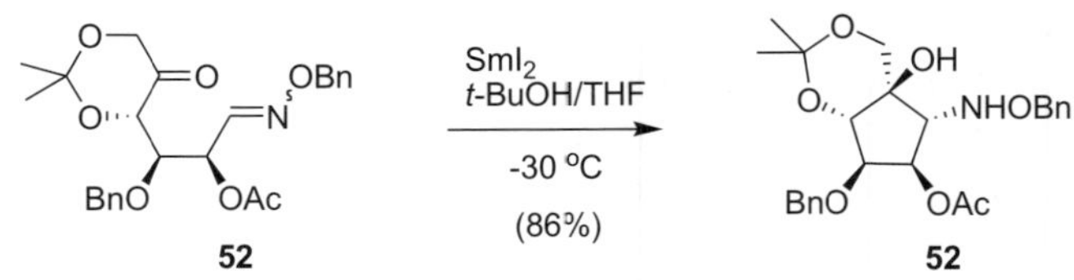

Scheme 39 An example of a SmI_2-mediated intramolecular pinacol coupling

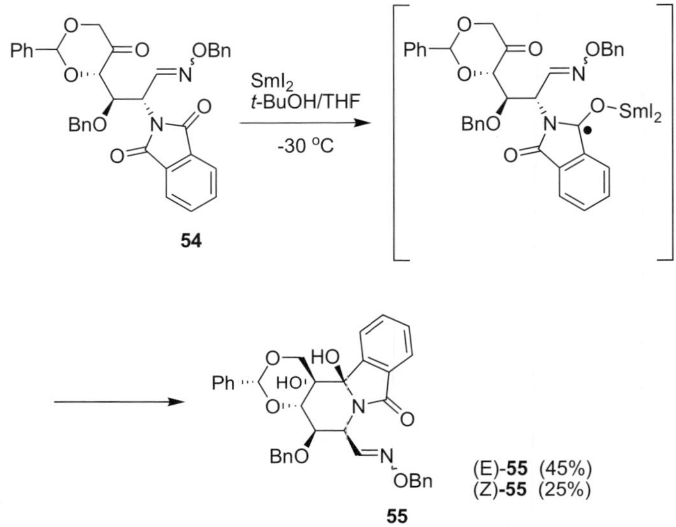

Scheme 40 Participation of the phthalimido group in an SmI_2-promoted intramolecular pinacol coupling

ability of keto-oximes such as **52** to undergo efficient SmI_2-promoted pinacol cyclization to the cyclopentitol **53** (Scheme 39) [64].

Attempts to extend this work to the keto-oxime substrate **54** derived from D-glucosamine with an *N*-phthalimido group resulted in the formation of a completely different product (Scheme 40). In this case, cyclization was initiated by reduction of the phthalimido carbonyl group to its corresponding ketyl radical anion followed by cyclization onto the ketone, providing an α-hydroxylactam **55** which was proposed to be a potentially useful scaffold for diversity-oriented synthesis.

4 Conclusion

As illustrated by the number of examples given in this review, radical chemistry presents itself as an interesting and often mild method for the structural modification of Nature's most important biomolecules. The applications include both selective side-chain and backbone modifications of amino acids and peptides, as well as transformations of carbohydrate substituents and cyclization reactions. Undoubtedly, many more applications and developments in radical chemistry will follow in the near future, considering the rapid developments in recent years. These will provide synthetic chemists with new opportunities to perform even more delicate and selective structural changes.

References

1. Easton CJ (1997) Chem Rev 97:53
2. Giese B, Wettstein P, Stähelin C, Barbosa F, Neuburger M, Zehnder M, Wessig P (1999) Angew Chem Int Ed 38:2586
3. Fuji K, Fuji K, Kawabata T (1998) Chem Eur J 4:373
4. Sinicropi A, Barbosa F, Basosi R, Giese B, Olivucci M (2005) Angew Chem 117:2442
5. Jonas M, Blechert S, Steckhan E (2001) J Org Chem 66:6896
6. Zhang X, Yeh SR, Hong S, Freccero M, Albini A, Falvey DE, Mariano PS (1994) J Am Chem Soc 116:4211
7. Yoon UC, Jin YX, Oh SW, Park CH, Park JH, Campana CF, Cai X, Cuesler EN, Mariano PS (2003) J Am Chem Soc 125:10664
8. Balraju V, Reddy DS, Periasamy M, Iqbal J (2005) Tetrahedron Lett 46:5207
9. Andrukiewicz R, Loska R, Prisyahnyuk V, Stalinski K (2003) J Org Chem 68:1552
10. Stork G, Sher PM (1986) J Am Chem Soc 108:303
11. Ricci M, Blakskjær P, Skrydstrup T (2000) J Am Chem Soc 122:12413
12. Blakskjær P, Gavrila A, Andersen L, Skrydstrup T (2004) Tetrahedron Lett 45:9091
13. Jacobsen MF, Turks M, Hazell R, Skrydstrup T (2002) J Org Chem 67:2411
14. Zard SZ (1997) Angew Chem Int Ed Engl 36:672
15. Blakskjær P, Pedersen L, Skrydstrup T (2001) J Chem Soc Perkin Trans I 910
16. Liard A, Quiclet-Sire B, Zard SZ (1996) Tetrahedron Lett 37:5877
17. Quiclet-Sire B, Seguin S, Zard SZ (1998) Angew Chem Int Ed Engl 37:2864
18. Blakskjær P, Høj B, Riber D, Skrydstrup T (2003) J Am Chem Soc 125:4030
19. Stojanovic A, Renaud P (1997) Synlett 181
20. Jensen CM, Lindsay KB, Taaning RH, Karaffa J, Hansen AM, Skrydstrup T (2005) J Am Chem Soc 127:6544
21. Hansen AM, Lindsay KB, Skrydstrup T, unpublished results
22. Matsumura Y, Shirakawa Y, Satoh Y, Umimo M, Tanaka T, Maki T, Onomura O (2000) Org Lett 2:1689
23. Boto A, Hernández R, León Y, Suárez E (2001) J Org Chem 66:7796
24. Boto A, Hernández R, Montoya A, Suárez E (2002) Tetrahedron Lett 43:8269
25. Easton CJ, Ivory AJ, Smith CA (1997) J Chem Soc Perkin Trans 2:503
26. Gavrila A, Andersen L, Skrydstrup T (2005) Tetrahedron Lett 46:6205
27. Spantulescu MD, Jain RP, Derksen DJ, Vederas JC (2003) Org Lett 5:2963,
28. Jain RP, Vederas JC (2003) Org Lett 5:4669
29. Postema MHD (1995) C-glycoside synthesis, 1st edn. CRC, Boca Raton
30. Levy DE, Tang C (1995) The chemistry of C-glycosides, 1st edn. Elsevier, Oxford
31. Keck GE, Enholm EJ, Yates JB, Wiley MR (1985) Tetrahedron 41:4079
32. Postema MHD, Piper JL, Liu L, Shen J, Faust M, Andreana P (2003) J Org Chem 68:4748

33. Mikkelsen LM, Skrydstrup T (2003) J Org Chem 68:2123
34. Chung WJ, Jeon GH, Yoon JY, Kim S (2002) Bull Korean Chem Soc 23:1187
35. Kim S, Lee IY, Yoon JY, Oh DH (1996) J Am Chem Soc 118:5138
36. Kim S, Lee IY, Yoon J (1997) Synlett 475
37. Kim S, Yoon JY (1997) J Am Chem Soc 119:5982
38. Bertrand F, Quiclet-Sire B, Zard SZ (1999) Angew Chem Int Ed 38:1943
39. Ollivier C, Renaud P (2001) J Am Chem Soc 123:4717
40. Dötz KH, Gomes da Silva E (2000) Tetrahedron 56:8291,
41. Parrish JD, Little RD (2002) Org Lett 4:1439
42. Gansauer A, Bluhm H (2000) Chem Rev 100:2771
43. Gansauer A, Lauterbach T, Narayan S (2003) Angew Chem Int Ed 42:5556
44. de Greef M, Zard SZ (2004) Tetrahedron 60:7781
45. Cordero-Vargas A, Quiclet-Sire B, Zard SZ (2004) Tetrahedron Lett 45:7335
46. Francisco CG, González Martin C, Suárez E (1998) J Org Chem 63:2099
47. Boto A, Hernández D, Hernández R, Suárez E (2003) J Org Chem 68:5310
48. Boto A, Hernández D, Hernández R, Suárez E (2004) Org Lett 6:3785
49. Alonso-Cruz CR, León EI, Ortiz-López FJ, Rodríguez MS, Suárez E (2005) Tetrahedron Lett 46:5265
50. Francisco CG, León EI, Martin A, Moreno P, Rodríguez MS, Suárez E (2001) J Org Chem 66:6967
51. Martín A, Pérez-Martín I, Suárez E (2002) Tetrahedron Lett 43:4781
52. Crich D, Yao Q (2003) Org Lett 5:2189
53. Roberts BP, Smits TM (2000) Tetrahedron Lett 42:3663
54. Hanessian S, Plessas NR (1969) J Org Chem 34:1035
55. Hanessian S, Plessas NR (1969) J Org Chem 34:1045
56. Hanessian S, Plessas NR (1969) J Org Chem 34:1053
57. Spencer RP, Schwartz J (2000) Tetrahedron 56:2103
58. Hansen T, Daasbjerg K, Skrydstrup T (2000) Tetrahedron Lett 41:8645
59. Gansauer A (1998) Synlett 801
60. Gansauer A, Bluhm H (2000) Chem Rev 100:2771
61. Gansauer A, Lauterbach T, Narayan S (2003) Angew Chem Int Ed 42:5556
62. Francisco CG, Herrera A, Suárez E (2002) J Org Chem 67:7439
63. Chiara JL, García Á, Cristóbal-Lumbroso G (2005) J Org Chem 70:4142
64. de Gracia IS, Bobo S, Martín-Ortega MD, Chiara JL (1999) Org Lett 1:1705

Top Curr Chem (2006) 264: 163–200
DOI 10.1007/128_021

Published online: 13 December 2005

Unusual Radical Cyclisations

John C. Walton

School of Chemistry, EaStCHEM, University of St. Andrews, Purdie Building,
St. Andrews, Fife KY16 9ST, UK
jcw@st-and.ac.uk

Abstract Methods for solving the problems associated with reverse ring opening of small-ring intermediates are described. For three- and four-membered rings particular patterns of substituents are required. Rapid methods of trapping small cyclic structures and extracting them from unfavourable equilibria depend on special new reagents. The new methods enable mono- and polycyclic cyclopropanes and cyclobutanes containing a range of functionality to be readily prepared. Furthermore, several of the newer reagents allow a good measure of stereocontrol to be exercised. Molecules containing the azetidinone ring system (β-lactam) have been made by no less than four different radical-based disconnections. Syntheses of medium-sized rings invoke *endo*-type cyclisations and are favoured by precursors containing electron-withdrawing groups at the terminus of the radical acceptor group. Promising routes to seven- and eight-membered lactams are described as well as to 11- and 12-membered lactams. Alternative radical-based strategies for making rings by intramolecular cyclo-dimerisations of diradicals and by intramolecular cyclo-coupling of radicals with carbanionic centres are included.

Keywords Synthetic methods · Radical cyclisations · Cyclopropanes · Lactams · $S_{RN}1$ reactions

Abbreviations

AIBN	Azoisobutyronitrile
cpm	Cyclopropylmethyl
C^{3x}	3-*exo* Cyclisation
C^{4n}	4-*endo* Cyclisation
Coll	Collidine
DEPO	Diethylphosphine oxide
DFT	Density functional theory
DTBP	Di-*tert*-butyl peroxide
EPHP	*N*-Ethylpiperidine hypophosphite
FGI	Functional group interconversions
HMPA	Hexamethylphosphoramide
LDA	Lithium diisopropylamide
MAP	4-Methoxyacetophenone
PTOC	*N*-Hydroxypyridine thione
TMM	Trimethylenemethane
trityl	Triphenylmethyl

1 Introduction

Three important global strategies for making molecular rings can be distinguished. The first, and probably most popular, relies on intramolecular nucleophilic substitution or addition and enables high yields of many five-, six- and seven-membered rings to be obtained. The second strategy employs pericyclic reactions of the Diels–Alder, 1,3-dipolar cycloaddition and related types and promotes ready access to five- and six-membered rings. The third emerging strategy is based on intramolecular free radical addition to unsaturated acceptor groups. With this third strategy, five-, six- and seven-membered rings are again easily prepared. Three-membered rings are also comparatively easily made by all three strategies, but reverse ring opening is often a problem. Four-membered rings and larger rings (> 7) are comparatively difficult to make.

A spectacular increase in the development and application of radical-based synthetic methodology for ring closures has taken place over the last 20 years or so [1–6]. By far the most popular radical-based strategy relies on intramolecular addition of a radical (carbon or heteroatom centred) to an unsaturated bond [Scheme 1 (1)]. Cyclisations of this type giving three-, four-, eight- and larger-membered rings are considered as unusual [7] and are covered in this review.

A second strategy, analogous to nucleophilic substitution, involves intramolecular homolytic substitution at atom A with loss of a new radical L [S_Hi, Scheme 1, (2)]. In this process the atom undergoing substitution (A) cannot be carbon, except for a few special situations, and hence the strategy is limited in scope [8–10]. The central element in the third strategy is an

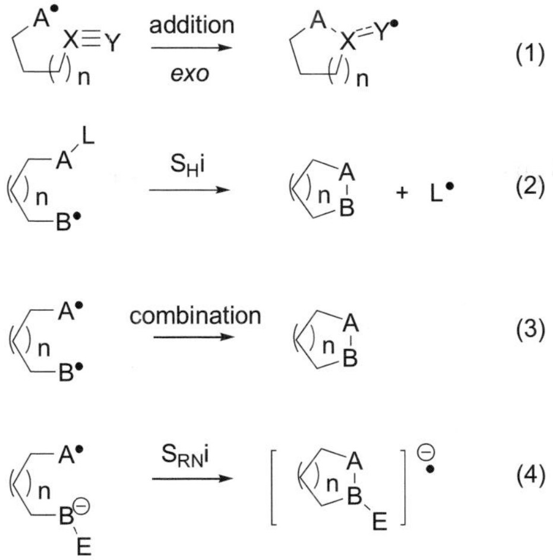

Scheme 1 Radical-mediated ring closure strategies

intramolecular dimerisation reaction of a diradical [Scheme 1, (3)]. The radical centre is destroyed in this process and hence this cyclo-combination does not lend itself to chain propagation. The fourth and final strategy involves creation of radical and anionic centres within a single species. Intramolecular coupling (cyclo-coupling) of these centres then leads to ring formation. This is the intramolecular version (S_{RN}i) of the radical nucleophilic substitution reaction [Scheme 1, (4)] and it can be incorporated into a radical chain process [11, 12]. Strategies (3) and (4) can also be classified as unusual cyclisations and recent examples will be reviewed in this chapter.

In planning the preparation of a novel cyclic molecule the synthetic chemist can, of course, rely on an encyclopaedic knowledge of literature reactions. Recently, however, several valuable tools have become available to aid synthetic design. Rate constants for many model radical reactions in solution have become available thanks to applications of the "free radical clocks" concept. Since most radical reactions are insensitive to solvent effects [13], these data can be freely applied. Cyclisation methodology via intramolecular addition will usually involve competition between the first-order cyclisation step (k_c) and bimolecular trapping (k_H) of the intermediate radical by some added trap such as Bu_3SnH, $(Me_3Si)_3SiH$ or cyclohexadiene. The ratio of the rates of cyclisation to trapping will therefore be given by: k_c/k_H[trap]. The rate constants k_H for H-atom transfer by these reaction partners are 25×10^5, 4×10^5 and 0.5×10^5 $M^{-1}s^{-1}$, respectively, at 300 K [14, 15]. In most synthetic protocols the trap concentration will be in the range 0.01 to 0.1 M and hence k_c values of $\geq 10^3$ s^{-1} at 300 K are needed for cyclisations to be synthetically useful. Experimental rate constants relevant to unusual cyclisations of model radicals are shown in Scheme 2. It is apparent that reverse ring opening processes dynamically oppose both 3-*exo* and 4-*exo* cyclisa-

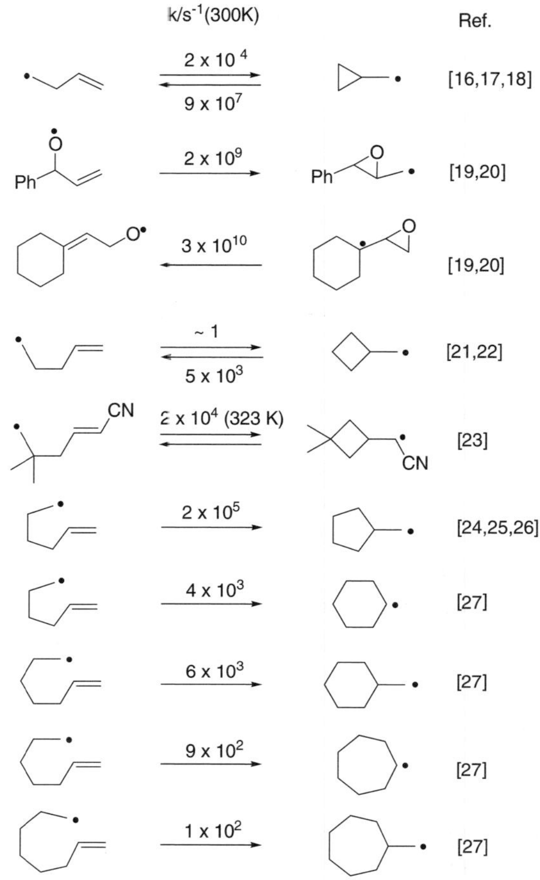

Scheme 2 Rate constants (k_c) at 300 K relevant to unusual radical cyclisations

tions. Preparations of three- and four-membered rings can, therefore, only be achieved by this intramolecular addition route for precursors containing substituents, or structural factors, which favour cyclisation and/or disfavour ring opening. The rate constants for cyclisation rapidly diminish as ring size increases. Furthermore, although *exo* cyclisation is usually faster than *endo* cyclisation, the difference gets smaller as ring size increases. Special substituents and/or structural features are needed for clean syntheses of ring sizes greater than seven-membered.

A second important tool stems from improved computational quantum mechanical methods, particularly DFT, which are now capable of predicting the molecular properties of small- to medium-size organic structures (< about 30 first row atoms) with a precision approaching that of experiment. Good quality computed electronic energies, coupled with improved force constants (and hence good vibrational analyses), enable molecular bond dissociation energies, ionisation potentials, electron affinities and proton affinities to be acquired with good precision. In this way reliable information about the structures of precursors, transition states and products, as well as thermodynamic and kinetic aspects of a proposed reaction channel, can be arrived at in advance and used in the design of precursor molecules and as a guide to experimental conditions. Several studies of this kind will be referred to in this chapter.

2 Intramolecular Free Radical Additions to Acceptor Groups

2.1 3-*exo* Cyclisation Reactions: Preparations of Functionalised Cyclopropanes

In principle, the archetype 3-butenyl radical could cyclise in the 4-*endo* mode (C^{4n}) to give a cyclobutyl radical. In practice, however, this 4-*endo* cyclisation is strongly disfavoured and preparations of four-membered rings using this reaction are essentially unknown. The adverse equilibrium between the 3-*exo* ring closure of 3-butenyl (C^{3x}) and the β-scission of the cyclopropylmethyl radical (cpm) can become favourable in certain compounds. A long-established example involves norbornenyl and related systems, in which the cyclisation rate is increased to about equal the reverse ring-opening rate thanks to the molecular architecture [28]. Srikrishna and co-workers have established useful preparations of tricyclooctan-4-ones **2** via organotin hydride treatment of bicyclo[2.2.2]oct-5-en-2-yl bromides **1** and related compounds [29, 30] (Scheme 3). The favourable structure and the presence of the aryl substituent on the double bond combined to shift the equilibrium in favour of the cyclised form. With the analogous norbornene system **3**, the extent of 3-*exo* cyclisation depended on the substituents at C-7. For example,

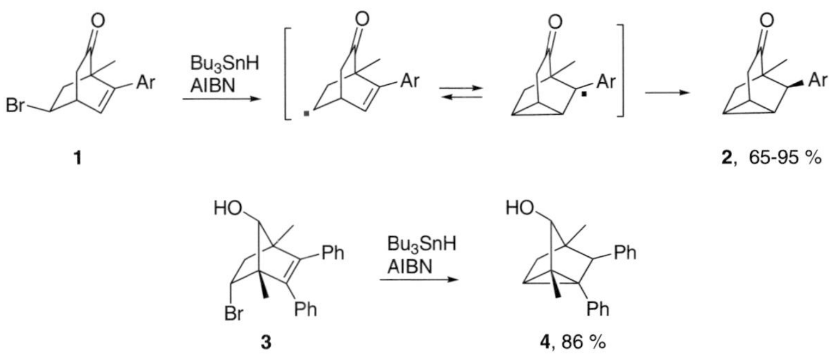

Scheme 3 3-*Exo* cyclisations of bicyclo[2.2.2]oct-5-enyl and norbornenyl derivatives

the *syn* alcohol **3** furnished predominantly the tricyclo product **4**, whereas with the *anti* alcohol, no 3-*exo* cyclisation took place [31]. These and related rearrangements have been reviewed recently [7].

An ingenious tactic to prevent opening of the cpm radical involves positioning a leaving group such as thiophenyl or bromine at its β position. Rapid loss of this leaving group, after 3-*exo* cyclisation, ensures that the three-membered ring is preserved. Saičić and co-workers employed this method to make several vinylcyclopropanes from 2-thioxo-thiazole ester precursors [32, 33]. Subsequently, they expanded the scope of the reaction to enable bicyclo[3.1.0]hex-2-ene derivatives **7** to be prepared via an addition, double cyclisation cascade [34, 35] (Scheme 4). The precursors were thioxothiazoles **5** that released 1-thiophenylbut-3-enyl radicals on photolysis in benzene. The butenyl radicals added to electron-deficient alkynes, generating vinyl-type radicals that underwent 5-*exo* cyclisations to afford cyclopentenylcarbinyl radicals. 3-*Exo* cyclisation of the latter was made permanent by loss of the phenylsulfanyl radical with production of the bicyclo[3.1.0]hexene **7**. The released PhS radical then attacked another molecule of precursor **5**, thus propagating the chain. Precursors such as **8** containing cycloalkene units underwent a similar annulation sequence to afford tricyclic products **9**, albeit in low yields [35].

An exciting recent development has been the discovery that titanocene-mediated reactions of ω-carbonyl epoxides [36] and ω-alkenyl epoxides [37] can produce functionalised cyclopropane and cyclobutane derivatives. For example, Fernández-Mateos and co-workers discovered that treatment of

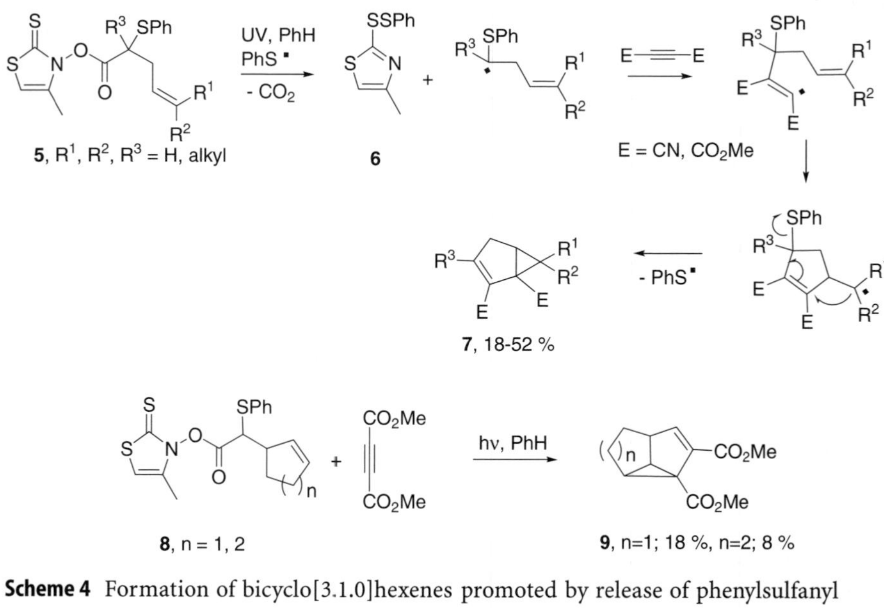

Scheme 4 Formation of bicyclo[3.1.0]hexenes promoted by release of phenylsulfanyl

β-formyl epoxide **10**, derived from α-cyclocitral, with a stoichiometric amount of titanocene(III) chloride led to formation of the bicyclic diol **12** in excellent yield [36].

Bicyclo[4.1.0]heptanols were also obtained from analogous epoxyketones and α,β-unsaturated ketones [38], and epoxynitriles reacted in a similar fashion [39]. The likely mechanism (Scheme 5) involves opening of the epoxide ring by Ti(III) to generate a carbon-centred radical that undergoes stereoselective 3-*exo* cyclisation onto the carbonyl acceptor to afford bicyclic alkoxyl radical **11**. It is probably the rapid trapping of **11** by H-atom abstraction from the THF solvent that prevents ring opening and stabilises the three-membered-ring product.

Gansäuer et al. discovered a related process in which epoxides bearing β-alkene groups with electron-withdrawing substituents (COR, CO_2R,

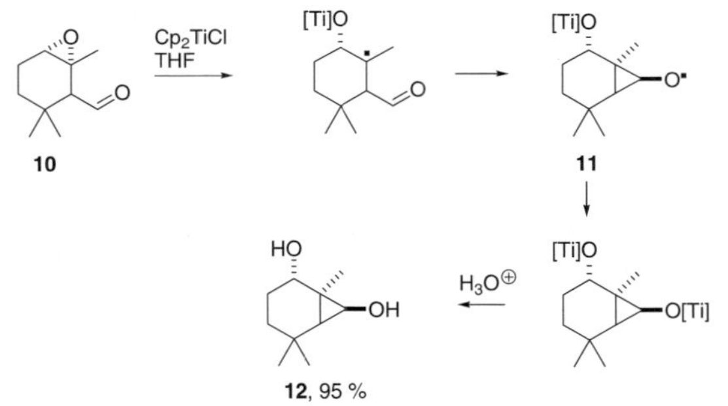

Scheme 5 Titanocene-mediated rearrangement of β-formyl epoxides

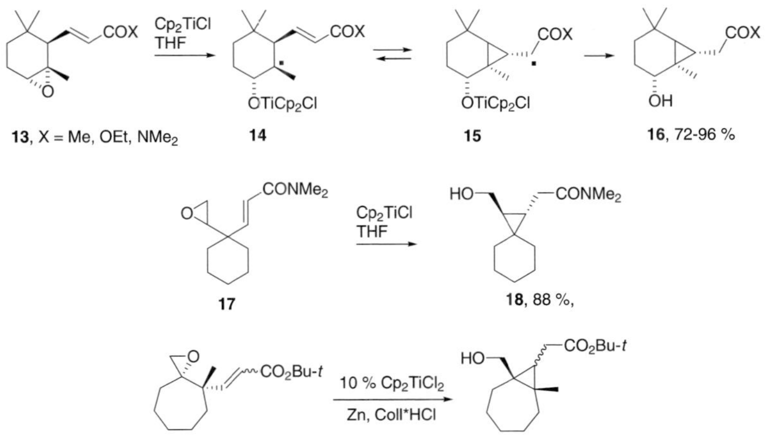

Scheme 6 Preparation of bicyclic and spiro cyclopropanes by titanocene-catalysed reactions of alkenyl epoxides

$CONR_2$), on treatment with catalytic amounts of Ti(III) complexes as electron-transfer agents, gave high yields of functionalised cyclopropanes [40, 41]. Several examples are shown in Scheme 6. The success of the method was due to chemoselectivity in the radical reductions. Cyclised enol radicals such as **15** were electrophilic and were quickly reduced by the titanocene catalyst, whereas the initial butenyl-type radicals **14** were nucleophilic and were unaffected. Epoxides containing alkenes with Ph or vinyl substituents did not yield cyclopropanes, even though the cyclised radicals would have been resonance stabilised. A DFT computational investigation of this system, for a range of model radicals, showed that the 3-*exo* cyclisations were exothermic with low activation energies ($< 16\,\mathrm{kJ\,mol^{-1}}$). The ring-opening reactions were also predicted to be swift and comparable in rate to the parent butenyl radical. The computations suggested that the cyclisation process would be reversible and under thermodynamic control. The correctness of this prediction was supported by the experimental diastereoselectivities and by the fact that for precursor **19** the product d.r. was independent of concentration and of the alkene geometry. These titanocene-mediated reactions show great promise as a high-yielding synthon for mono- and polycyclic cyclopropanes containing a range of functionality useful for further functional group interconversions (FGI).

On treatment with SmI_2 and *t*-BuOH in THF, δ-oxo-α,β-unsaturated esters **21** cyclised to give *anti*-cyclopropanols with complete diastereocontrol [42] (Scheme 7). Electron transfer to the carbonyl group gave intermediate **22** that underwent 3-*exo* cyclisation and reduction to enolate **23**. For most substituents, proton transfer from the *t*-BuOH and quenching with acid gave a cyclopropanol **24** as the major product, accompanied by a minor amount of the corresponding bicyclic lactone **25**. For R^1 = isopropyl the lactone was the major product [43]. The complete stereocontrol giving *anti*-cyclopropanols was attributed to the predominance of the *anti* rotamer **22** in the cyclisation step.

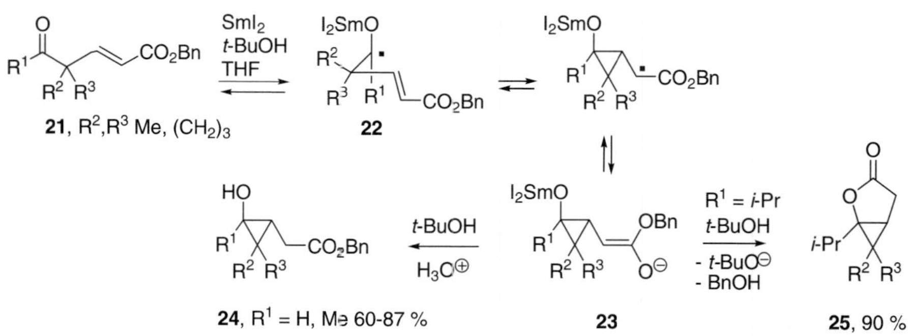

Scheme 7 Samarium(II)-mediated 3-*exo* ring closures of β,γ-unsaturated ketones

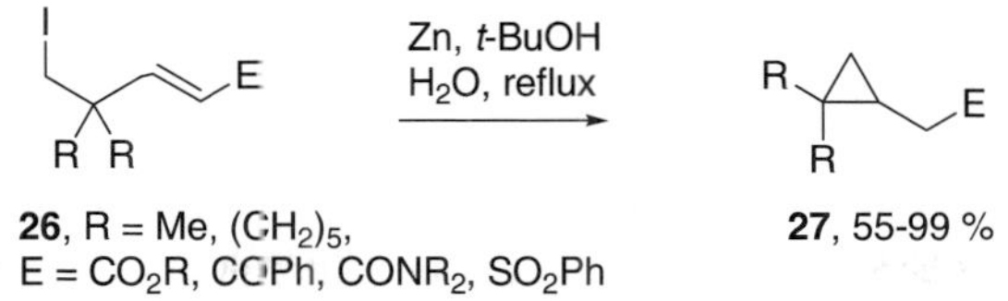

Scheme 8 Zinc-mediated preparation of cyclopropanes from functionalised iodobutenes

Simple but-3-enyl halides cannot be cyclised by treatment with organotin hydrides. However, Togo and Sakuma developed an environmentally friendly methodology for related substrates [44]. When but-3-enyl iodides containing electron-withdrawing substituents at the terminus of the double bond in **26** were refluxed with zinc powder in a mixture of *t*-butanol and water, the corresponding cyclopropanes **27** were formed. Good yields were only obtained for substrates containing bismethyl or analogous substitution in the chain (Scheme 8).

The 3-*exo* cyclisation of ester-substituted 3-butenyl radicals is important in the rearrangement of 2-methyleneglutamate to 3-methyl itaconate catalysed by α-methyleneglutamate mutase. Newcomb and co-workers have applied laser flash photolysis to cleverly designed precursors to show that an ester group at the 1-position of 3-buten-1-yl accelerates the 3-*exo* cyclisation by a factor of about 3, but that the same substituent at the 3-position slows the process by a factor of about 50 [45].

Recent study of aziridinylmethyl radicals has supported earlier reports [46] that they ring open preferentially by C–N cleavage to give N-centred radicals [47, 48]. No preparative routes to aziridines have been reported, but aziridinylmethyl radicals having pentenyl substituents cyclise to pyrrolizidines [49].

2.2 4-*Exo* Cyclisation Reactions

Cyclisation of the archetype 4-pentenyl radical in the 4-*exo* mode (C^{4x}) to give the cyclobutylmethyl radical is comparatively slow, due to the strain in the four-membered ring, and the reverse ring-opening process is considerably faster (Scheme 2). However, the equilibrium can be shifted in favour of cyclobutane formation by rapid trapping of the cyclised radical, by 2,2-dimethyl substitution and by electron-withdrawing substituents at the terminus of the alkene. Cyclisation in the 5-*endo* mode to give the cyclopentyl radical (C^{5n}) is disfavoured on stereoelectronic grounds, but numerous examples have been reported mainly for 4-pentenyl-type radicals containing Ph or other substituents at the 4-position that stabilise the cyclopentyl radical by electron delocalisation [7, 28].

2.2.1 Preparations of Cyclobutanes, Cyclobutanols and Cyclobutanones

The cyclobutane ring is an important structural motif found in many natural products. Until recently, this ring was usually made by some variant of the photochemical [2 + 2] cycloaddition reaction. The first report of a preparative radical 4-*exo* cyclisation appears to be that of Piccardi et al. who made a tetrafluorocyclobutane derivative by adding the elements of carbon tetrachloride to 3,3,4,4-tetrafluorohexa-1,5-diene [50]. The kinetic study of Newcomb and co-workers of the cyclisation of the 5-cyano-2,2-dimethylpentenyl radical (Scheme 2) [23] highlighted the importance of having a quaternary C-2 atom in the chain and an electron-withdrawing group at the alkene terminus. Jung and co-workers showed that this reaction could be used preparatively by slow addition of organotin hydride to 1-bromopent-4-enes with an ester group at C-5 and a quaternary C-2 atom [51–53].

The discovery of general, stereoselective syntheses of cyclobutanols and cyclobutanones, involving samarium(II) reductions of unsaturated aldehydes and titanium(III) reductions of unsaturated epoxides, ranks as a major synthetic advance. Building on an earlier lead [54], Procter and co-workers showed that the samarium(II)-mediated reaction of γ,δ-unsaturated aldehydes **28** provided cyclobutanols **29** in good yields [55, 56]. They found that disubstitution in the chain was crucial to cyclisation and that the esters **28** ($E = CO_2Et$) gave better yields than the sulfones ($E = SO_2Ph$). An attractive feature was that the reaction occurred with complete *anti* stereoselectivity (Scheme 9). By way of contrast, unsaturated ketones **28** (R = Me), when treated under the same conditions with SmI_2 in MeOH/THF, produced cyclopentanols **30** by a non-radical, reduction–intramolecular aldol sequence. However, if SmI_2 in HMPA/*t*-BuOH/THF was employed instead, the corresponding cyclobutanols were again obtained, albeit in low yields [57]. The same research group carried out a partial synthesis of the caryophyllene-type sesquiterpene, pestalotiopsin A (**34**) [58]. 4-*Exo* ring closure of precursor aldehyde **31** gave cyclobutanol **32** in high yield. The latter was transformed to the core pestalopsin A unit, i.e. **33**, in a series of steps.

Williams and Blann developed a route to cyclobutanols starting from carbohydrate-derived precursors [59]. The ketyl intermediates formed on treatment of precursors **35** with SmI_2 in THF/HMPA underwent 4-*exo*-ketyl-olefin cyclisations to afford cyclobutanols **36** as isomer pairs (major isomer shown). Good yields were obtained for R = *t*-butyldimethylsilyl and trityl (triphenylmethyl), but for R = pivaloyl the reaction was diverted into an elimination pathway.

Fernandez-Mateos and co-workers showed that γ,δ-epoxyaldehydes and nitriles could be transformed into the corresponding cyclobutanols and cyclobutanones, respectively, by 4-*exo* cyclisations of the intermediate radicals mediated by titanocene complexes [36, 38, 39]. For example, treatment of

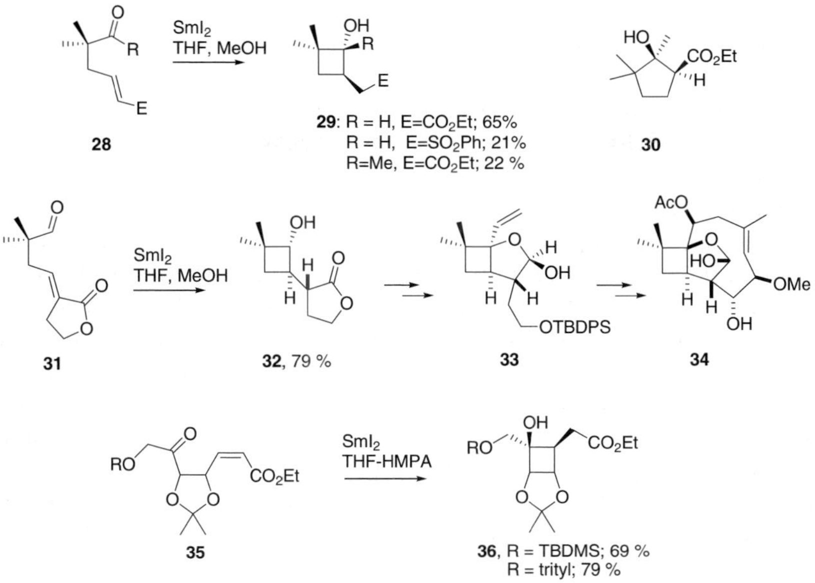

Scheme 9 Preparation of cyclobutanols by samarium(II)-promoted ring closures of γ,δ-unsaturated aldehydes

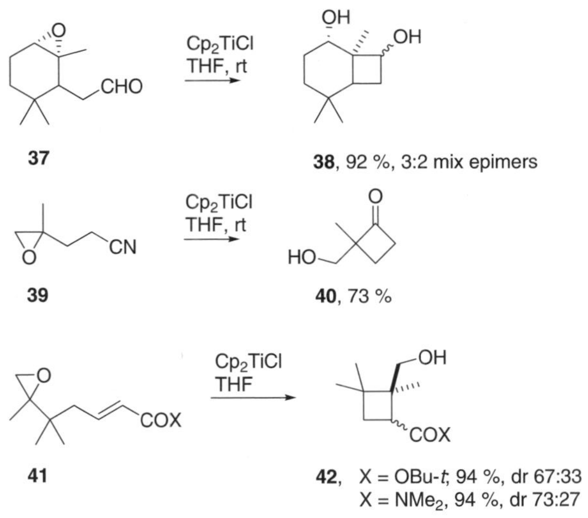

Scheme 10 Preparation of cyclobutane derivatives by titanocene-mediated reactions of epoxides

aldehyde **37** with titanocene furnished bicyclobutanol **38** in 92% yield as a mixture of two epimers (Scheme 10). Similarly, epoxynitrile **39** yielded cyclobutanone **40**. Analogous epoxyketones could also be used although product yields tended to be lower. The mechanism of the transformation was similar to that shown for the related 3-*exo* process (Scheme 5). Gansäuer et al. obtained high yields of cyclobutylmethanols **42** from the titanocene-catalysed reaction of unsaturated epoxy esters and amides **41** [40].

2.2.2 Preparations of β-Lactams

The azetidinone ring system **43** is an important structural feature of the powerful β-lactam families of antibiotics and appears in many other natural products such as clavulanic acid. Free radical-based routes to this ring system are remarkable for their variety and range. Four distinct radical-based disconnections have been investigated for azetidinone preparation. Disconnection **a** implies C^{4x} closure of a carbamoyl-type radical onto an unsaturated acceptor group. Disconnection **b** implies a C^{4x} closure of an amidoalkyl (α-carbamoyl) radical onto an enamide acceptor. Disconnection **c** points to an amidyl radical ring closure onto an alkene acceptor. Finally, disconnection **d** connotes C^{4x} ring closure of an acyl radical onto an imine acceptor (Scheme 11).

Amidoalkyl radicals for use with disconnection **b** have been prepared in several ways. Belletire and co-workers made them by slow addition of tributyltin hydride to bromoenamides **44** [60]. The C^{4x} closure was favoured by the presence of two phenyl groups at the terminus of the double bond and *trans*-β-lactams **45** were obtained in good yields (Scheme 12). Ikeda, Ishibashi and co-workers made a thorough study of substituent effects on γ-enamidoalkyl radical ring closures [61–65]. 5-*Endo* cyclisation was favoured by precursors containing a phenyl or alkyl substituent on the internal C atom of the alkene, or for when the alkene was contained in a ring. Model radicals lacking the carbonyl in the chain gave only uncyclised prod-

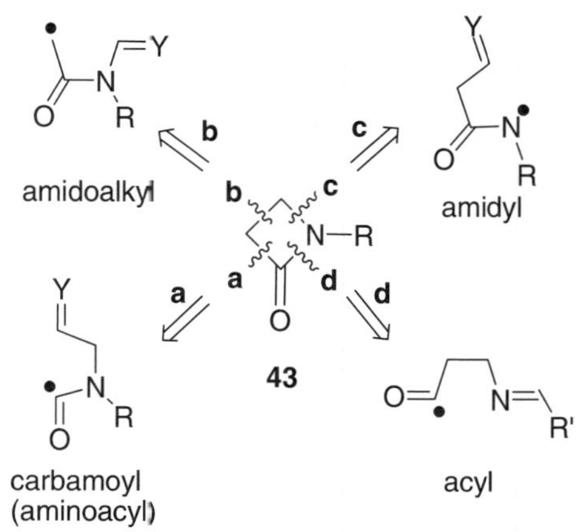

Scheme 11 Four radical-based disconnections for the azetidinone ring

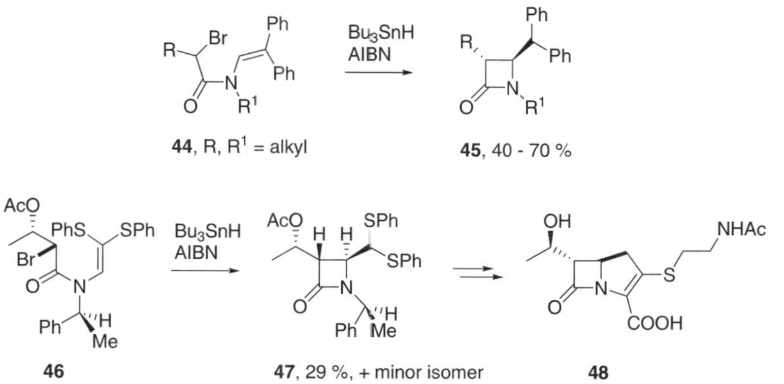

Scheme 12 Preparation of (+)-thienamycin from a bromoenamide

uct. 4-*Exo* cyclisation was favoured by bromoenamides containing one or more thiophenyl substituents at the terminus of the alkene. Ikeda, Ishibashi et al. carried out numerous successful preparations of β-lactams using this methodology, and also developed syntheses of antibiotics PS-5 and thienamycin **48** (Scheme 12) [62, 66]. Studies with chiral substrates showed that it was necessary to use the enamide with the (*S*)-acetoxy group. Furthermore, a chiral auxiliary on nitrogen with matched chirality **46** led to improved diastereoselectivity.

Zard and co-workers prepared a series of *N*-alkenyl trichloroacetamides such as **49** and **51** [67, 68]. On treatment with nickel powder and acetic acid in refluxing 2-propanol, amidodichloroalkyl radicals were generated (Scheme 13). In the absence of a trap, or if the trap was not efficient enough, 5-*endo* cyclisation was observed yielding functionalised γ-lactams. The amidoalkyl radicals underwent reversible 4-*exo* cyclisations, and β-lactams could be isolated, provided a good trap was used. The most efficient method was to include a fragmentation step. Thus, compounds **49** and **51** afforded β-lactams in good yields, after elimination of the phenylsulfanyl radical (Scheme 13). The Parsons group examined CuCl and $RuCl_2(PPh_3)_3$ as catalysts for similar ring-closure processes [69–71]. They found that both the C^{4x} and C^{5n} processes were reversible and that the 5-*endo* cyclisation was favoured except when a bulky radical-stabilising group was present on the β position of the enamide double bond. Enamides **53** and **55**, on treatment with the copper or ruthenium reagents, gave good yields of the corresponding azetidinones **54** and **56** (Scheme 13). An oxidative route to amidoalkyl radicals employed ceric ammonium nitrate treatment of α-ethoxycarbonylacetamides and yielded various functionalised β-lactams [72, 73].

Zard and his group have also applied their xanthate methodology for preparation of azetidinones [74]. For example, treatment of xanthate **57** with lauroyl peroxide afforded β-lactam **58**. The reversible C^{4x} process was helped

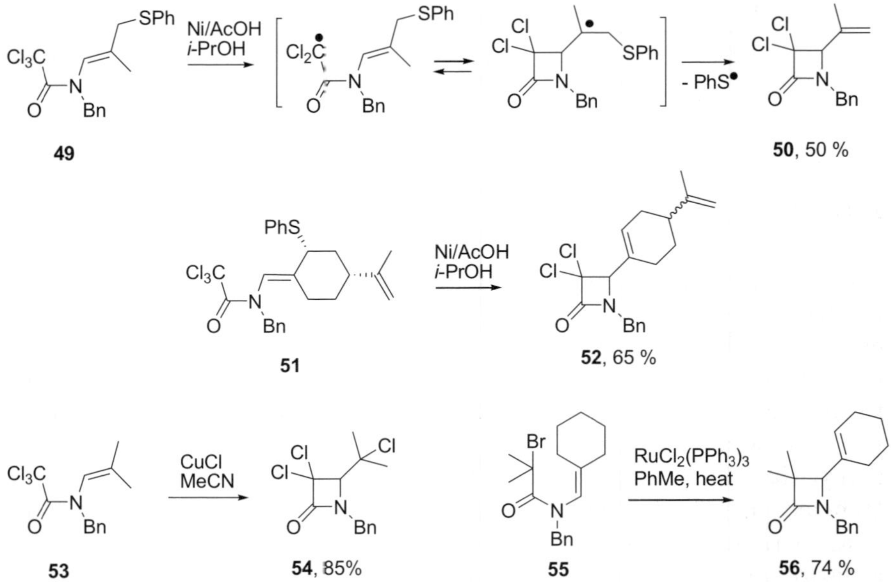

Scheme 13

Scheme 13 Metal-mediated cyclisations of haloenamides

to completion by elimination of the appropriately placed thiophenyl group (Scheme 14).

Quite a range of β-lactams have been made by methodologies following disconnection **a** with carbamoyl radicals (aminoacyls) as intermediates. Pattenden and co-workers made carbamoyl cobalt salophen complexes and showed that on photolysis carbamoyl radicals were released and underwent 4-*exo* cyclisations [75–77]. For example, carbamyl chloride derivative **59** was converted to cobalt complex **60**, which on photolysis yielded the cyclised cobalt-azetidinone complex **61**. The free azetidinone **62** was released by heating the cobalt complex in toluene and was transformed into thienamycin in several subsequent steps (Scheme 15).

A number of metal-free routes to carbamoyl radicals have also been developed recently. Hydrogen abstraction by incoming radicals is regioselec-

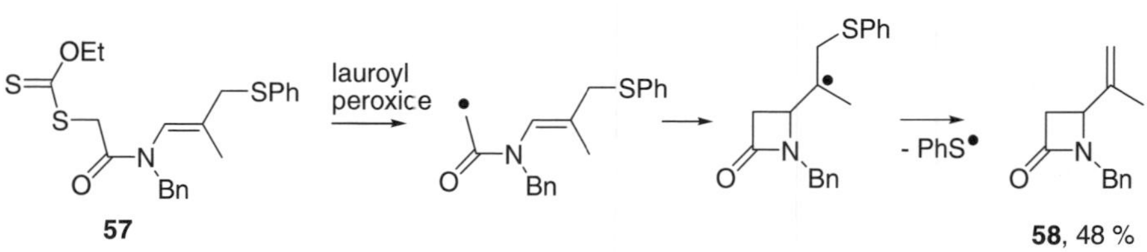

Scheme 14 Xanthate precursors for azetidinone preparations

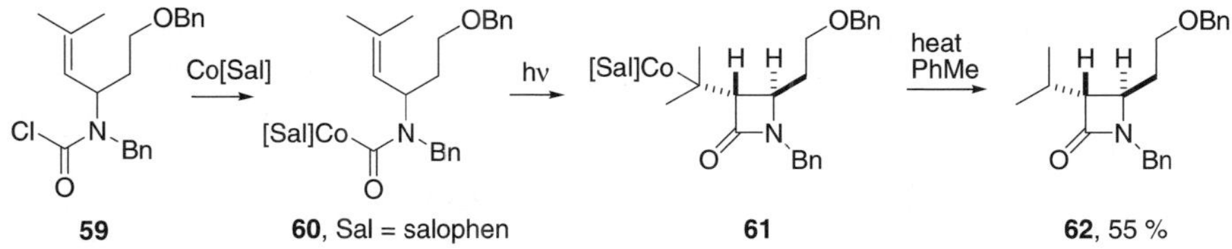

Scheme 15 β-Lactam preparations via cobalt salophen complexes

tive for the bisallylic site of amidocyclohexadienes such as **63**. Treatment of precursor **63** with a peroxide initiator led to generation of cyclohexadienyl radical **64** which readily dissociated to release *N*-allylcarbamoyl radical **65**. The process is thermodynamically favourable because an aromatic sextet coalesces during traversal of the reaction coordinate such that toluene is a co-product. C^{4x} ring closures of carbamoyl radicals with allyl and cinnamyl side chains were achieved giving azetidinones in yields of 30–45% (Scheme 16) [78–80]. Minor amounts of formamides derived from direct H-atom transfer to carbamoyl radicals **65** were also formed. However, no aromatic amides were observed in any of the reactions, indicating that the undesired fragmentation of intermediate **64** to $Me^{\bullet}$ radicals did not compete. As expected, a carbamoyl radical with a propargyl side chain cyclised very inefficiently to a 3-methylene-azetidinone (ca. 7%) and no C^{4x} ring closure at all was observed onto the CN group of the cyanomethyl-containing analogue. For these two triply bonded substrates, the main products were just the formamides derived from reduction of the carbamoyl radicals.

The cinnamyl-substituted amide **63** (R = Ph) was selected to enhance the difficult C^{4x} step, because resonance stabilisation in the cyclised radical **66** (R = Ph) should favour this. However, initiation with di-*tert*-butyl peroxide (DTBP) or benzoyl peroxide gave low yields, probably because of slow H-atom abstraction by the delocalised radical **66**. Use of lauroyl peroxide as initia-

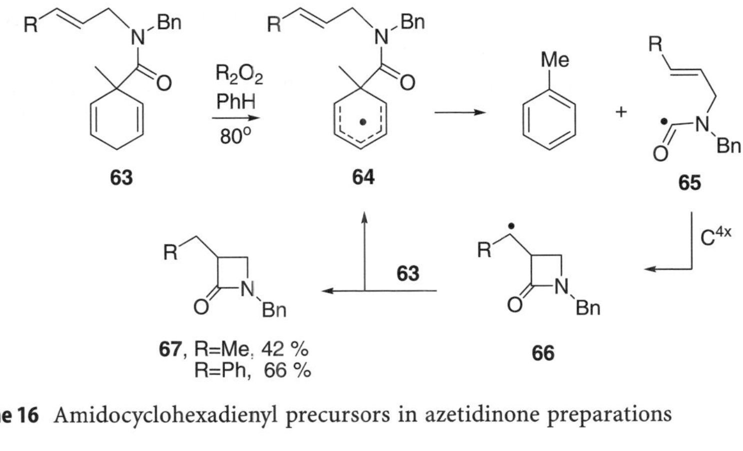

Scheme 16 Amidocyclohexadienyl precursors in azetidinone preparations

tor, together with methyl thioglycolate (RSH) as H donor, led to a greatly improved yield of **67** (R = Ph, 66%). It is likely that polarity reversal catalysis [81, 82] played a part in enhancing this yield. The azetidinylbenzyl radical **66** (R = Ph) will be nucleophilic. Hence, a polar effect should favour H-atom abstraction from the electronegative RSH. The electrophilic sulfanyl radical (RS$^{\bullet}$) generated in this way will, in turn, abstract H atoms more readily from the bisallylic site of **63**, thus regenerating RSH and continuing the chain.

Oxime oxalate amides such as **68** are easily prepared from an oxime, oxalyl chloride and an *N*-alkenylamine. They also function as novel, clean sources of carbamoyl radicals [83]. The best conditions for preparations of lactams involved photolyses of dilute oxime oxalate amide solutions in toluene, with a threefold excess of 4-methoxyacetophenone (MAP). In these photosensitised reactions sufficient energy was transferred to the oxime oxalate amide to break the weak N – O bond and release the phenyliminyl radical together with a carbamoyl radical **69** (Scheme 17). The iminyl radicals simply abstracted hydrogen from the toluene solvent and the resulting imine was hydrolysed to benzaldehyde during work-up. Using this methodology, C^{4x} cyclisation of carbamoyl **69** gave azetidinylmethyl radical **70** which abstracted an H atom from the solvent to afford a modest yield of *N*-benzyl-3-methylazetidin-2-one

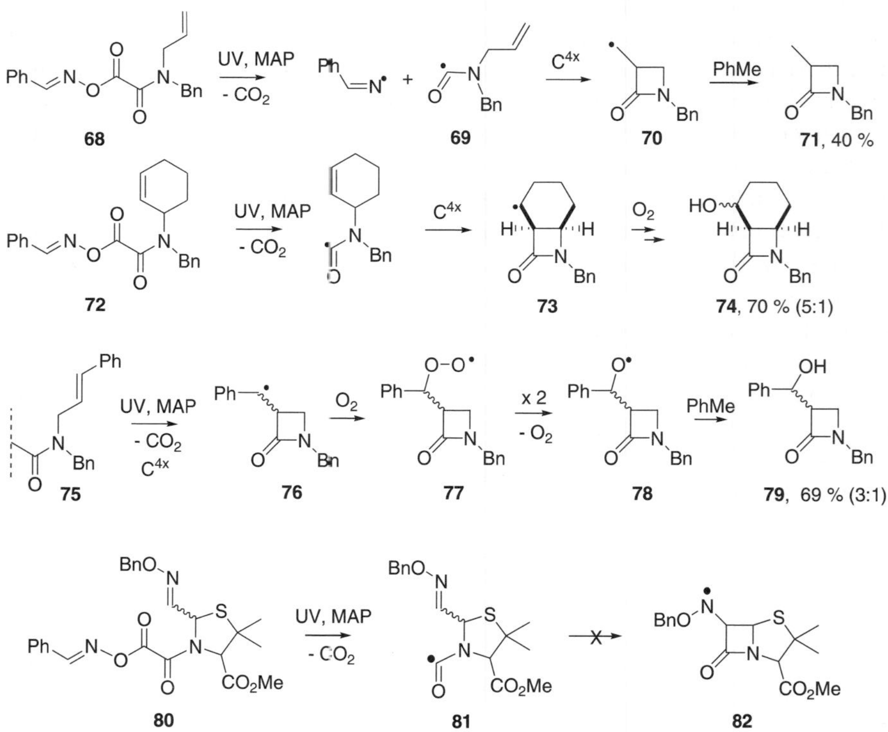

Scheme 17 Oxime oxalate amides as β-lactam precursors

71. Ring closures of the carbamoyl radicals derived from the cyclohexenyl- and phenyl-substituted oxime oxalate amides **72** and **75** afforded the bicyclic β-lactam **74** and 3-benzyl-substituted β-lactam **79**, respectively [83, 84]. In both cases the cyclised radicals were more stabilised than **70** (one was secondary and the other secondary-benzylic) and this favoured 4-*exo* ring closure. An interesting feature was that both **74** and **79** were obtained as hydroxyl derivatives, the former as a 5 : 1 mixture of *anti* and *syn* isomers and the latter as a pair of diastereoisomers (3 : 1) (Scheme 17). Probably hydrogen abstraction from the toluene solvent was slower for both the secondary precursor radical **73** and the benzylic precursor radical **76**, so that addition of dissolved dioxygen supervened. The peroxyl radicals formed in this way, e.g. **77**, would be converted to more reactive oxyl radicals **78** (by self-coupling and O_2 loss) that did abstract hydrogen, hence affording the hydroxyl derivatives. As an alternative mechanism, radicals **73** and **76** might be oxidised to the corresponding carbocations and hence transformed to the hydroxyl derivatives via reaction with moisture.

Attempts were also made to access penicillin derivatives by this route. Several thiazolidine-containing oxime oxalate amides were prepared and photolysed under the same conditions, but without success [84]. It is known that C^{5x} and other conventional cyclisations onto oxime ether acceptors ($> C = NOR$) are faster than onto alkene acceptors [85]. In the hope that C^{4x} cyclisation onto an oxime ether acceptor would also be more efficient, oxime oxalate amide **80** containing both a thiazolidine ring and oxime ether acceptor was prepared. The photosensitised reaction of **80** did yield the desired carbamoyl radical **81**, as shown by EPR spectroscopy. However, the presence of the thiazolidine ring evidently inhibited 4-*exo* cyclisation because no significant amount of cyclisation to radical **82** took place and none of the penicillin derivative was isolated.

Oxime oxalate amides were found to be good precursors for EPR spectroscopic work. In several special cases both the initial carbamoyl radical and the cyclised azetidinyl-containing radical could simultaneously be observed. This enabled rate constants for 4-*exo* cyclisations of carbamoyl radicals to be measured by the EPR technique [86]. In this way the rate constant for C^{4x} closure of carbamoyl **83** onto an alkene acceptor to yield bicyclic radical **73** was found to be $5 \times 10^4\ s^{-1}$ at 300 K. The technique also indicated that the reverse ring-opening reaction was approximately an order of magnitude slower. By use of precursor **84** the cyclisation rate constant for carbamoyl **85** onto an oxime ether acceptor to give benzyloxyaminyl radical **86** was also determined [86] (Scheme 18). To gain further insight into the process, DFT computations were carried out to determine the barrier heights of cyclisation and ring opening for several carbamoyl radicals. Benchmarking with model radicals indicated that the B3LYP/6-31+G** basis set gave reasonable overall agreement with correlated ab initio results. The calculations indicated that the 4-*exo* cyclisations onto oxime ether type $C = N$ bonds were exothermic

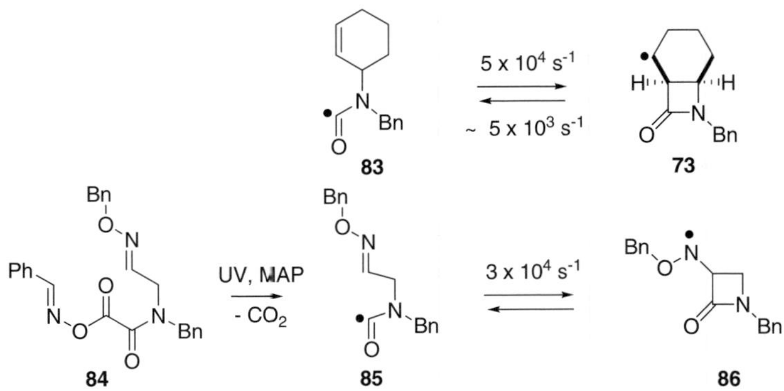

Scheme 18 Rate constants for 4-*exo* cyclisations of carbamoyl radicals

with all substituents. The barrier to cyclisation decreased with increasing size of N and O substituents and, in parallel with this, the barrier to ring opening increased. Therefore, the DFT computations supported the experimental data for comparatively slow reverse reactions.

As anticipated, the 4-*exo* cyclisations of **83** and **85** have smaller rate constants than the 5-*exo* cyclisations of 5-hexenyl-type radicals (Scheme 2). However, the 4-*exo* rate constants for carbamoyls are significantly greater than for 4-butenyl radicals. This is in good agreement with the deduction from preparative work that azetidinone rings form more easily than cyclobutyl-type rings. Interestingly, carbamoyl cyclisations were found to be less reversible than C^{4x} cyclisations of amidoalkyls for which there is much evidence that azetidinone formation is reversible (see above and [71]). 4-*Exo* cyclisation onto a $>C=NOR$ bond was expected to be faster than onto a $C=C$ bond. The error limits are large on both sets of kinetic data so that although Scheme 18 shows that the k_c value for **83** (onto a $C=C$ bond) is marginally greater than k_c for **85** (onto a $C=NO$ bond), the two are the same within the error limits. The two specific examples are stereoelectronically dissimilar so that a general conclusion of this kind cannot safely be made.

Grainger and Innocenti have shown that *N*,*N*-dialkyl dithiocarbamates such as **87** can be prepared in high yields by treatment of the corresponding carbamoyl chlorides with commercially available sodium diethyldithiocarbamate trihydrate [87]. Initiation with light or peroxides led to release of carbamoyl radicals which were shown to take part in chain cyclisation reactions. For example, photolysis of dithiocarbamate **87** in refluxing cyclohexane released carbamoyl **88** and hence gave bicyclic β-lactam **90** in high yield (Scheme 19). The method also worked well for lactams with larger rings.

Ryu and co-workers have pioneered a host of elegant radical carbonylation reactions mediated by tin, silicon or germanium hydrides and gaseous CO [88]. When an azaenyne was used as the substrate, the process led to aze-

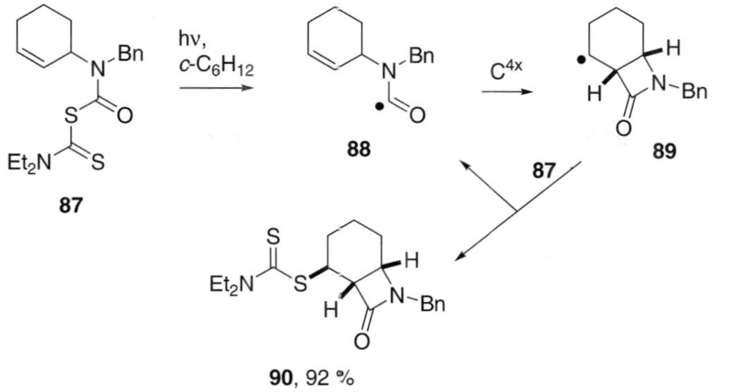

Scheme 19 Preparation of a bicyclic β-lactam from an *N*,*N*-dialkyl dithiocarbamate

tidinone formation via disconnection **d**. The reactions were carried out with the precursor and Bu_3SnH, or $(Me_3Si)_3SiH$ or other hydride, under 90 atm of CO at 90 °C and with azoisobutyronitrile (AIBN) as initiator [89, 90]. For example, with azaenyne **91** addition of an organometallic radical gave vinyl radical **92** which picked up CO to generate acyl radical **93**. The latter underwent an unusual C^{4x} cyclisation onto the N atom of the imine yielding the azetidinylalkyl radical **94**, which was trapped by H-atom transfer with the metal hydride (Scheme 20). Good yields of a number of tin-containing β-lactams were obtained in this way as mixtures of *E* and *Z* isomers **95**. The products were quantitatively converted to the corresponding iodides by treat-

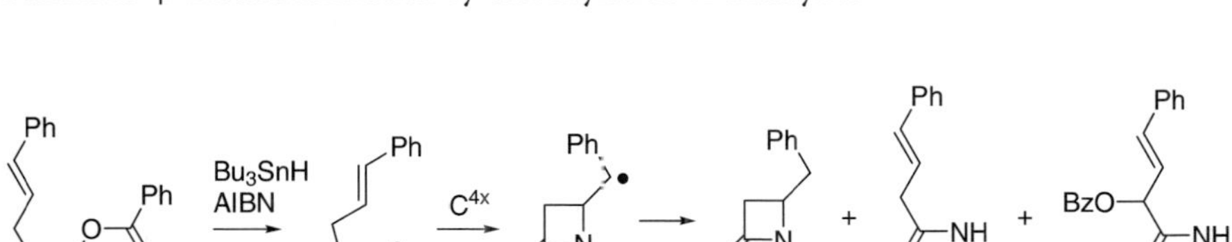

Scheme 20 β-Lactam formation by carbonylation of azaenynes

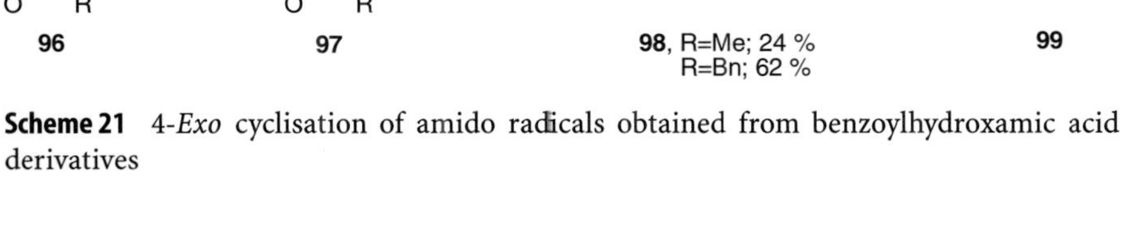

Scheme 21 4-*Exo* cyclisation of amido radicals obtained from benzoylhydroxamic acid derivatives

ment with iodine in ether at 0 °C. Use of tin hydride led to a preponderance of the *Z* isomer, whereas the reverse was the case for $(Me_3Si)_3SiH$.

The final disconnection c, projected to give β-lactams by amidyl radical cyclisations, was investigated by Clark et al. [91, 92]. Treatment of *O*-benzoylhydroxamic acid derivatives **96** with Bu_3SnH and AIBN using slow addition furnished the azetidinones **98**, via 4-*exo* cyclisation of the corresponding amidyl radicals **97**, accompanied by the straight reduction product and a benzoyloxy group migration product **99** (Scheme 21).

2.2.3 Preparation of β-Lactones

Recently, it has been demonstrated by Sweeney and co-workers that β-lactones can also be prepared by radical 4-*exo* cyclisation methodology [93]. The intermediate *O*-(alkenoyl)oxyalkyl radicals **101** could be prepared by tin hydride treatment of the corresponding (1-bromopropyl)cinnamate. However, best results were obtained with *N*-hydroxypyridine thione (PTOC) esters **100**. In this case a toluene solution of the PTOC ester and Bu_3SnH was slowly added by syringe pump to a heated flask irradiated with a tungsten lamp (Scheme 22). For precursors containing only one substituent at the terminus of the double bond the direct reduction was the main process. For precursors **100** with two phenyl or two arylthio substituents, 4-*exo* cyclisation of the intermediate C-centred radicals **101** was efficient and good yields of β-lactones **103** were isolated, accompanied in each case by minor amounts of the direct reduction product. The bis-arylthio functionality was particularly appropriate for further FGI.

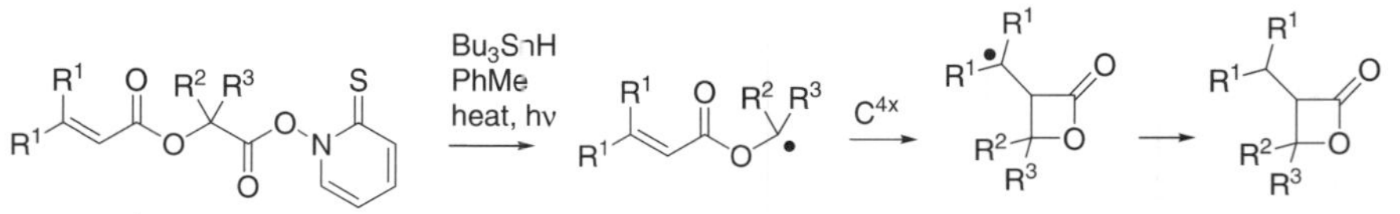

Scheme 22 Preparation of β-lactones by cyclisation of *O*-(alkenoyl)oxyalkyl radicals

2.3 Preparations of Medium and Larger Rings

The kinetic data in Scheme 2 illustrate the fact that radical cyclisation rates decrease as the chain length of the unsaturated radical increases. Furthermore, the chains become more flexible, and consequently the regioselectivity between the *exo* and *endo* cyclisation modes also decreases. From a synthetic perspective, the consequences of these factors are first, that uncyclised materials often appear as by-products and second, that it is difficult to cleanly obtain a desired medium ring, particularly in reactions mediated by good donors such as organotin hydrides. These conclusions were nicely exemplified by Bailey and Longstaff's study of methyl-substituted hept-6-enyl radicals [94]. To minimise premature reduction, they used $(Me_3Si)_3SiH$ as coreactant with various methyl-substituted 1-iodohept-6-enes. However, they still obtained methylcycloheptane along with dimethylcyclohexanes and uncyclised material. The relative proportion of 6-*exo* to 7-*endo* cyclisation varied from 8.4 for the 2-methylheptenyl radical to 3.3 for the 5-methylheptenyl radical. Similarly, for each methylheptenyl, the stereoselectivity amongst the possible dimethylcyclohexanes was also low, although the major isomers could be satisfactorily predicted in each case from consideration of Hannesian-type [95] chair transition states.

In spite of the foregoing constraints, many compounds with rings larger than the conventional five and six members have been obtained. Yet [96] and Srikrishna [7] have provided comprehensive reviews of such cyclisations including literature up to the year 2000, so this section focuses on more recent material. The presence of one or more electron-delocalising groups (Ph, CO_2R, $CONR_2$) at the terminus of the double bond in the unsaturated radical favours *endo* cyclisation. *Exo* cyclisations are often favoured when the chain contains two sp^2 atoms (e.g. an aromatic ring).

2.3.1
Preparations of Seven- and Eight-Membered Lactams

Several of the radical-based methods outlined for β-lactam preparations have been applied not only to syntheses of γ- and δ-lactams but also for larger-ring lactams. For example, Liu et al. generated amidomethyl radicals from *N*-alkenyliodoacetamides by treatment with $BF_3 \cdot OEt_2$. They found that the *N*-(4-pentenyl) precursor **104** cyclised regioselectively in favour of the eight-membered-ring lactam **105**, whereas the *N*-(2-allylphenyl)-substituted precursor **106** gave a mixture of products from both 7-*exo* and 8-*endo* cyclisation [97] (Scheme 23). Murphy and co-workers generated amidoalkyl radicals by treatment of bromoamides with *N*-ethylpiperidine hypophosphite (EPHP) or with diethylphosphine oxide (DEPO) [98]. When the precursors **107** were refluxed in toluene with VA20 initiator, mixtures of seven-membered- and eight-membered-ring lactams were obtained from the 7-*exo*- and 8-*endo* cyclisation processes (Scheme 23). Similar results were obtained with either EPHP or DEPO and, with the latter reagent, water could be used as solvent. Premature reduction of the intermediate radicals was not a problem for either reagent, and in the DEPO-mediated reactions there was no need for slow addition with a syringe pump. Analogous bromoesters **108** were also examined and found to give solely eight-membered lactones from 8-*endo* cyclisations.

Ryu and co-workers applied their organotin-mediated carbonylation method (see Sect. 2.2.2, Scheme 20) to long-chain azaenynes and prepared

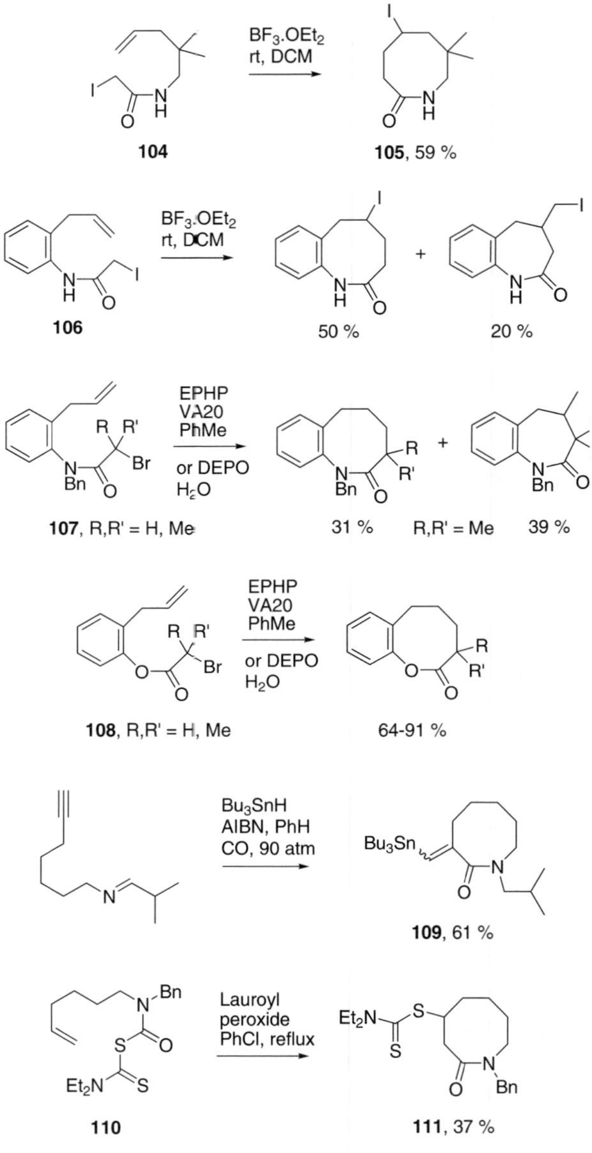

Scheme 23 Synthetic routes to eight-membered lactams and lactones

eight-membered-ring α-stannylmethylene-lactam **109** in good yield [90] (Scheme 23). Grainger's dithiocarbamate group transfer methodology was also applied to the synthesis of larger-ring lactams [87]. For example, *N*,*N*-diethyldithiocarbamate **110** with an *N*-hexenyl substituent cyclised cleanly in the *endo* mode to afford eight-membered-ring lactam **111**. The next lower homologue, with an *N*-pentenyl substituent, gave a mixture of

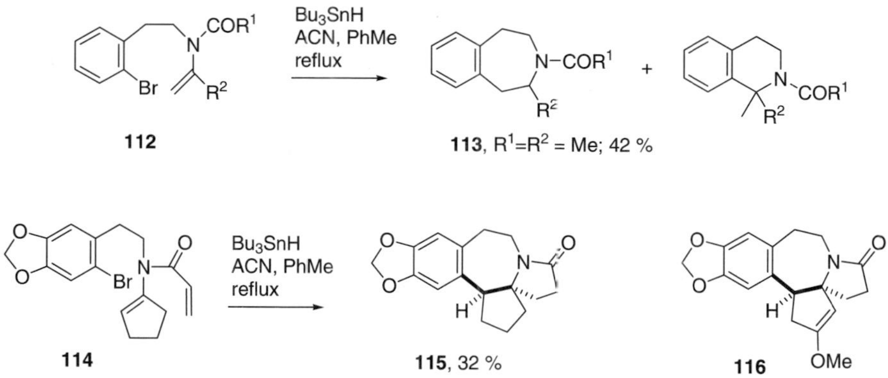

Scheme 24 Organotin-mediated syntheses of tetrahydrobenzoazepines

the corresponding six- and seven-membered-ring lactams, with the former predominating.

The organotin-mediated cyclisations of 2-(2-bromophenyl)-*N*-alkenylacetamides **112** were investigated by Taniguchi et al. [99] (Scheme 24). The cyclisation mode depended on alkenyl substituent R^2 and, as expected, for R^2 = Me the *endo* product **113** predominated, whereas for R^2 = H a mixture of *exo* and *endo* products resulted. They applied this methodology to a short construction of the skeleton of the antileukaemic cephalotaxine **116**. Treatment of precursor **114** with tributyltin hydride was followed by a double cyclisation of the resulting aryl radical. A C^{7n} cyclisation onto the cyclopentene acceptor was followed by a C^{5n} cyclisation onto the vinyl group giving tetracyclic product **115** with a stereochemistry identical to that of natural cephalotaxine.

2.3.2 Preparations of 11- and 12-Membered Lactams and Cyclic Peptides

Porter's general guideline for radical macrocyclisations, i.e. that "*endo* cyclisation modes are favoured" [100], has been receiving additional support as data have accumulated. Recently, Prado and co-workers published an interesting investigation of macrocyclisations of 2-iodobenzamides bearing allyloxy groups in their side chains [*N*-4-(allyloxyalkyl)-2-iodobenzamides **117**] [101–104] (Scheme 25). Their method employed slow addition of a solution of tributyltin hydride and dibenzoyl peroxide to the substrate. Precursors **117a, b** and **c**, containing progressively longer allyloxyalkyl substituents, were designed to test for 10-, 11- and 12-*endo* cyclisations of the corresponding aryl radicals. Only the direct reduction product **118a** was obtained from **117a**, i.e. 10-*endo* cyclisation was negligible. However, for the substrates bearing allyloxypropyl and allyloxybutyl substituents **117b** and **117c**, low yields of the products from 11-*endo* and 12-*endo* cyclisations **119b** and **119c**, respectively,

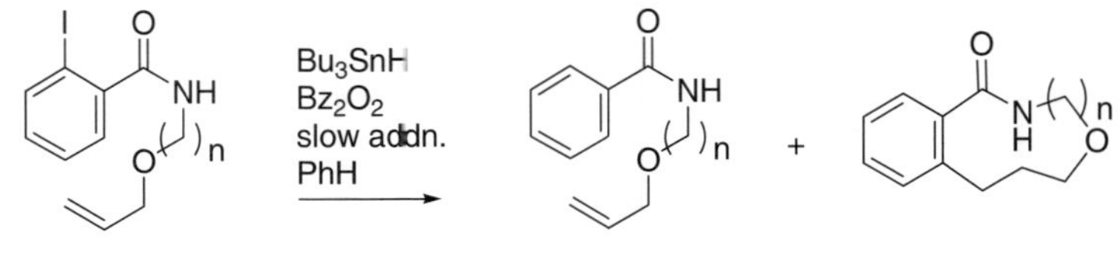

117 a; n = 2; b; n = 3; c; n = 4

10-*endo*; 118a, 76 %; 11-*endo*; b, 85 %; 12-*endo*; c, 33 %

119a, 0 %; b, 14 %; c, 19 %

Scheme 25 Preparations of 11- and 12-membered lactams

were isolated along with the products of direct reduction. No macrocyclic lactam from an *exo* cyclisation was obtained in any case. These results showed that a marked *endo* preference applied in these systems and, in agreement with literature data, showed that ten-membered rings are very difficult to make by radical methodology.

In an intriguing parallel study, Prado and co-workers also examined analogous reactions of glucopyranose derivatives containing allyloxy and 2-iodobenzamido substituents [101, 105] (Scheme 26). Treatment of 2-iodobenzamido precursor **120** with organotin radicals generated the corresponding aryl radical which underwent direct reduction yielding **122**. However, an impressive 40% yield of the macrocyclic lactam **121** was also isolated. It appears that conformational restraints imposed by the glucose unit favoured

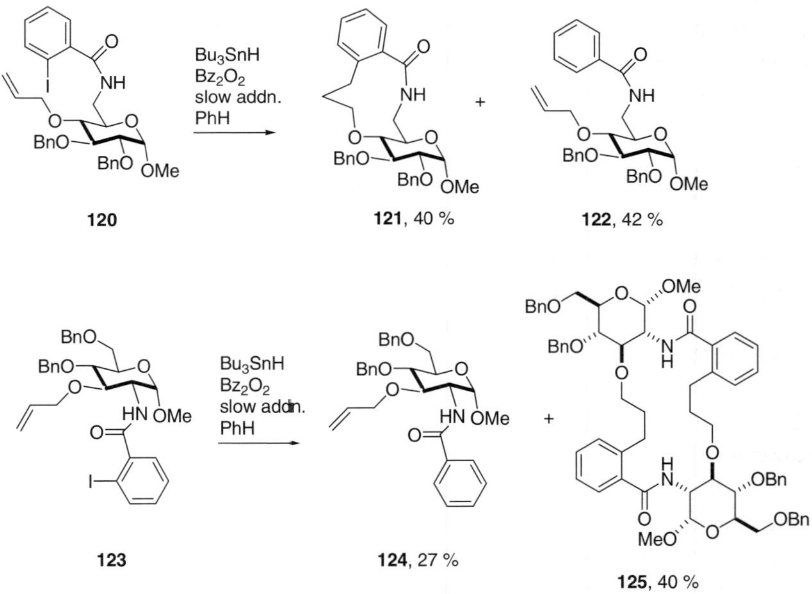

Scheme 26 Macrocyclisations of 2-iodobenzamides bearing allyloxy groups in their side chains

126a : X = NH, R = Bn, R' = *i*-Bu
b: X = NH, R = *i*-Bu, R' = Me
c: X = O , R = Bn, R' = *i*-Bu

127a, 47 %
b, 59 %
c, 0 %

Scheme 27 Radical-mediated syntheses of cyclic di- and tripeptides

the 11-*endo*-type ring closure onto the allyloxy radicophile. The related glucopyranose substrate **123** containing a different substitution pattern was also examined. As expected, a significant amount of the direct reduction product **124** was obtained but none of the product from the potential 9-*exo* or 10-*endo* cyclisations was formed. Instead, the benzomacrodilactam **125** was isolated in 40% yield. This was accounted for by a mechanism involving intermolecular addition of the initial aryl radical onto another precursor molecule, followed by a subsequent unique 20-*endo* intramolecular ring closure to give **125.** It was proposed that the intermolecular addition was favoured because closure of the initial aryl radical by a 10-*endo* process was highly unfavourable.

Iqbal and co-workers studied the radical-mediated ring closures of di- and tripeptides having 3-bromobenzyl C-termini and acryloyl groups at their *N*-termini [106]. For example, aryl radicals, generated on treatment of dipeptide **126** with Bu_3SnH/AIBN, cyclised in the 14-*endo* mode to afford cyclic dipeptides **127** in good yield (Scheme 27). Similar quality results were obtained for analogous functionalised acyclic tripeptides from which the corresponding cyclic tripeptides were obtained. By way of contrast, cyclisation of the ester-containing analogue **126c** did not occur. The efficiency of these macrocyclisations was believed to be due to the presence of an intramolecular H bond (γ/β-turn) in the acyclic peptide precursors which pre-organised them. The method evidently holds significant promise as a general route for making small cyclic peptides constrained with 3-(3-aminomethylphenyl)propionic acid and related linkers.

2.3.3
Preparations of Dibenzocyclooctadienols

Building on their earlier research into 8-*endo* cyclisations of SmI_2-generated ketyl-type radicals, Molander and co-workers [107, 108] developed this process as a key step in the syntheses of several naturally occurring lignans of the dibenzocyclooctadiene type. For example, they prepared the biarylchromium tricarbonyl complex **128** containing *ortho*-formyl and butenolide

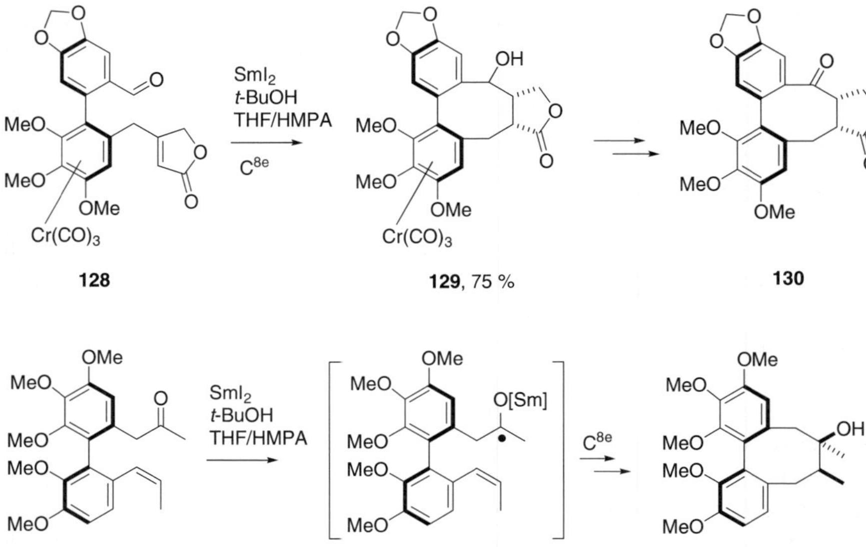

Scheme 28 Preparations of dibenzocyclooctadienols from functionalised biaryls

substituents in adjacent aromatic rings and treated it with SmI_2 and *t*-BuOH in THF/HMPA [109] (Scheme 28). Under these conditions the corresponding ketyl radical was generated and it cyclised onto the butenolide acceptor in the 8-*endo* mode to afford dibenzocyclooctanol **129** in good yield. A similar but less stereoselective cyclisation took place with the uncomplexed precursor, but the $Cr(CO)_3$ complex gave a single stereoisomer of **129**. This cyclisation enabled (±)-steganone **130** to be obtained in six steps from commercial 3,4,5-trimethoxybenzyl alcohol. Similarly, the SmI_2-promoted stereocontrolled 8-*endo* cyclisation of ketyl radical **132**, generated from precursor **131**, enabled (+)-isoschizandrin **133** to be obtained with excellent stereo- and regioselectivity [110].

3 Intramolecular Dimerisations of Diradicals

The classic precursors used for generating diradicals are cyclic, bicyclic and polycyclic diazenes. However, diradicals have also been made by Norrish type I photochemical extrusion of CO from cyclic ketones, by thermal cleavage of vinyl and divinylcyclopropanes, by pinacol reactions of diketones, by Bergman-type cyclisations of endiynes, by several types of photoelectron transfer reactions and in other ways. Most synthetic applications have started with a derivative of 2,3-diazabicyclo[2.2.1]hept-2-ene which on heating or photolysis gives a cyclopenta-trimethylenemethane (TMM) diradical. A review of earlier work with these compounds leading to small, strained polycyclic compounds has appeared [111]. The Little group have explored the synthetic chemistry stemming from an impressive variety of such precursors, and recently gave a fascinating account of many striking syntheses of polycyclics and natural products [112]. The mechanism of the nitrogen extrusion from these and related molecules has received much experimental and theoretical study. The dissociation normally takes place by both one-bond and two-bond processes but substituents can modify this [113]. Similarly the electronic state of the initial diradical depends on the substituent pattern [114, 115].

3.1 Thermal and Photochemical Dissociations of Bicyclic Diazenes

Thermal dissociations of 2,3-diazabicyclo[2.2.1]heptenes with unsaturated substituents at the 7-position (**134**) generate TMM diradicals of type **135** (Scheme 29). The most important products from intramolecular cycloadditions are the bridged compounds **136** (octahydro-4,8-methanoazulenes) and the linear tricycles **137** (octahydro-1*H*-cyclopenta[*a*]pentalenes). Switching between the two reaction modes could be achieved by changing the siting and electronic characteristics of substituents. Large substituents R^2 favoured the bridged product **136**, but when R^2 was an electron-withdrawing group, the linear product **137** was preferred. Little and co-workers gave a useful rule of thumb that "intramolecular diyl trapping reactions selectively afford linear cycloadducts from the singlet state of the diyl when an electron-withdrawing group is appended to either carbon of the diylophile; selective formation of bridged cycloadducts occurs from the triplet state of the diyl when a large alkyl group is appended to the internal carbon of the diylophile" [112].

Recently, Little and co-workers have examined the thermal dissociations of several diazabicyclo[2.2.1]heptenes with vinylcyclopropane substituents and found they gave access to eight-membered rings [116–118]. Simple reflux of **138** in benzene afforded bicyclotrienes **139** together with compounds **140** (Scheme 30). The ketone **138b** was found to undergo a facile Cope rearrange-

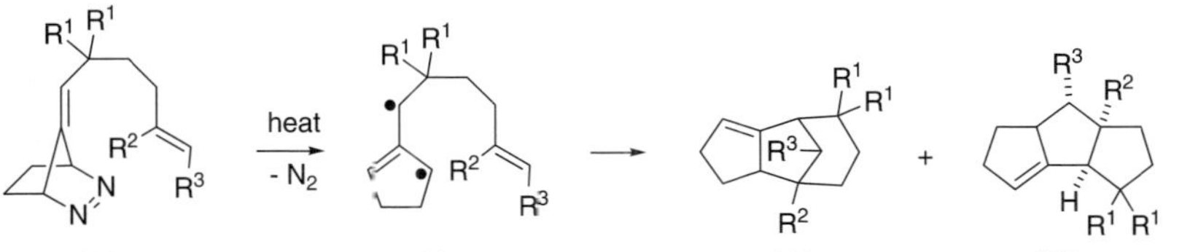

134 135 136 137

Scheme 29 Cyclodimerisation modes of diradicals derived from functionalised diazabicyclo[2.2.1]heptenes

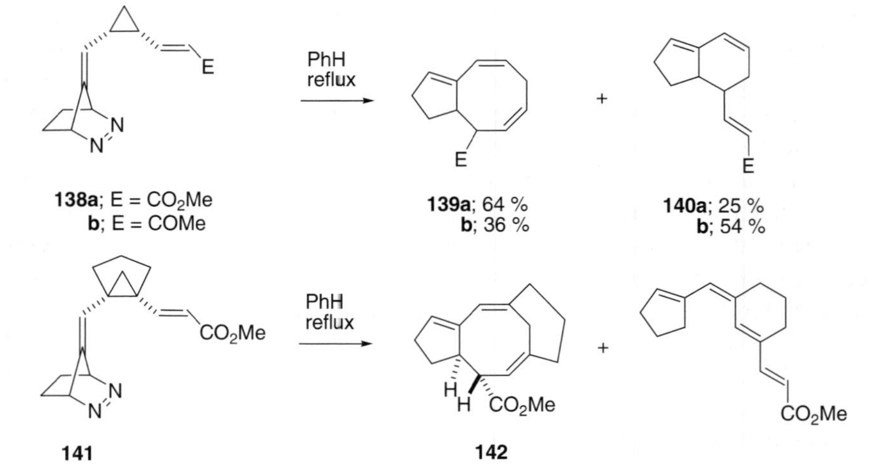

Scheme 30 Thermal dissociations of bicyclic diazenes with vinylcyclopropane substituents

ment whereas the ester **138a** did not, due to different conformations adopted by the precursors, and this accounted for the different product ratios. Remarkably, thermolysis of the more strained precursor **141** also gave tricyclic compound **142**, containing an eight-membered ring, as a single stereoisomer, albeit in low yield.

The thermal, photochemical and Lewis acid catalysed dissociations of diazabicyclo[2.2.1]heptenes with side chains of various lengths terminated by aldehyde groups **143** have also been examined [119] (Scheme 31). The course of the reaction depended on the length of the tether. For the precursor with n = 1, tetrahydroindenol **144** was the main isolated product. Thermolysis of

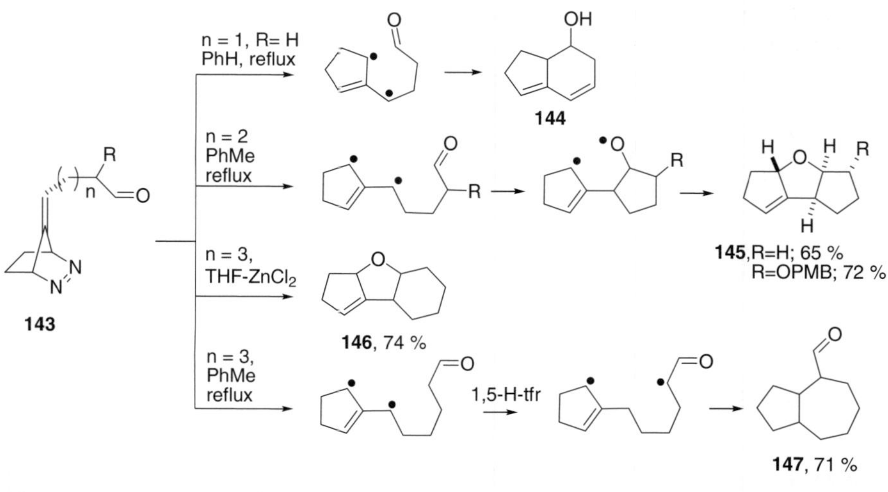

Scheme 31 Preparations of octahydrodicyclopentafuran and decahydroazulene derivatives

the precursor with n = 2 generated the expected cyclopenta-TMM diradical, which underwent C^{5x} ring closure onto the carbonyl carbon followed by intramolecular radical dimerisation to yield the octahydrodicyclopentafuran **145**. Somewhat higher yields of **145** were obtained in reactions initiated photochemically in CH_3CN and in reactions mediated by $ZnCl_2$ in THF. The precursor having tether length n = 3 underwent a similar sequence of reactions when treated with THF/$ZnCl_2$, yielding analogous product **146**. However, on thermolysis in toluene, the diradical derived from this precursor underwent 1,5-hydrogen transfer, followed by intramolecular combination, to yield decahydroazulene-4-carbaldehyde **147**. In general it was found that in the absence of a Lewis acid, diyl combination occurred if the tether was too short to allow it to attain the requisite geometry for H-atom transfer (n = 1). In the presence of a Lewis acid, cycloadditon occurred when the five-membered ring could be formed via addition to the carbonyl carbon. Atom-transfer cyclisation occurred when the tether was long enough to accommodate a "nearly linear" atom-transfer angle and the resulting radical was stabilised [119].

3.2 Miscellaneous Cyclo-Dimerisations of Diradicals

Rüedi et al. discovered a novel and efficient two-carbon ring expansion of vinylcycloalkanones **148** (Scheme 32). Precursors with n = 9 to 15 were thermolysed at 600–630 °C to generate diradical intermediates, which took part in intramolecular recombinations to yield ring-expanded γ,δ-unsaturated cycloalkanones **149** [120]. In each case a mixture of *E* and *Z* isomers was obtained. Thermolysis of the 12-membered-ring ethynyl-substituted cycloalkanone **149** yielded a novel 14-membered-ring allenyl-cycloalkanone **150** via an analogous mechanism.

The SmI_2-mediated intramolecular pinacol coupling reactions of carbonyls **151** were found to afford heterocyclic diols **152** in diastereoselective

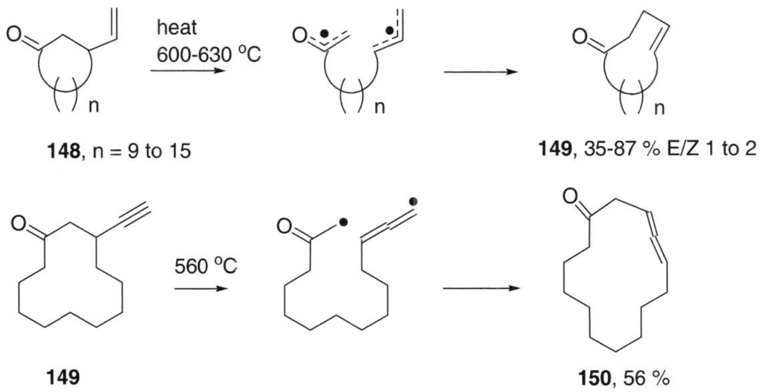

Scheme 32 Two-carbon ring expansions of vinylcycloalkanones

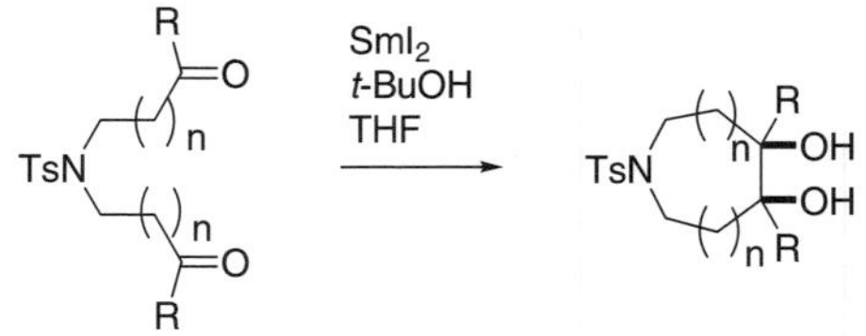

151, R=H, Me **152**, 62-73 %

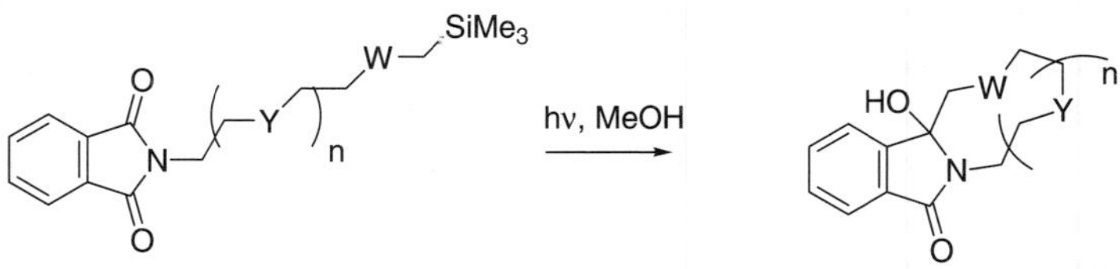

153

154, Y=W=NMs, n=2; 91 %
Y=S, W=NMs, n=2; 97 %
Y=W=O, 50-99 %
Y=W=S, 72-80 %

Scheme 33 Cyclodimerisations via ketyl diradicals and zwitterionic diradicals

reactions [121]. Good yields and selectivity were observed for five- and six-membered rings but more *trans*-diols were produced for larger rings. When mediated by titanocene (Cp_2TiPh), diols with the opposite stereochemistry were obtained.

Diradicals have also been obtained in several ways from photoelectron transfer reactions of suitable substrates [122]. Yoon and co-workers have published noteworthy examples of the production of macrocyclic polyamides, polyethers and polythioethers from photochemical reactions of functionalised phthalimides and naphthalimides [123, 124]. Photolysis in methanol of a phthalimide or naphthalimide, *N*-functionalised with an alkyl, polyamide, polyether or polythioether side chain containing a terminal trimethylsilyl substituent, generated a zwitterionic diradical intermediate that underwent a macrocyclisation with extrusion of the Me_3Si moiety. For example, substrates **153** gave high yields of the ring-closed derivatives **154** (Scheme 33). The efficiency of these ring closures was attributed to electrostatic attraction between the oppositely charged ends of the zwitterionic intermediates.

4 Intramolecular Radical-Carbanion Cyclo-Coupling

Coupling of a carbanion with a free radical can also form a carbon–carbon bond. The overall result is substitution of some leaving group by a nucleophile, but the process can take place as a chain reaction ($S_{RN}1$) involving radical-anion intermediates. Intermolecular radical-carbanion coupling is extremely fast and often diffusion controlled [12, 125–129]. The intramolecular version of the reaction [Scheme 1, (4)] is therefore also expected to be extremely fast. Potentially, cyclisation by this means should be applicable for a different range of functionality, should permit ready formation of *quaternary* bonds, and should also enable a range of ring sizes to be accessed. Currently, reports of ring closures by radical-carbanion cyclo-coupling (S_{RN}i) are comparatively scant in number. The research on S_{RN}i ring closures up to 2003 has been ably reviewed by Rossi et al. along with comprehensive coverage of intermolecular processes [12]. In some systems an intermolecular $S_{RN}1$ reaction precedes a conventional cyclisation, and in some useful syntheses a conventional cyclisation is followed by an $S_{RN}1$ reaction [12, 130, 131]. However, this section will focus on recent examples in which the ring closure step is the cyclo-coupling reaction.

Almost all known intramolecular radical-anion cyclo-coupling reactions involve the production of an aryl-type radical by electron transfer to an aryl halide. The mechanism of the process is outlined in Scheme 34. Suitable precursors **155** contain an *ortho*-halogen X in the aromatic ring as a leaving group and an electron-withdrawing group E in the chain to facilitate carbanion formation. Treatment with base (MNH_2 in liquid NH_3, lithium diisopropylamide (LDA) in THF, or KOBu-*t*) provides the carbanion and, simultaneously, UV photolysis is required. Initiation occurs by photoelectron transfer from the solvent or from some added reagent. The intermediate **156** loses halide and is transformed into the radical anion **157**. Internal cyclo-coupling of **157** produces the ring-closed radical anion **158** which transfers an electron to more substrate **155**, thus propagating the chain and yielding the neutral product **159**.

Many of the known examples involve base-promoted carbanion formation from aromatic amides. For example, indol-2-one derivatives **161** have been made by treatment of *N*-acyl-*o*-haloanilines **160** with LDA in THF, or KNH_2 in liquid ammonia [132–134] (Scheme 35). Similarly, α,β-unsaturated anilides **162** afford the ring-closed 3-alkylideneoxindoles **163** on irradiation and treat-

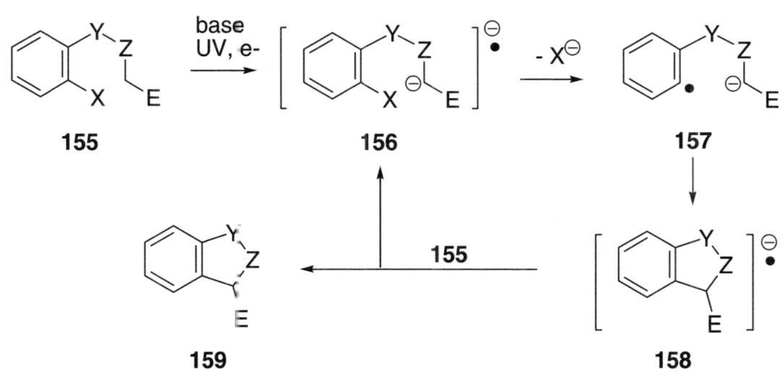

Scheme 34 S_{RN}i chain mechanism of radical-carbanion cyclo-coupling

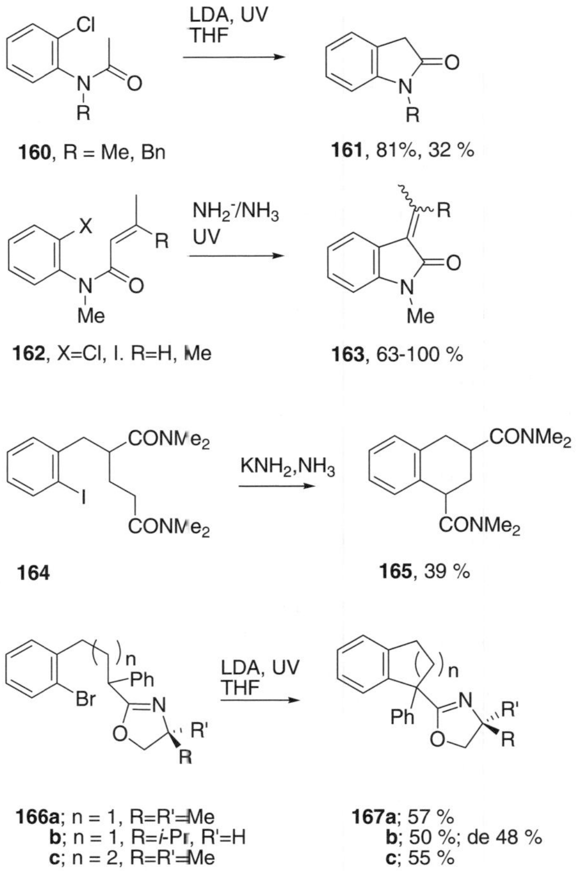

Scheme 35 Synthetic applications of radical-carbanion cyclo-coupling

ment with KNH_2/NH_3 [132]. Dihydroisoquinolin-2-one derivatives [132] and 2-phenyl-1,3-benzothiazoles [135] have been prepared by analogous routes. Six-membered and larger rings have also been made in good yields using this methodology. For example, Wolfe and co-workers treated bis-amide **164** with KNH_2/NH_3 and obtained cyclo-diamide **165** [136]. Several alkaloids such as eupoulauramine, (±)-tortuosamine [137, 138] and an ergot-type alkaloid [139, 140] have been made by routes in which radical-anion cyclo-couplings were the key steps.

Recently it has been demonstrated that radical-anion cyclo-coupling reactions can also take place with haloaromatics containing 2-oxazoline side chains. ω-(2-Halophenyl)alkyl-2-oxazolines **166**, when treated with LDA in THF and subjected to UV irradiation, gave good yields of the corresponding 1-phenyl-1-oxazolino-indan **167a** or -tetralin **167c** derivatives (Scheme 35) [141]. The cyclo-coupling enabled *quaternary* centres to be

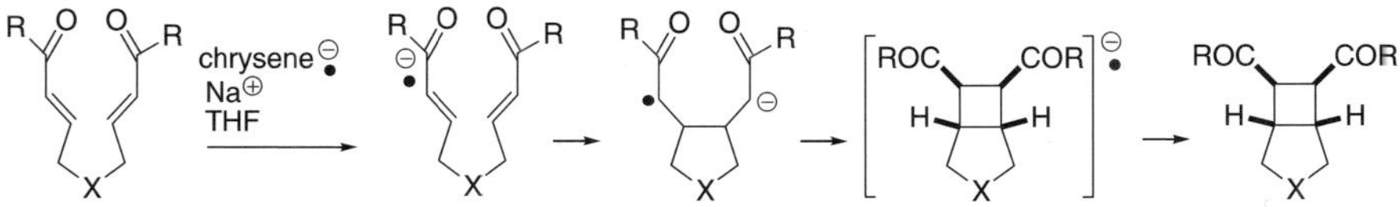

168, X= CH_2, O
R=Ph, 4-PhPh, 2-nap
169
170
171
172, 30-62 %

Scheme 36 Cyclobutanation reaction of bis-enones

readily obtained and showed that both five- and six-membered rings were accessible by this base-promoted route. Chiral 2-oxazolines are well known as effective chiral auxiliaries, so the reaction of **166b**, containing a chiral centre in the 4-(*S*)-isopropyl-2-oxazoline unit, was also investigated. Moderate stereocontrol of the cyclo-coupling was achieved (**167b**; *de* = 48%) but this needs considerable improvement to be synthetically useful. An interesting aspect of this study was that the intermediate radical anions were directly observed by EPR spectroscopy, thus providing good support for the proposed S_{RN}i mechanism of Scheme 34.

A somewhat different approach has been followed by Bauld and coworkers who treated bis-enones **168** with stoichiometric amounts of aromatic radical anions, particularly the chrysene radical anion in THF [142]. The main products were bicyclo[3.2.0]heptanes (or 3-oxabicyclo[3.2.0]heptanes) **172** obtained as single stereoisomers, along with minor amounts of reductive aldolisation and reductive cyclisation products. The mechanism was believed to involve radical-carbanion cyclo-coupling of the intermediate **170** (Scheme 36). The cyclobutanation products were favoured by R substituents such as 2-naphthyl or 4-phenylphenyl that provided extra stabilisation of the product radical anion **171** due to electron delocalisation. The pronounced *cis* stereoselectivity was attributed to a strong electrostatic interaction between the sodium cation and both carbonyl groups in the transition state for cyclisation. Similar cyclobutanations were accomplished electrochemically but this methodology yielded mixtures of *cis*- and *trans*-cyclobutane isomers. Analogous reactions involving metal-complexed anion radicals were also studied [143].

5
Concluding Remarks

During the last decade radical-mediated synthetic methodologies for preparing compounds containing small and medium rings have begun to take a more prominent place. Several ingenious ways and means of solving the problems associated with reverse ring opening of small rings have been discovered. For three- and four-membered rings the patterns of substituents favouring cyclisation have been established. Rapid methods of trapping small cyclic structures and extracting them from unfavourable equilibria have been devised. The new methods enable mono- and polycyclic cyclopropanes and cyclobutanes containing a range of functionality to readily be prepared. Furthermore, several of the newer reagents allow a good measure of stereocontrol to be exercised. The azetidinone ring system (β-lactam) has played a key role in these developments and has received special attention. Molecules containing this ring system have been made by no less than four different disconnections and via an impressive variety of radical types.

Recent research into preparative methods for compounds containing medium rings has steadily built on established principles. *Endo*-type cyclisations are favoured and precursors containing electron-withdrawing groups at the terminus of the radical acceptor group are very advantageous. Numerous efficient preparations of seven- and eight-membered lactams have been described and promising routes to 11- and 12-membered lactams have been explored. Radical-based methodology for making cyclic di- and tripeptides, constrained by particular linkers, appears to benefit from intramolecular hydrogen bonding which pre-organises the desired structures.

Alternative radical-based strategies for making rings by intramolecular cyclo-dimerisation of diradicals and by intramolecular cyclo-coupling of radicals with carbanionic centres have been gradually developing alongside the classic intramolecular addition strategy. The delightful variety of polycyclic structures and natural products obtainable from thermolyses of bicyclic diazenes continues to expand. Advantage is also being taken of the rapidity of radical-carbanion cyclo-coupling processes in that ring closures forming quaternary centres are facilitated. There is clear promise that cyclisations hitherto considered unusual will steadily enter into the mainstream of radical-based synthetic methods.

References

1. Giese B (1986) Radicals in organic synthesis: formation of carbon–carbon bonds. Pergamon, Oxford
2. Motherwell WB, Crich D (1992) Free radical chain reactions in organic synthesis. Academic, New York
3. Giese B, Kopping B, Göbel T, Dickhaut J, Thoma G, Kulicke KJ, Trach F (1996) Radical cyclization reactions. Org Reactions 48:301
4. Renaud P, Sibi M (eds) (2001) Radicals in organic synthesis, vols 1 and 2. Wiley, Weinheim
5. Zard SZ (2003) Radical reactions in organic synthesis. Oxford University Press, Oxford
6. Togo H (2004) Advanced free radical reactions for organic synthesis. Elsevier, Amsterdam
7. Srikrishna A (2001) Unusual cyclizations. In: Renaud P, Sibi M (eds) Radicals in organic synthesis, vol 2. Wiley, Weinheim, p 151

8. Schiesser CH, Wild LM (1996) Tetrahedron 52:13265
9. Walton JC (1998) Acc Chem Res 31:99
10. Bowman WR, Bridge CF, Brookes P (2000) J Chem Soc Perkin Trans 1:1
11. Savéant JM (2000) Adv Phys Org Chem 35:117
12. Rossi RA, Pierini AB, Peñéñory AB (2003) Chem Rev 103:71
13. Snelgrove DW, Lusztyk J, Banks JT, Mulder P, Ingold KU (2001) J Am Chem Soc 123:469
14. Newcomb M (1993) Tetrahedron 49:1151
15. Studer A, Amrein S, Schleth F, Schulte T, Walton JC (2003) J Am Chem Soc 125:5726
16. Bowry VW, Lusztyk J, Ingold KU (1990) Chem Commun 923
17. Newcomb M, Choi SY, Horner JH (1999) J Org Chem 64:1225
18. Newcomb M, Glenn AG, Williams WG (1989) J Org Chem 54:2675
19. Ziegler FE, Petersen AK (1995) J Org Chem 60:2666
20. Krishnamurthy V, Rawal VH (1997) J Org Chem 62:1572
21. Beckwith ALJ, Moad G (1980) J Chem Soc Perkin Trans 2:970
22. Walton JC (1989) J Chem Soc Perkin Trans 2:173
23. Park SU, Varick TR, Newcomb M (1990) Tetrahedron Lett 31:2975
24. Chatgilialoglu C, Ingold KU, Scaiano JC (1981) J Am Chem Soc 103:7739
25. Beckwith ALJ, Easton CJ, Lawrence T, Serelis AK (1983) Aust J Chem 36:545
26. Beckwith ALJ, Schiesser CH (1985) Tetrahedron Lett 26:373
27. Beckwith ALJ, Moad G (1974) Chem Commun 472
28. Beckwith ALJ, Ingold KU (1980) Free radical rearrangements. In: de Mayo P (ed) Rearrangements in ground and excited states, vol 1. Academic, New York, p 161
29. Srikrishna A, Sharma GVR (1997) J Chem Soc Perkin Trans 1:177
30. Srikrishna A, Danieldos S (1997) J Org Chem 62:7863
31. Srikrishna A, Viswajanani R, Reddy TJ, Vijaykumar D, Kumar PP (1997) J Org Chem 62:5232
32. Čeković Ž, Saičić RN (1990) Tetrahedron Lett 31:6085
33. Saičić RN, Čeković Ž (1992) Tetrahedron 48:8975
34. Ferjančić Z, Saičić RN, Čeković Ž (1997) Tetrahedron Lett 38:4165
35. Ferjančić Z, Čeković Ž, Saičić RN (2001) C R Acad Sci Ser IIC 599
36. Fernández-Mateos A, de la Nava EM, Coca GP, Silva AR, González RR (1999) Org Lett 1:607
37. Gansäuer A, Lauterbach T, Zarayan S (2003) Angew Chem Int Ed 42:5556
38. Fernández-Mateos A, Burón LM, de la Nava EM, Clemente RR, González RR, González FS (2004) Synlett 2553
39. Fernández-Mateos A, Burón LM, Clemente RR, Silva AR, González RR (2004) Synlett 1011
40. Gansäuer A, Lauterbach T, Geich-Gimbel D (2004) Chem Eur J 10:4983
41. Friedrich J, Dolg M, Gansäuer A, Geich-Gimbel D, Lauterbach T (2005) J Am Chem Soc 127:7071
42. Villar H, Guibé F (2002) Tetrahedron Lett 43:9517
43. Bezzenine-Lafollée S, Guibé F, Villar H, Zriba R (2004) Tetrahedron 60:6931
44. Sakuna D, Togo H (2004) Synlett 2501
45. Newcomb M, Miranda N (2003) J Am Chem Soc 125:4080
46. Murphy JA, Dickinson JM (1992) Chem Commun 434
47. Smith DM, Nicolaides A, Golding BT, Radom L (1998) J Am Chem Soc 120:10223
48. Wang Y-M, Fu Y, Liu L, Guo Q-X (2005) J Org Chem 70:3633
49. Van Speybroek V, De Kimpe N, Waroquier M (2005) J Org Chem 70:3674
50. Piccardi P, Massardo P, Modena M, Santoro E (1973) J Chem Soc Perkin Trans 1:982

51. Jung ME, Gervay J (1991) J Am Chem Soc 113:224
52. Jung ME, Marquez R, Houk KN (1999) Tetrahedron Lett 40:2661
53. Jung ME (1999) Synlett 843
54. Weinges K, Schmidbauer SB, Schick H (1994) Chem Ber 127:1305
55. Johnston D, McCusker CM, Procter DJ (1999) Tetrahedron Lett 40:4913
56. Johnston D, McCusker CM, Muir K, Procter DJ (2000) J Chem Soc Perkin Trans 1:681
57. Hutton TK, Muir K, Procter DJ (2002) Org Lett 4:2345
58. Johnson D, Francon N, Edmonds DJ, Procter DJ (2001) Org Lett 3:2001
59. Williams DBG, Blann K (2004) Eur J Org Chem 3286
60. Fremont SL, Belletire JL, Ho DM (1991) Tetrahedron Lett 32:2335
61. Ishibashi H, Nakamura N, Sato T, Takenuchi M, Ikeda M (1991) Tetrahedron Lett 32:1725
62. Ishibashi H, Kameoki C, Kodama K, Ikeda M (1996) Tetrahedron 52:489
63. Ishibashi H, Kodama K, Kameoki C, Kawanami H, Ikeda M (1996) Tetrahedron 52:13867
64. Ishibashi H, Kameoki C, Kodama K, Kawanami H, Hamada M, Ikeda M (1997) Tetrahedron 53:9611
65. Ikeda M, Ohtani S, Yamamoto T, Sato T, Ishibashi H (1998) J Chem Soc Perkin Trans 1:1763
66. Ishibashi H, Kameoki C, Kodama K, Sato T, Ikeda M (1995) J Org Chem 60:1276
67. Quiclet-Sire B, Saunier JB, Zard SZ (1994) Tetrahedron Lett 35:1397
68. Cassayre J, Quiclet-Sire B, Saunier JB, Zard SZ (1998) Tetrahedron 54:1029
69. Bryans JS, Chessum NEA, Parsons AF, Ghelfi F (2001) Tetrahedron Lett 42:2901
70. Ghelfi F, Parsons AF (2000) J Org Chem 65:6249
71. Bryans JS, Chessum NEA, Huther N, Parsons AF, Ghelfi F (2003) Tetrahedron 59:6221
72. D'Annibale A, Pesce A, Resta S, Trogolo C (1997) Tetrahedron Lett 38:1829
73. Cerreti A, D'Annibale A, Trogolo C, Umani F (2000) Tetrahedron Lett 41:3261
74. Boiteau L, Boivin J, Quiclet-Sire B, Saunier JB, Zard SZ (1998) Tetrahedron 54:2087
75. Gill GB, Pattenden G, Reynolds SJ (1989) Tetrahedron Lett 30:3229
76. Pattenden G, Gill GB, Reynolds SJ (1994) J Chem Soc Perkin Trans 1:369
77. Pattenden G, Reynolds SJ (1994) J Chem Soc Perkin Trans 1:379
78. Jackson LV, Walton JC (2000) Chem Commun 2327
79. Bella AF, Jackson LV, Walton JC (2002) J Chem Soc Perkin Trans 2:1839
80. Bella AF, Jackson LV, Walton JC (2004) Org Biomol Chem: 2421
81. Roberts BP (1999) Chem Soc Rev 28:25
82. Fielding AJ, Roberts BP (2001) Tetrahedron Lett 42:4061
83. Scanlan EM, Walton JC (2002) Chem Commun 2086
84. Scanlan EM, Slawin AMZ, Walton JC (2004) Org Biomol Chem 2:716
85. Fallis AG, Brinza IM (1997) Tetrahedron 53:17543
86. DiLabio GA, Scanlan EM, Walton JC (2005) Org Lett 7:155
87. Grainger RS, Innocenti P (2004) Angew Chem Int Ed 43:3445
88. Ryu I (2001) Radical carbonylations mediated by tin, germanium and silicon reagents. In: Renaud P, Sibi M (eds) Radicals in organic synthesis, vol 2. Wiley, Weinheim, p 22
89. Tojino M, Otsuka N, Fukuyama T, Matsubara H, Schiesser CH, Kuriyama H, Miyazato H, Minakata S, Ryu I (2003) Org Biomol Chem 1:4262
90. Ryu I, Miyazata H, Kuriyama H, Matsu K, Tojino M, Fukuyama T, Minakata S, Komatsu M (2003) J Am Chem Soc 125:5632
91. Clark AJ, Peacock JL (1998) Tetrahedron Lett 39:1265
92. Clark AJ, Filik RP, Peacock JL, Thomas GH (1999) Synlett 441

93. Castle K, Hau CS, Sweeney JB, Tindall C (2003) Org Lett 5:757
94. Bailey WF, Longstaff SC (2001) Org Lett 3:2217
95. Hanessian S, Dhanoa DS, Beaulieu PL (1987) Can J Chem 65:1859
96. Yet L (1999) Tetrahedron 55:9349
97. Liu L, Chen Q, Wu YD, Li C (2005) J Org Chem 70:1539
98. Lang S, Corr M, Muir N, Khan TA, Schonebeck F, Murphy JA, Payne AH, Williams AC (2005) Tetrahedron Lett 46:4027
99. Taniguchi T, Ishita A, Uchiyama M, Tamura O, Muraoka O, Tanabe G, Ishibashi H (2005) J Org Chem 70:1922
100. Porter NA, Chang VHT (1987) J Am Chem Soc 109:4976
101. Prado MAF, Alves RJ, Souza-Filho JD, Alves RB, Pedrosa MTC, Prado RF, Faraco AAG (2000) J Chem Soc Perkin Trans 1:1853
102. Binatti I, Prado MAF, Alves RJ, Souza-Filho JD (2002) J Braz Chem Soc 3:570
103. Olivera RB, Prado MAF, Alves RJ, Souza-Filho JD (2003) J Braz Chem Soc 14:442
104. Faraco AAG, Prado MAF, Alves RJ, Souza-Filho JD, Alves RB, Faraco RFP (2003) Synth Commun 33:463
105. Faraco AAG, Prado MAF, Alves RB, Faraco RFP, Alves RJ, Souza-Filho JD, Meurer EC, Eberlin MN (2004) Tetrahedron Lett 45:3317
106. Balraju V, Reddy DS, Periasamy M, Iqbal J (2005) Tetrahedron Lett 46:5207
107. Molander GA, McKie JA (1994) J Org Chem 59:3186
108. Dinesh CU, Reissig HU (1999) Angew Chem Int Ed 35:789
109. Monovich LG, Le Huérou Y, Rönn M, Molander GA (2000) J Am Chem Soc 122:52
110. Molander GA, George KM, Monovich LG (2003) J Org Chem 68:9533
111. Walton JC (1997) Cyclopropylmethyl to but-3-enyl radical rearrangements. In: de Meijere A (ed) Carbocyclic three-membered-ring compounds. Thieme, Stuttgart, p 2438
112. Parrish JD, Little RD (2001) Diradicals in synthesis. In: Renaud P, Sibi M (eds) Radicals in organic synthesis, vol 2. Wiley, Weinheim, p 383
113. Khuong KS, Houk KN (2003) J Am Chem Soc 125:14867
114. Adam W, Garcia H, Didering M, Martí V, Olivucci M, Palomares E (2002) J Am Chem Soc 124:12192
115. Abe M, Ishihara C, Kawanami S, Masuyama A (2005) J Am Chem Soc 127:10
116. Mikesell PJ, Little RD (2001) Tetrahedron Lett 42:4095
117. Carroll GL, Harrison R, Gerken JB, Little RD (2003) Tetrahedron Lett 44:2109
118. Carroll GL, Harrison R, Gerken JB, Little RD (2003) Tetrahedron Lett 44:3433
119. Maiti A, Gerken JB, Masjedizadeh MR, Mimieux YS, Little RD (2004) J Org Chem 69:8574
120. Ruedi G, Nagel M, Hansen HJ (2003) Org Lett 5:4211
121. Handa S, Kachala MS, Lowe SR (2005) Tetrahedron Lett 45:253
122. Cossy J (2001) Photoinduced electron transfer in radical reactions. In: Renaud P, Sibi M (eds) Radicals in organic synthesis, vol 2. Wiley, Weinheim, p 229
123. Yoon UC, Mariano PS (2001) Acc Chem Res 34:523
124. Yoon UC, Kwon HC, Hyung TG, Choi KH, Oh SW, Yang S, Zhao Z, Mariano PS (2004) J Am Chem Soc 126:1110
125. Savéant JM (2000) Adv Phys Org Chem 35:117
126. Bowman WR (1988) Chem Soc Rev 17:283
127. Rossi RA, Pierini AB, Palacios SM (1990) Adv Free Radical Chem 1:193
128. Albini A, Fasani E, Mella M (1990) Top Curr Chem 168:143
129. Kornblum M (1990) Aldrichim Acta 23:71
130. Vaillard SE, Postigo A, Rossi RA (2002) J Org Chem 67:8500

131. Barolo SM, Lukach AE, Rossi RA (2003) J Org Chem 68:2807
132. Goehring RR, Sachdeva YP, Pisipati JS, Sleevi MC, Wolfe JF (1985) J Am Chem Soc 107:435
133. Wolfe JF, Sleevi MC, Goehring RR (1980) J Am Chem Soc 102:3646
134. Wu GS, Tao T, Cao JJ, Wei XL (1992) Acta Chimi Sin 50:614
135. Bowman WR, Heaney H, Smith PHG (1982) Tetrahedron Lett 23:5093
136. Dandekar SA, Greenwood SN, Greenwood TD, Mabic S, Merola JS, Tanko JM, Wolfe JF (1999) J Org Chem 64:1543
137. Goehring RR (1992) Tetrahedron Lett 33:6045
138. Goehring RR (1994) Tetrahedron Lett 35:8145
139. Martin SF, Liras S (1993) J Am Chem Soc 113:10450
140. Lyras S, Lynch CL, Fryer AM, Vu BT, Martin SF (2001) J Am Chem Soc 123:5918
141. Roydhouse MD, Walton JC (2005) Chem Commun 4453
142. Yang JK, Felton GAN, Bauld NL, Krische MJ (2004) J Am Chem Soc 126:1634
143. Wang LC, Jang HY, Roh Y, Lynch V, Schultz AJ, Wang X, Krische MJ (2002) J Am Chem Soc 124:9448

Top Curr Chem (2006) 264: 201–236
DOI 10.1007/128_029

Published online: 19 January 2006

The Degenerative Radical Transfer of Xanthates and Related Derivatives: An Unusually Powerful Tool for the Creation of Carbon–Carbon Bonds

Béatrice Quiclet-Sire · Samir Z. Zard (✉)

Laboratoire de Synthèse Organique associé au CNRS, Ecole Polytechnique, 91128 Palaiseau, France
zard@poly.polytechnique.fr

Abstract This review summarises recent work on the dithiocarbonyl group transfer reaction. Xanthates, in particular, have proved to be extremely useful for both inter- and intra-molecular additions. The broad applicability of the intermolecular addition to unactivated olefins opens tremendous opportunities for synthesis, since various functional groups can be brought together under mild conditions and complex structures can be rapidly assembled. The presence of the xanthate in the product is also a powerful asset for further modifications, by both radical and non-radical pathways. Of special importance is the access to highly substituted aromatic and heteroaromatic derivatives and the synthesis of block polymers through a controlled radical polymerisation mediated by various dithiocarbonyl agents (RAFT and MADIX processes).

Keywords Block polymers · Radical addition · Radical cyclisation · Radical-polar-crossover · Xanthates

Abbreviations

AIBN azo-bis(isobutyronitrile)
ACCN azo-bis(cyclohexanecarbonitrile)
DCE 1,2-dichloroethane
PMB *p*-methoxybenzyl

1 Introduction

Only a handful of reactions are capable of creating carbon–carbon bonds in an *intermolecular* fashion starting with simple, *un-activated* olefins, and hardly any is of truly broad generality. Even though radicals are capable of reacting with simple olefins, most of the developments and applications in this area have been confined to intramolecular transformations or additions to activated olefins, with the notable exception of allylations using allyl stannanes and analogous reagents. For recent books on radicals in synthesis, see [1, 2]. Intermolecular additions to *un-activated* olefins are simply too slow to compete with other pathways open to the radical intermediate. The rate constants between an intermolecular addition to a simple olefin (Scheme 1) and hydrogen abstraction from a stannane (a silane, an organomercury hydride, etc.), for example, generally differ by a factor greater than 10^3. This cannot therefore be overcome by playing on concentration effects through the use of syringe pump techniques, high dilution, mode of addition of the reagents, etc [1, 2].

One simple solution to this longstanding problem in radical chemistry is based on the degenerative radical exchange of dithiocarbonyl derivatives [3–6]. The principal, unwanted competing pathway is eliminated and, as a consequence, the intermediate radicals acquire a sufficient lifetime to allow them to add to un-activated olefins or, more generally, to undergo slow inter- or intra-molecular processes.

This key property can be easily understood by examining the reaction manifold displayed in Scheme 2. The initiation step generates a very small concentration of radicals R$^•$, which will rapidly add to the thiocarbonyl group of the starting xanthate **1** (path **A**). This addition is fast but the stabilised adduct radical **2** is too hindered to dimerise (or does so reversibly) and cannot disproportionate. The only pathway left is fragmentation by scission of the C – O (path **B**) or the C – S bond (path **C**). The former is quite difficult, since it involves a particularly strong bond and generates a high-energy ethyl radical. The latter (path **C**) leads simply to the starting xanthate and the same radical R. In other words, *the reaction of the initial radical R$^•$ with its xanthate precursor is reversible and degenerate.* This point is crucial to understanding the properties of the system: since R$^•$ is continuously regenerated, its effective lifetime in the medium increases considerably. It is therefore able to add even to an un-activated olefin **3** (path **D**) to give a new radical **4**, which in turn reacts *reversibly* with the starting xanthate to furnish ultimately, via intermediate radical **5**, adduct **6** as well as the initial radical R$^•$ to propagate the chain.

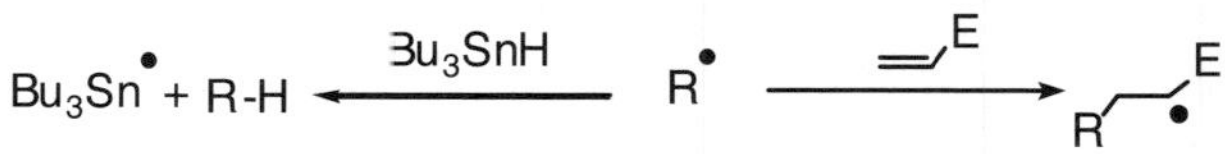

Scheme 1 Two competing pathways

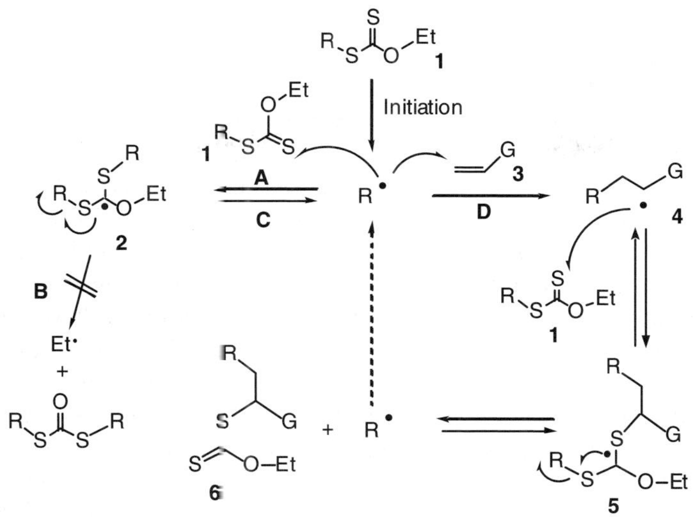

Scheme 2 The degenerate xanthate transfer manifold

The overall mechanism is relatively simple and corresponds to the addition of the elements of the xanthate across the double bond of the olefinic trap. Xanthates are used in this scheme since most of the transformations discussed below involve this group, but the same applies to other related derivatives such as dithioesters, dithiocarbamates, trithiocarbonates, etc.

There are, however, several subtleties embodied in the process, which need to be appreciated. The thiocarbonyl group is usually vastly more radicophilic than a simple olefin. This means that reactive radicals (namely R, **4**, and any radicals arising from the initiator) are rapidly removed from the medium and stored as stabilised adducts of type **2** and **5**. These adducts fragment to liberate preferentially the most stabilised radical (stabilisation implies thermodynamic effects and this may be used as a rule of thumb, but polar factors speeding up the fragmentation step can also have a significant influence). The xanthate group exerts in fact a powerful regulating influence on the concentration of the various radicals present in the medium. Furthermore, because intermediate **5** can fragment both ways, it is important to bias the fragmentation in the desired direction by making adduct radical **4** less "stable" (with the caveat noted above) than the initial radical R. The chain will otherwise be slowed down causing the appearance of unwanted side reactions. This simple consideration ensures that as long as the starting xanthate **1** is present, the product xanthate **6** is "protected" by being in a sense prevented from undergoing radical additions to the olefin resulting in telomerisation. The greater the difference in stability between R$^{\bullet}$ and **4** the better is the dichotomy in the reactivity of the two xanthates and the easier it is to control the process. This is a key point that is important to keep in mind, especially when dealing with intermolecular reactions.

There are many advantages attached to the use of the reversible addition-fragmentation on dithiocarbonyl derivatives.

- The reagents are cheap, generally stable, easy to handle, and readily available.
- No heavy metals are involved.
- The reactions are usually convergent and atom economical.
- The radical additions can be advantageously run under high concentration (reduced waste).
- The processes are self-regulating, safe, and easily scalable.
- Although peroxides are usually the preferred initiators for triggering the chain reaction, other initiators such as diazo derivatives, a combination of triethylborane and oxygen etc., can also be used in some cases. Initiation may equally be performed photochemically.
- There is a remarkable tolerance for most functional groups, allowing easy access to a very wide diversity of structures.
- A classical chain process is not strictly necessary. In some cases, the adduct radical **4** can be oxidised by the peroxide causing a crossover from a radical to a polar manifold. This will have its importance when dealing with the synthesis of aromatic and heteroaromatic derivatives. The peroxide thus behaves both as an initiator and a reagent and needs to be used in stoichiometric amounts.

2 Some Examples

Xanthates have been used most frequently because they offer the best combination in terms of reactivity, stability, and accessibility. Potassium *O*-ethyl xanthate is commercially available and cheap. It is an excellent nucleophile and many xanthates can be made trivially by displacement of a suitable leaving group. The examples of radical additions displayed in Scheme 3 are representative and underscore many of the desirable features discussed above. As can be seen, the xanthate reagent is readily made from the hemiacetal of trifluoroacetaldehyde and adds efficiently to a large assortment of olefins [7]. 1,2-Dichloroethane was used as the solvent but many other solvents can be used. Several related trifluoromethylated xanthates have been prepared and can be used to introduce a trifluoromethyl group into a complex molecule or to prepare various fluorinated starting materials [8–10].

The second set of examples pictured in Scheme 4 illustrates a route to a *S*-trifluoromethyl xanthate **9**, which cannot be made by a simple displacement reaction [10]. In this case the corresponding *S*-acyl xanthate **8** is first prepared

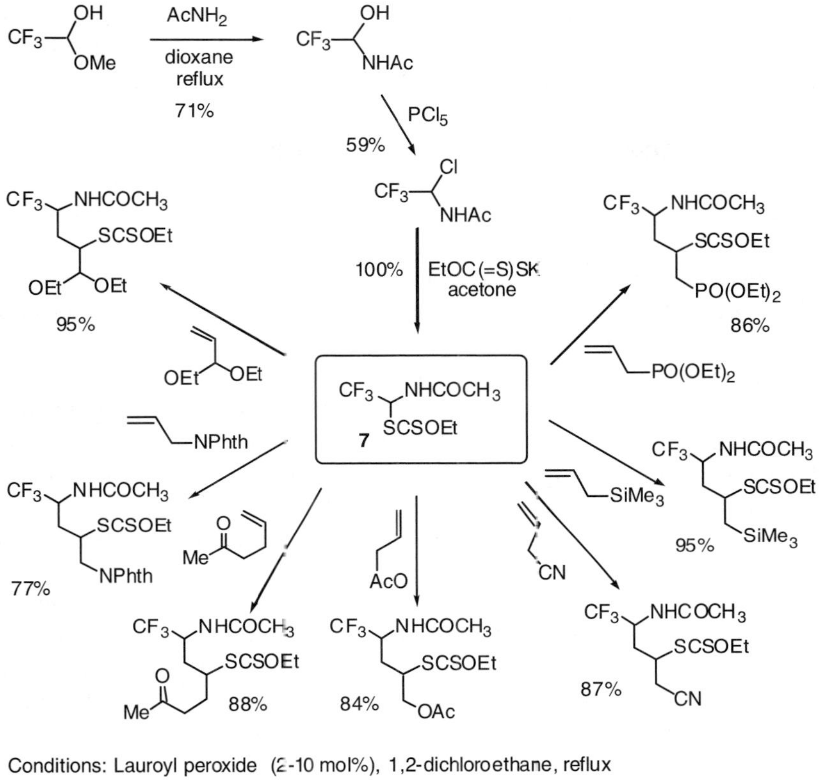

Scheme 3 Examples of radical addition to variously functionalised olefins

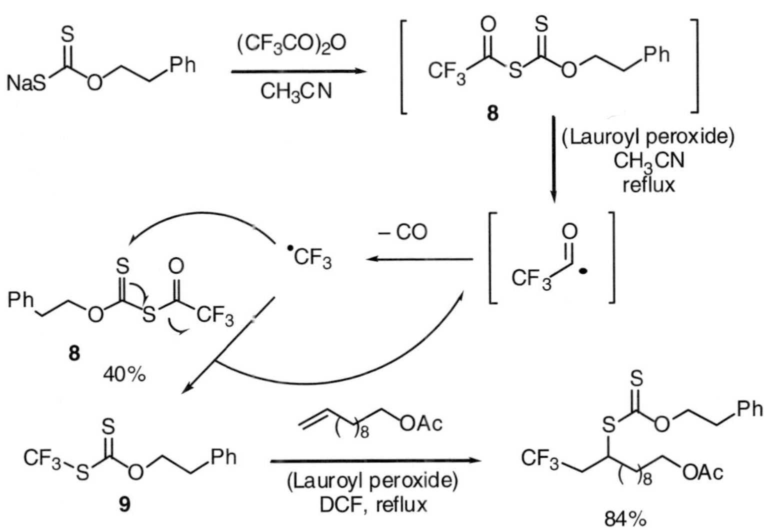

Scheme 4 Introduction of a trifluoromethyl group

and allowed to extrude carbon monoxide. The ethyl group in the xanthate was replaced by the heavier phenethyl in order to avoid the handling of a possibly volatile fluorinated molecule.

Yet another method to access xanthates consists of reacting an anion with a bis-xanthate [11]. Three such examples are displayed in Scheme 5. In the first, bisphosphonate xanthate **10** was prepared in this manner and found to add efficiently to 10-undecenyl pivalate [12]. As with xanthate **7** in Scheme 3, numerous other olefinic traps can be used besides the one shown, giving a plethora of new *geminal* bis-phosphonates. The second transformation corresponds to an annelation sequence devised by Saicic and co-workers [13] starting from xanthate **11**, itself prepared by reaction of an ester enolate with the bis-xanthate. The last example illustrates the possibility of using xanthate chemistry to generate and capture amidyl radicals [14]. The preparation of precursor **12**, belonging to a hitherto unknown family of xanthates, also relies on the substitution reaction of the amidyl anion with the bis-xanthate. The use of potassium hydride instead of sodium hydride is sometimes advantageous, as the potassium salt of the amide exhibits enhanced nucleophilicity towards the bis-xanthate.

There is a radical analogy to the above method, which involves reacting a radical with a bis-xanthate. If the radical precursor is a diazo derivative, an interesting situation occurs, since in principle nothing is wasted except for one molecule of nitrogen. This approach is summarised in Scheme 6 for the synthesis of tertiary xanthate **13** by the reaction of a bis-xanthate with an

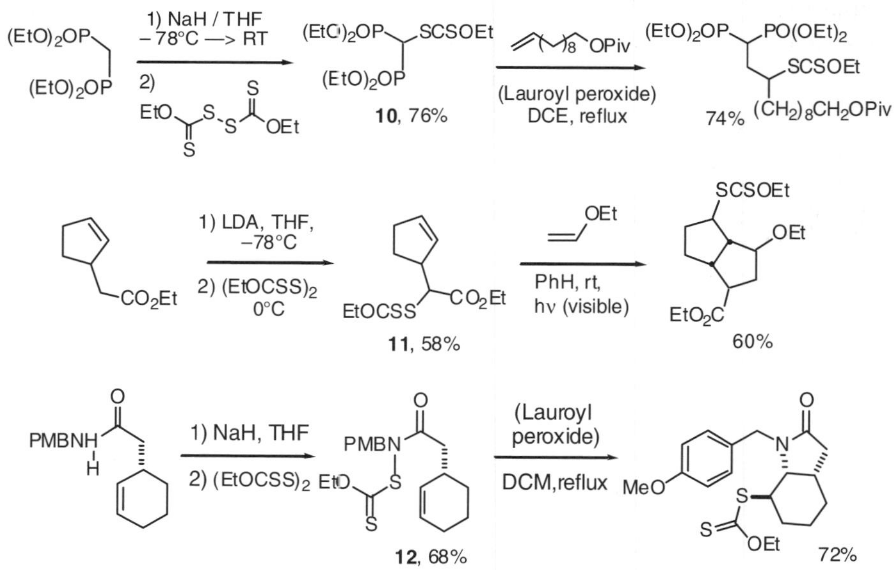

Scheme 5 Xanthates prepared by nucleophilic substitution on a bis-xanthate

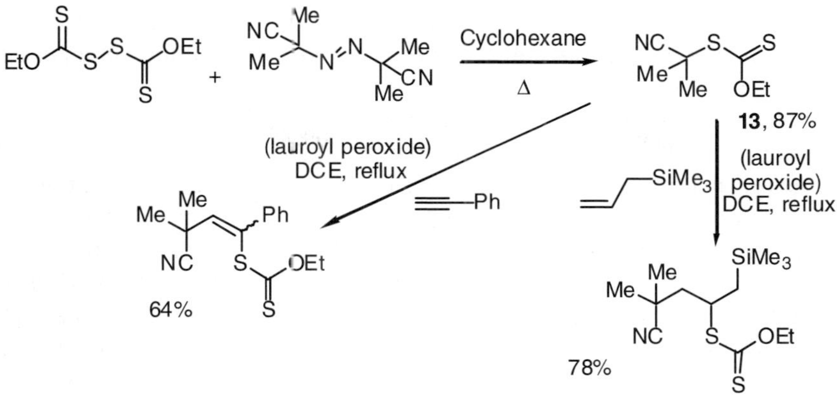

Scheme 6 Additions of a xanthate derived from AIBN

equimolar amount of AIBN. Various other hindered xanthates, dithioesters, dithiocarbamates etc., can be prepared in the same way [15, 16]. Addition of such xanthates to various olefins or to activated terminal alkynes such as phenylacetylene represents a simple method for generating quaternary centres [15].

Xanthate salts add reversibly to electrophilic olefins, but this conjugate addition is reversible [17]. By operating under acidic conditions, the reverse reaction can be prevented and a high yield of the desired xanthates can be secured. In this way, a number of otherwise inaccessible xanthates can be prepared and used in radical addition reactions [18]. Of particular interest are tertiary xanthates, such as compound **14** shown in Scheme 7, since the intermolecular addition leads to the formation of a quaternary centre. Further-

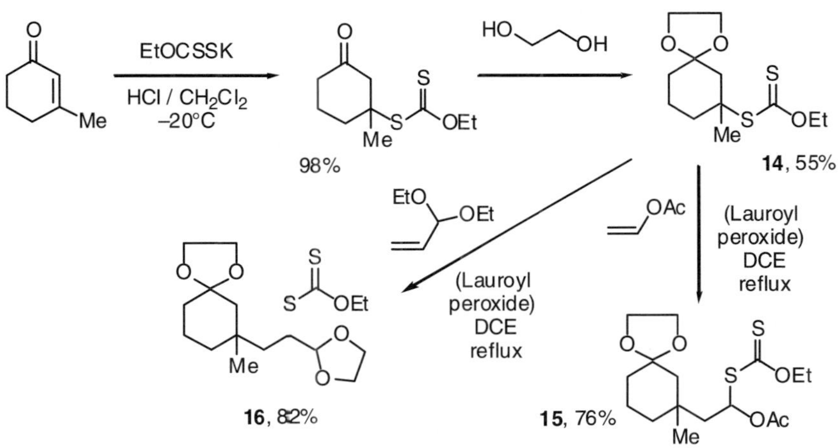

Scheme 7 Generation of quaternary centres

more, the addition product **15** to vinyl acetate contains a masked aldehyde function, and both **15** and **16** can be engaged in principle in Robinson-type annelations. The need to protect the ketone after the conjugate addition of the xanthate is not usually necessary, but in the present case the retro-Michael is particularly easy and has to be prevented by masking the ketone temporarily.

One unusual route to xanthates is the Leuckart reaction involving the reaction of a xanthate anion with a diazonium salt to give *S*-aryl xanthates. This reaction is reputed to lead sometimes to serious explosions and is therefore little used in practice. In the light of the general mechanism of Scheme 2, it is now possible to write a coherent chain mechanism for this transformation and expand greatly its synthetic utility as well as improve its safety aspects [19]. In the earlier procedures, the xanthate salt is added to the aqueous diazotation mixture. Combination of the diazonium and xanthate salts gives rise to a reactive diazo xanthate **17** (Scheme 8) which, being normally non polar, separates from the aqueous medium. Spontaneous initiation by light or metallic impurities can trigger an exothermic chain reaction within the *neat* diazo xanthate and lead to an explosion. All the reactions described in the present account do not present a safety hazard and can be safely scaled up. In contrast, the Leuckart reaction can pose a serious danger when performed under the original conditions, especially if the intermediate diazo xanthate **17** is a liquid and if the reaction is carried out on a large scale, for the process involves highly reactive aryl radicals and the rapid evolution of nitrogen gas. A simple, practical solution is to add an inert organic solvent to the diazotation mixture *before* portion-wise addition of the xanthate salt [19]. In

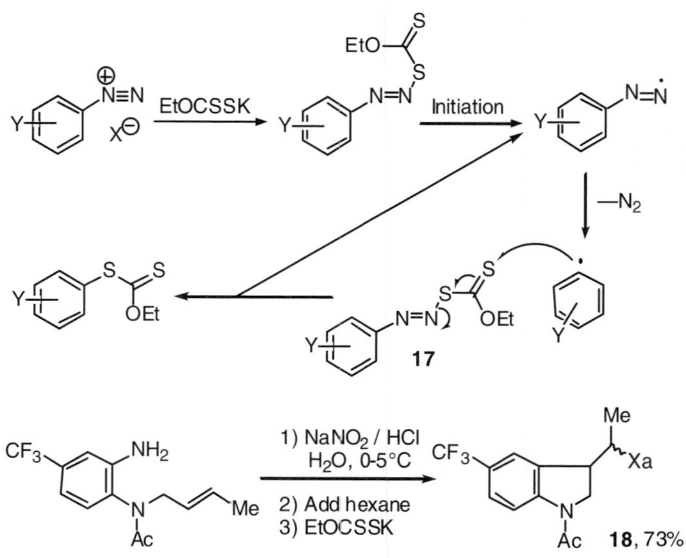

Scheme 8 A modification of the Leuckart reaction

this manner, the diazo-xanthate **17** is dissolved as it is formed and the chain reaction now takes place under dilute and well-controlled conditions. From a synthetic standpoint, capture of the intermediate aryl radical to produce cyclic structures such as **18** is especially interesting and can be done easily with the modified procedure.

3 Modifications of the Xanthate Group

The xanthate group in the adducts can be modified in many ways. The addition to vinyl pivalate shown in Scheme 9, like the addition to vinyl acetate mentioned in Scheme 7 above, provides a masked aldehyde because of the *geminal* disposition of the xanthate and pivalate group. It is therefore possible to take advantage of this fact to design a simple synthesis of pyrroles through a variation of the Paal–Knorr reaction (Quiclet-Sire et al, personal communication) [20]. By starting with an α-xanthyl ketone and vinyl pivalate, the radical addition leads efficiently to an intermediate **19** which undergoes condensation with benzylamine to give the corresponding pyrrole **20** (Scheme 9). Numerous pyrroles were analogously prepared by varying the different components. Interestingly, and unexpectedly, treatment of the same intermediate **19** with titanium tetrachloride produced a dithietanone **21** in good yield [21]. Dithietanones are very rare substances and their chemistry is still largely unexplored. Preliminary studies indicate them to be convenient precursors for thioaldehydes [Quiclet-Sire, B., Tétart, T., Zard, S. Z., unpublished results].

Another possibility open to α-xanthyl ketones is the formation of a dihydrothiophene by aminolysis of the xanthate group and ring closure of the

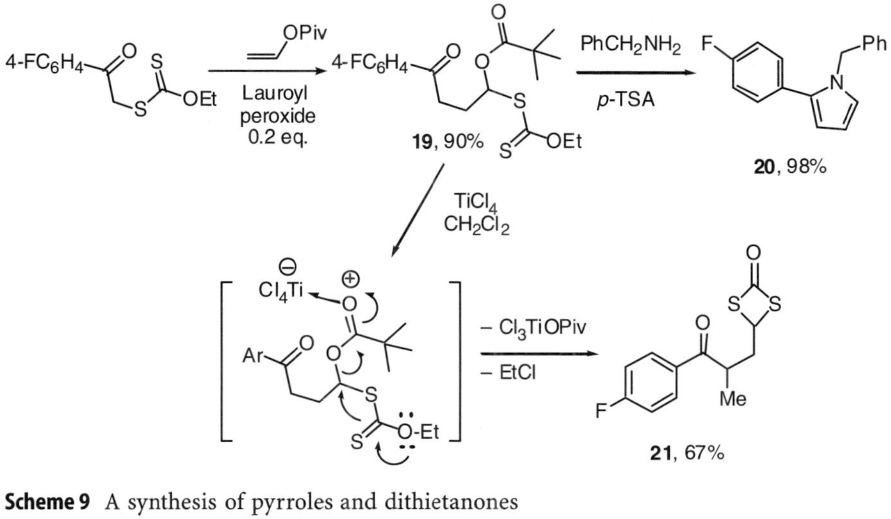

Scheme 9 A synthesis of pyrroles and dithietanones

resulting thiol onto the ketone, as indicated in Scheme 10. Oxidation to the sulfone followed by exposure to DBU in refluxing cyclohexane affords a diene through isomerisation of the olefinic bond and cheletropic elimination of sulfur dioxide [22]. Despite their deceptively simple structure, dihydrothiophenes and their corresponding sulfoxides and sulfones are tediously obtained by traditional routes and their tremendous synthetic potential still remains to be exploited.

The xanthate group, incorporated naturally into the addition products, represents a marvellous synthetic asset because it provides a simple, flexible point of entry into the unusually rich chemistry of sulfur. In many cases, however, the xanthate group is not needed and must be reductively removed. This task can be executed swiftly and efficiently using tributylstannane or tris-(trimethylsilyl)silane. Raney nickel or nickel boride are also often suitable reducing agents. Cheaper reagents include hypophosphorus reagents [23] or a combination of isopropanol and a stoichiometric amount of lauroyl peroxide [24]. The latter procedure relies on the fact that the primary, undecyl radical generated upon the thermal decomposition of the peroxide will preferentially react with the xanthate to liberate R, and the latter has no choice but to abstract a hydrogen from isopropanol because it cannot be consumed through the degenerate reaction with the xanthate. Abstraction of hydrogen from isopropanol leads to a ketyl radical. This species cannot propagate the chain because of its relative high stability but can be easily oxidised by electron transfer to the peroxide as indicated in the upper part of Scheme 11. The peroxide thus acts as both an initiator and an oxidant and must in consequence be used stoichiometrically. The modification of a complex molecule such as pleuromutilin without the need for protecting any of the functional

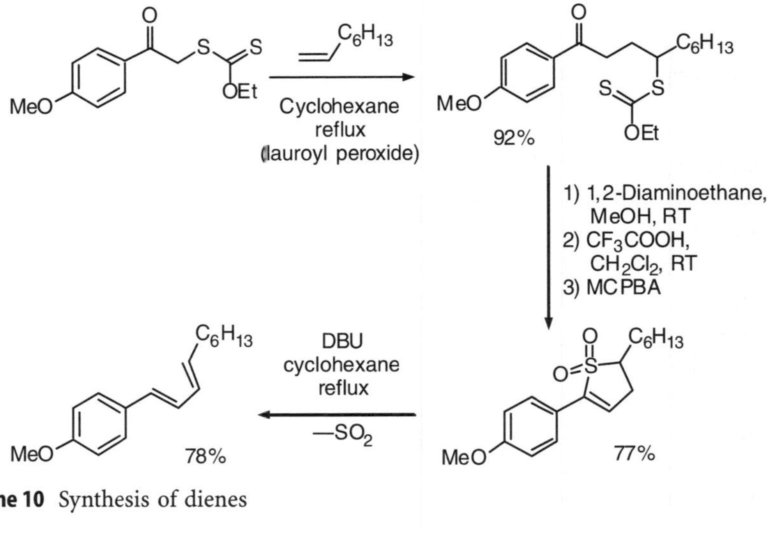

Scheme 10 Synthesis of dienes

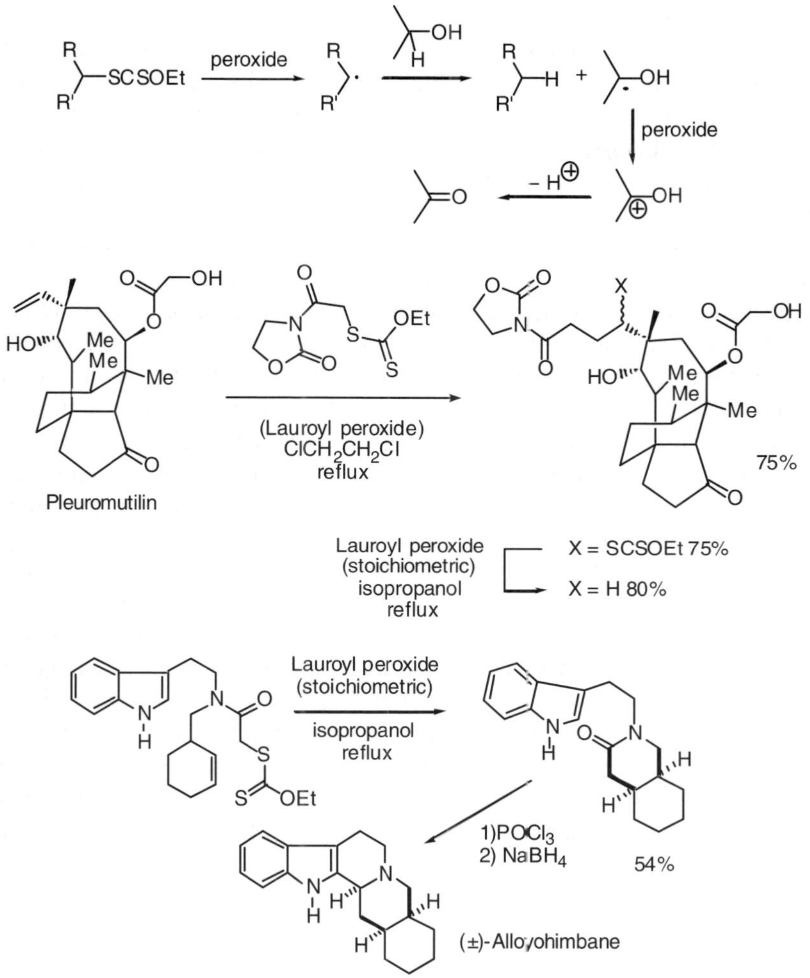

Scheme 11 Radical addition and reductive removal of the xanthate group

groups testifies to the mildness of both the radical addition and reduction steps [25]. Incidentally, the eight-membered ring in pleuromutilin can be constructed by a direct cyclisation of a suitable xanthate precursor [26]. The second example is a key step in the synthesis of alloyohimbane, where the 6-*exo* ring-closure is followed by reduction through a hydrogen atom abstraction from isopropanol [27]. It worth stressing that such cyclisations involving amides are inherently slow processes and cannot usually be accomplished efficiently using standard stannane-based methods.

Another transformation hinging on the use of stoichiometric amounts of peroxide is an exchange reaction that replaces the xanthate group with a bromide. This synthetically valuable functional group exchange is illustrated by the examples depicted in Scheme 12 [28]. Advantage is taken of the higher reactivity of a xanthate as compared with the bromine atom in ethyl α-bromoisobutyrate. The peroxide has to be used in equimolar amounts because the tertiary isobutyryl radical produced in the bromine transfer step is again too stabilised and incapable of propagating the chain. In some cases, as shown by the second transformation, the addition and exchange can be done directly in high yield.

More impressive transformations can be implemented by exploiting the properties of sulfonyl radicals. For instance, heating a xanthate in the presence of ethyl allyl sulfone and a small amount of a peroxide or a diazo initiator leads to an overall allylation reaction [29]; for a review see [30]. Two examples of this allylation procedure, including a simplified mechanistic rationale, are presented in Scheme 13. The key consideration is the fact that the ethylsulfonyl radical extrudes a molecule of sulfur dioxide to give a reactive ethyl radical that is now capable of propagating the chain. The ethyl group in the sulfone reagent can be replaced by a methyl or, for that matter, by any primary aliphatic group, as long as the derived radical is not stabilised. Many substituted allyl groups can be introduced. The fact that the method is tin-free makes it compatible with the presence of halogens. In the first transformation shown, base-induced elimination of bromide would lead to an alkyne resulting in an overall indirect radical propargylation. In the second example, a ring-closure precedes the allylation step.

The use of sulfones as relays can be extended to the introduction of a large variety of appendages. Vinylation is especially important because of the general difficulty of performing vinylations on sp^3 centres using transition metal chemistry. The examples assembled in Scheme 14 thus acquire a special significance [31]. It is also noteworthy that both the allylation and vinylation reactions, as well as some of the more exotic transformations discussed later in this section, all of which employ sulfone-based reagents, can be applied not just to aliphatic xanthates but also to iodides and tellurides [29, 37–44].

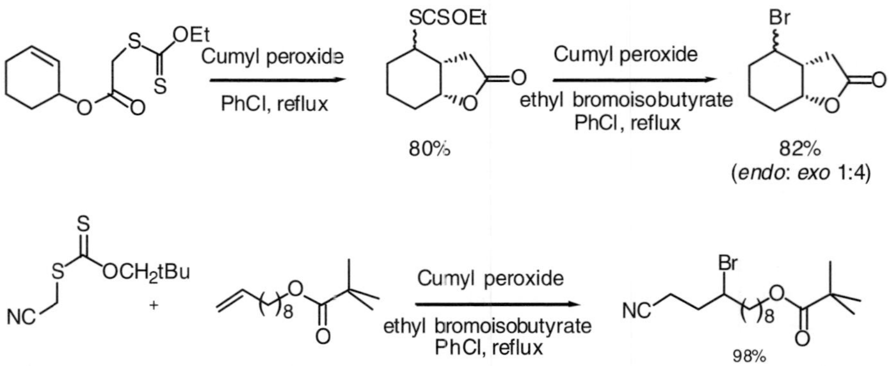

Scheme 12 Radical addition of xanthates and exchange with a bromine atom

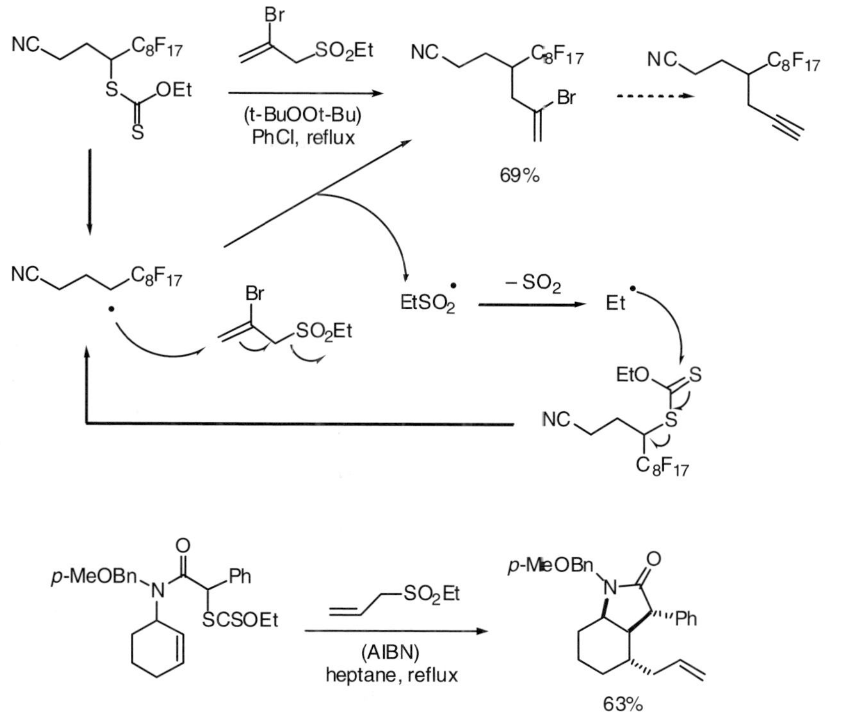

Scheme 13 Radical allylation of xanthates

Aliphatic iodides, and especially secondary and tertiary representatives, are subject to hydride elimination and are not generally useful substrates in transition metal catalysed coupling reactions. The last reaction in Scheme 14 [31], for instance, cannot be executed using current transition metal-based technology. In contrast, vinyl and aryl iodides, which are superb partners in many classical metal-induced coupling reactions, are very poor substrates in the present radical process because of the high energy of vinyl and aryl radicals. The two methods thus nicely complement each other.

One synthetic application of this methodology is highlighted in one of the two radical key steps in the formal total synthesis of (±)-lepadin B (Scheme 15) [32]. The 6-*exo* ring-closure leads, as in the case of the alloyohimbane, to the cis-decahydroquinoline system **23**. It is interesting to observe that even though the starting xanthate **22** is a mixture of diastereoisomers, the cyclisation leads to only one isomer, as far as the carbon bearing the methyl group is concerned. Epimerisation of the carbon in the position α- to the ketone occurs during the reaction, presumably induced by the small amount of lauric acid present in the medium. The xanthate in the cyclised product can then be exchanged for a styryl group by use of 2-methylsulfonylstyrene. The corresponding aldehyde, obtained by ozonoly-

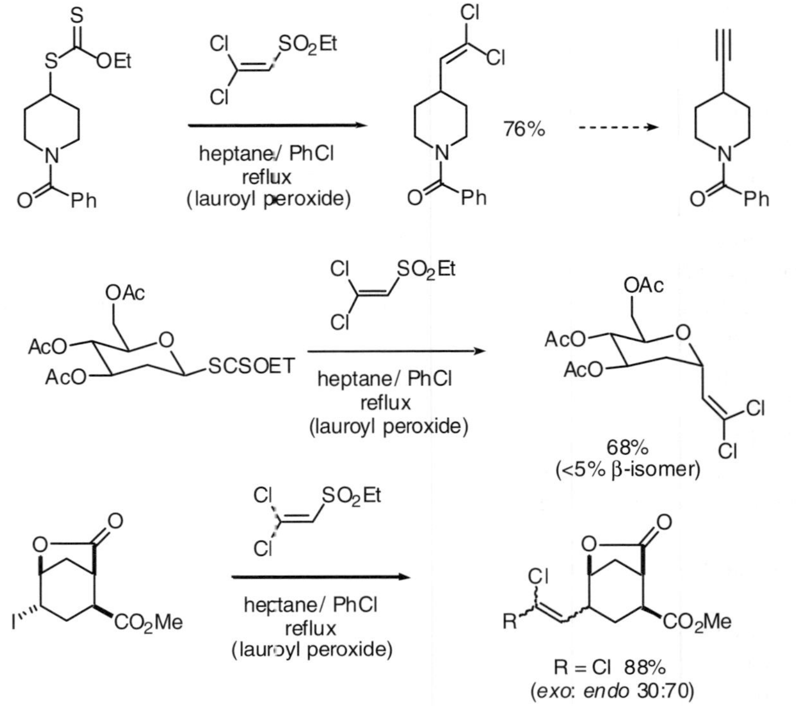

Scheme 14 Radical vinylation of xanthates and iodides

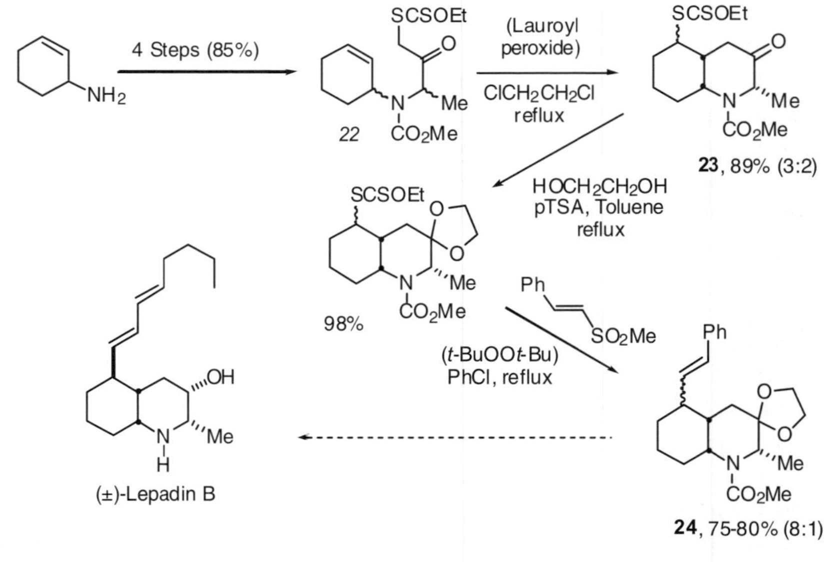

Scheme 15 Formal synthesis of (±)-Lepadin B

sis of the olefinic bond in **24**, serves to install the requisite octadienyl side chain via a Julia olefination reaction.

The leaving group and fragmentation abilities of the sulfonyl radicals can be exploited in many different ways. One interesting application is to force a given equilibrium to proceed in the desired direction by combining it with the β-scission of a sulfonyl radical. This contrivance was used for example to access 3-arylpiperidine structures that are of special interest to medicinal chemists and that are usually difficult to obtain by traditional ionic or organometallic chemistry [33–35]. As shown in Scheme 16, elimination of the sulfonyl group effectively drives the neophylic rearrangement towards the formation of the piperidine precursor **26**. Addition of ammonia or a primary amine followed by reduction of the intermediate cyclic imine (not shown) leads smoothly to various 3-arylpiperidines [33, 34]; for other approaches to piperidines and pyridines, see [35]. Structural diversity can be readily introduced by modifying the xanthate, the starting olefin **25** (which incidentally is easily prepared in two trivial steps from *p*-trifluoromethylbenzaldehyde), or the amine. Replacement of the reduction step with a dehydrogenation procedure would furnish the corresponding pyridine.

Yet another application concerns the synthesis of functionalised linear structures through the controlled translocation of the initial adduct radical. The principle of this approach is detailed in Scheme 17 for the first transformation, where the final elimination of a methylsulfonyl radical provides vinyl bromide **27** [36]. Internal hydrogen atom abstraction is not normally an intended pathway; it is often viewed as a complicating factor and a source

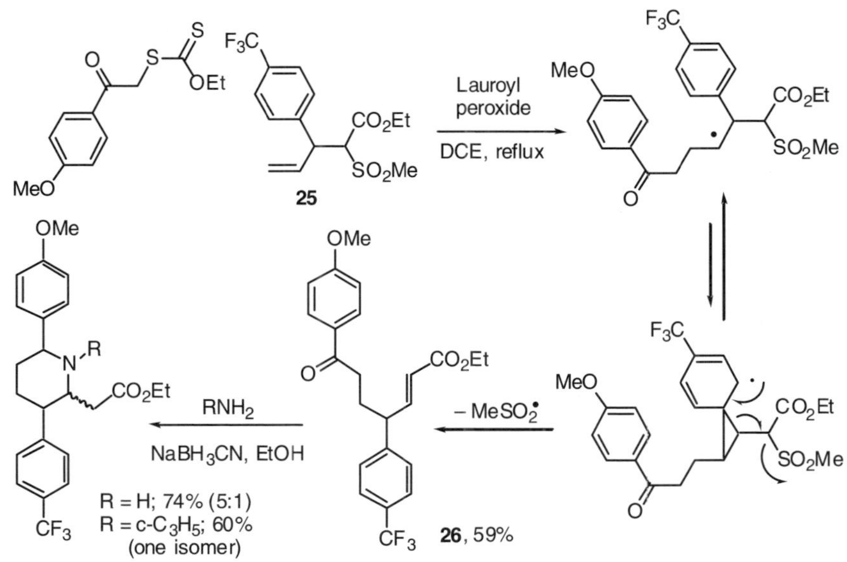

Scheme 16 The neophylic rearrangement in the synthesis of 3-aryl piperidines

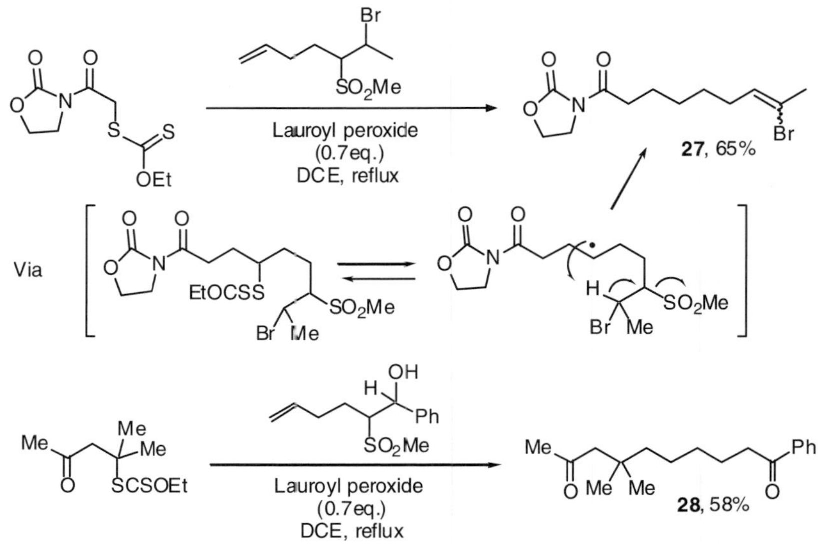

Scheme 17 Exploiting the 1,5-hydrogen atom shift

of unwanted side reactions. In the present case, the transfer allows the creation of a radical centre in a relatively inaccessible position and is ultimately harnessed in a productive manner. The second example in Scheme 17 constitutes an overall oxidation of the secondary alcohol through elimination of the sulfonyl radical to give ketone **28**.

The groups of Renaud and Kim introduced important extensions to the chemistry of sulfonyl radicals [37–44]. For instance, Renaud and his coworkers found that it is possible to use a sulfonyl azide to replace a xanthate or an iodide with an azide group [37–42], as illustrated by the two examples in Scheme 18. This reaction opens up numerous possibilities for the expeditious construction of complex alkaloid structures by combining the easy reduction of azides into amines with the convergence and flexibility of the intermolecular addition of xanthates. Azides have also recently gained high visibility in the context of "click" chemistry. In the lower part of Scheme 18 is given an example of a one-carbon elongation of a xanthate into the homologous aldehyde oxime taken from the work by Kim and his students [43, 44]. As a result of all these efforts, xanthates (and iodides and tellurides) can now be considered as springboards to a vast number of structures and functional groups.

Other motifs can be eliminated besides sulfonyls in order to introduce an unsaturation. Nitrogen dioxide is a good leaving group in a radical sense and, since nitroolefins are excellent Michael acceptors, it is easy to prepare β-nitro xanthates by conjugate addition of a xanthate salt. Exposure of these compounds to an equivalent amount of lauroyl peroxide furnishes the corres-

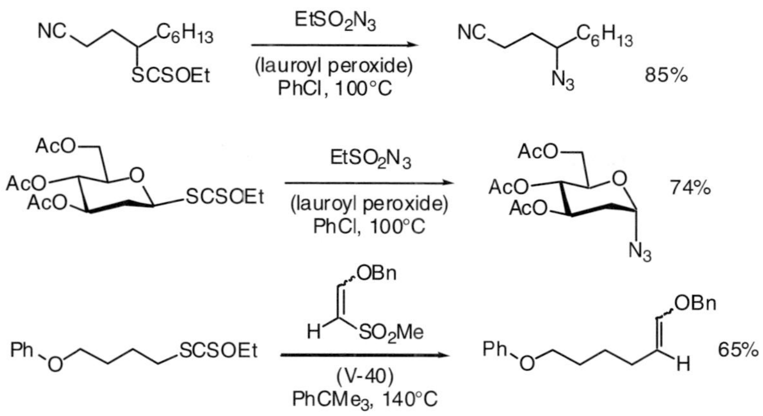

Scheme 18 Radical azidation and acylation

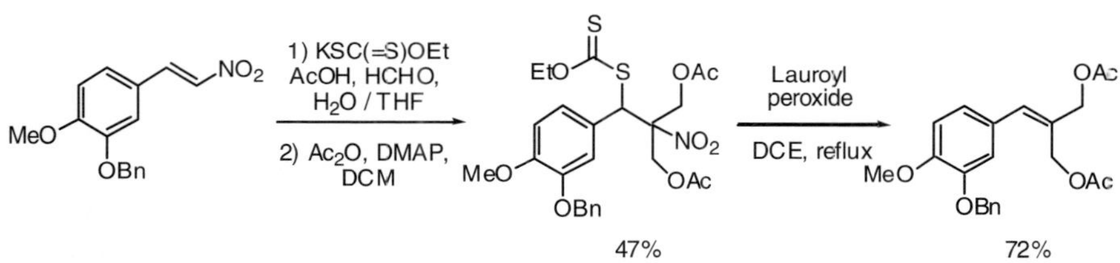

Scheme 19 Formation of alkenes from vicinal nitro xanthates

ponding alkenes [45]. The example in Scheme 19 is representative. It involves a conjugate addition of potassium *O*-ethyl xanthate to the nitrostyrene, followed by a double Henry reaction with formaldehyde and acetylation of the nitrodiol. The combination of the potent Henry reaction with the radical elimination can thus be used to advantage to assemble olefins that are not readily accessible by classical routes.

4 Some Further Synthetic Applications

The possibility of creating a carbon–carbon bond by an *intermolecular* addition to an *un-activated* olefin can be used to bring together functional groups in a manner that would otherwise be quite arduous to accomplish. These functional groups can then be made to react together by a modification of the reaction conditions. This conception is encapsulated by the two reactions in Scheme 20, where phosphonate-containing side-chains of different lengths can be readily attached to cyclobutanones [46, 47]. Treatment with base then induces an intramolecular Wittig–Horner–Emmons reaction to give either the cyclohexene or the cycloheptene derivative. It is worth stressing the fact

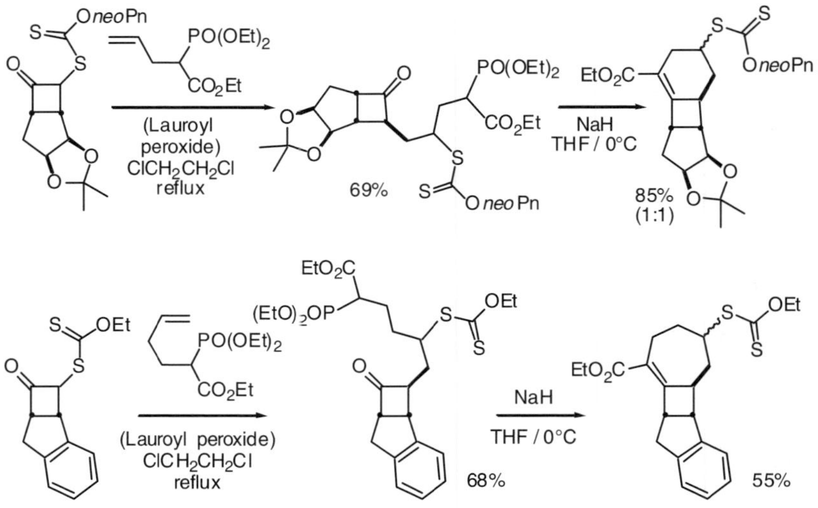

Scheme 20 Synthesis of cyclohexenes and cycloheptenes

that alkylation of cyclobutanones via the corresponding enolates is generally problematic and does not exhibit the generality expected with ordinary ketones. The use of the neopentyl xanthate in this case obviates an interesting but unwanted ionic side reaction. More generally, the neopentyl xanthate is preferred in cases where the substitution step used to prepare the xanthate starting material proves problematic with the simpler ethyl analogue. An ionic chain reaction is sometimes observed whereby xanthate RSC(= S)OEt is converted into an ethyl sulfide, R-SEt, with a net loss of carbon oxysulfide.

Exotic amino acids can be easily prepared, either by addition of a xanthate to an olefin containing a protected α-amino acid motif (or β-amino acid, or, for that matter, almost any disposition of an amine and carboxylic acid function) or by adding a xanthate bearing the amino acid function to various olefins. The former approach is illustrated by the first transformation in Scheme 21 involving the addition of xanthate **29** to a protected allylglycine (Heinrich and Zard, personal communication) [48]. The second example shows the addition of a protected glycine unit **30** to 1-hexene (Heinrich and Zard, personal communication) [48]. It was found that placing two Boc groups on the nitrogen enhanced significantly the reactivity of the corresponding radical and improved the yield of the reaction. The last, more complex example involving dipeptide **31**, is taken from the study of Skrydstrup and collaborators [49]. In this case, the addition product was not isolated but reduced using the combination of lauroyl peroxide and isopropanol discussed earlier.

Unusual and otherwise difficult to obtain boronates can be made by addition of a xanthate to allyl or vinyl boronates. The reaction proceeds in the same way as for ordinary olefins and compounds containing a boronate

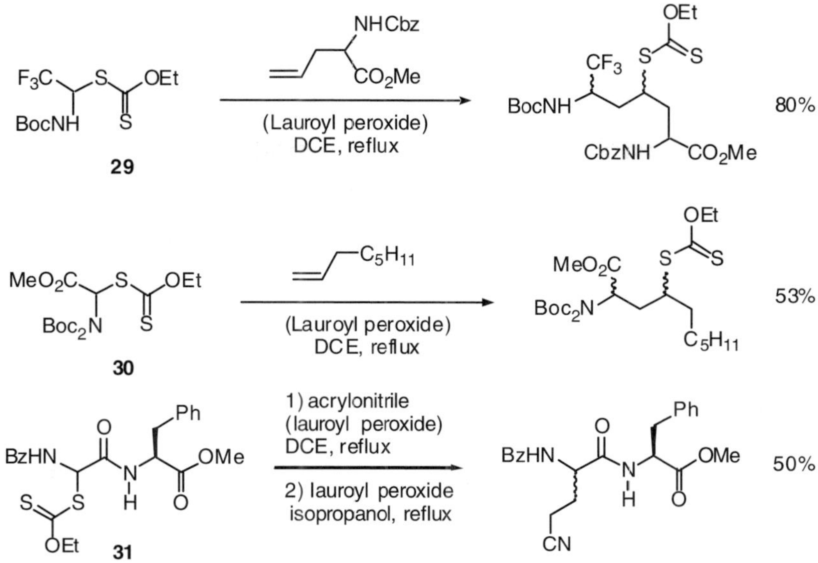

Scheme 21 Synthesis of amino acids

and various combinations of other functional groups can be readily prepared [50, 51]. The transformations sketched in Scheme 22 are typical. The distance separating the boronate and carbonyl groups can be modified by selecting a boronate partner with the correct side chain, as shown by the use of vinyl and allyl boronates in the first two examples [50]. Presumably higher homologues are also suitable. The addition of 1-phthalimidocyclopropylacyl xanthate **32** highlights the possibility of generating and capturing an acyl radical [51]. Interestingly, cyclopropyl acyl radical **33** undergoes neither loss of carbon monoxide, because the resulting cyclopropyl radical is in fact akin to a vinyl radical and therefore rather destabilised, nor opening of the three-membered ring [52]. The xanthate group in the resulting adduct **34** is now β- to the ketone and treatment with mild base causes elimination to give vinyl boronate **35** with an interesting and unusual array of functionalities. The role of the methyl iodide in this reaction is to intercept irreversibly the eliminated xanthate and to prevent it from undergoing undesired reactions.

Not only acyl, but also alkoxycarbonyl radicals can be generated from the corresponding xanthates, and their capture by addition to olefins produces esters or lactones, depending on whether the addition reaction is inter- or intra-molecular [53]. Grainger and Innocenti found that xanthates derived from carbamoyl chlorides were difficult to make and handle but, by replacing the xanthate salt by a dithiocarbamate, better precursors for the desired aminocarbonyl radicals were obtained [54]. Irradiation with a tungsten lamp proved more efficient than chemical initiation with lauroyl peroxide and lactams of various sizes could be readily obtained, as illustrated by the

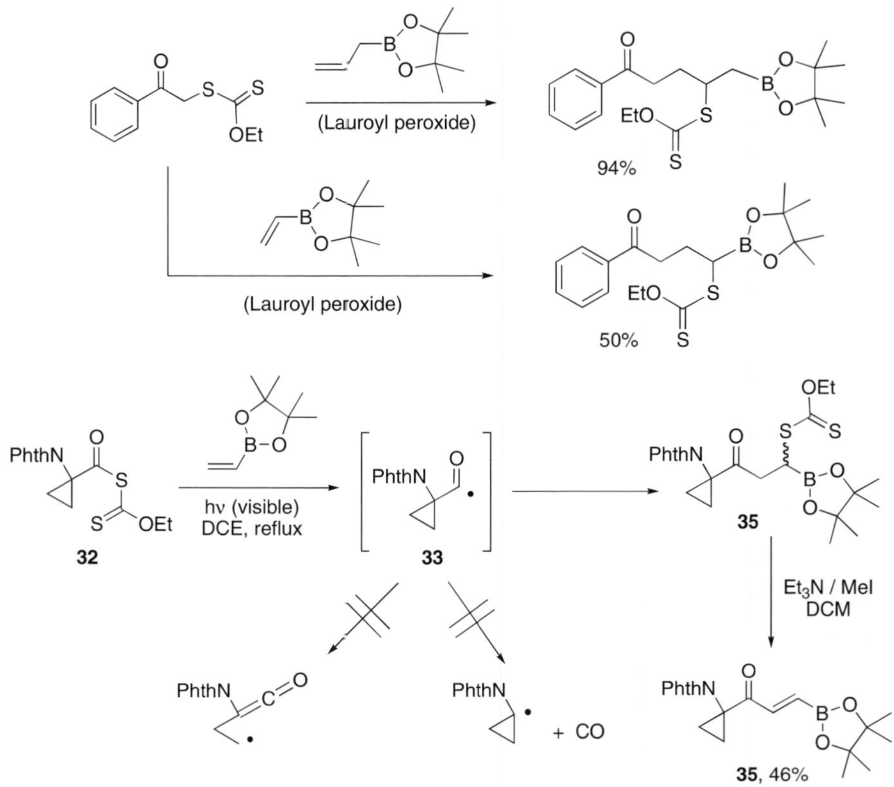

Scheme 22 Synthesis of novel boronates

transformations in Scheme 23. Most remarkable is the exceedingly efficient ring-closure to the strained bicyclic β-lactam. The lower yielding but nevertheless highly unusual formation of the eight-membered lactam is also worthy of note. With a chain on the nitrogen one carbon shorter, the reaction (not shown) furnished a mixture of δ- and ε-lactams resulting from competing 6-*exo* and 7-*endo* closures.

The introduction of a formyl group by a radical addition cannot be accomplished directly using the corresponding xanthate, which is too reactive to make or handle. A disguised form has to be employed, therefore, and a simple solution was found based on the chemistry of dithianes [55]. Although xanthate **36** could be prepared, its reaction with simple olefins failed to give the desired adducts. The corresponding dithianyl radical **37** is quite nucleophilic in character and only highly electrophilic alkenes, such as maleimide, were capable of capturing it in useful yield. Monosulfoxide **38**, in contrast, turned out to be a much more interesting reagent. Its derived radical **39** reacted efficiently with various olefins, as exemplified by the two reactions in Scheme 24 [55]. The dithiane *S*-oxide is in-

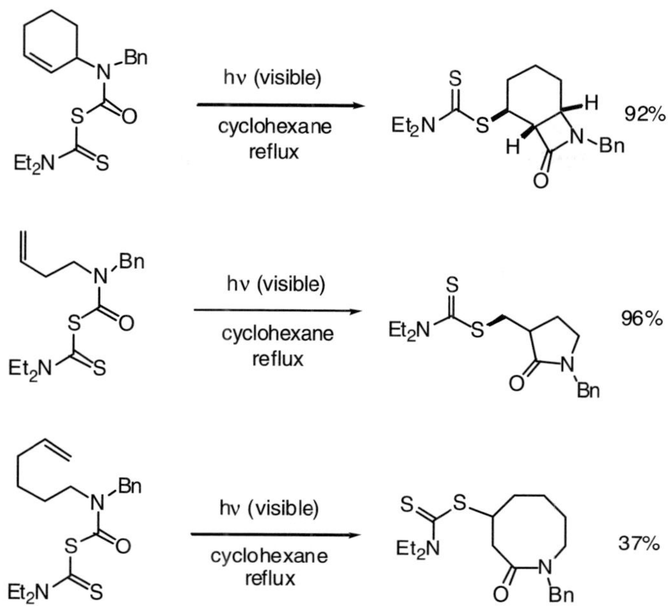

Scheme 23 Generation and capture of aminocarbonyl radicals

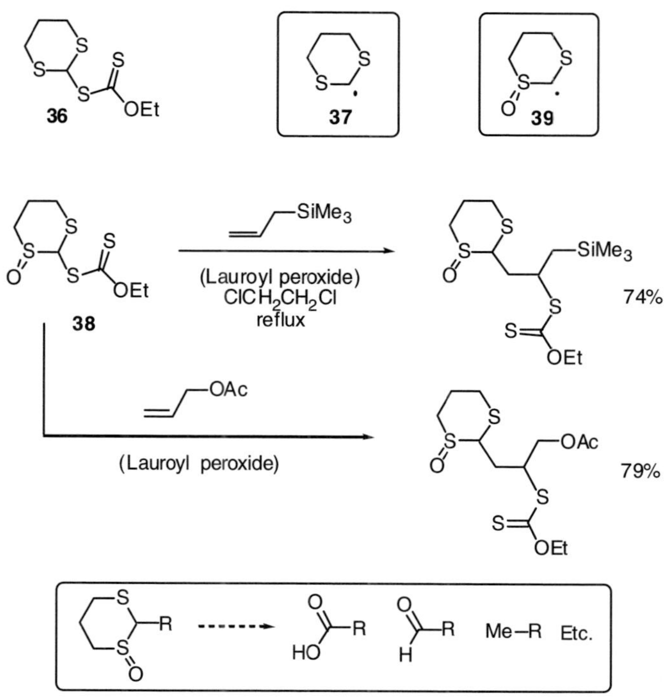

Scheme 24 A one carbon radical precursor

deed a true chameleon: it can be cleaved to an aldehyde, converted into a carboxylic acid through the Pummerer reaction, or reduced to a methyl group. Xanthate **38** is therefore a formal precursor of formyl, hydroxycarbonyl, or methyl radicals. None of these species can be generated directly or captured usefully, either because the corresponding xanthate cannot be made, as for the first two, or because the ensuing radical, as in the case of Me, is too high in energy and a chain process cannot in consequence be perpetuated.

5
Radical-Polar Crossover Reactions

As mentioned in the introduction, the propagation of the chain process depends to a large extent on the difference in stabilities between the initial and adduct radicals, the former having to be preferably more stable than the latter. It is therefore difficult to establish a chain reaction when the initial radical is too unstabilised (for example Me radical, as stated above, or vinyl and aro-

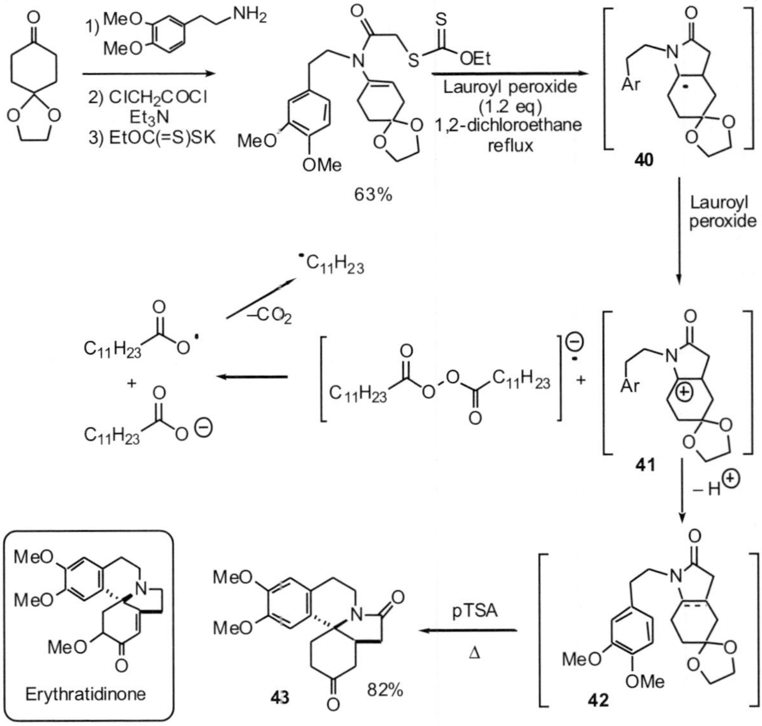

Scheme 25 Synthesis of the erythrina skeleton

matic radicals—see however the Leuckart reaction discussed earlier) or when the adduct radical is, *a contrario*, too stabilised. In the latter situation, if the stabilisation is due to an electron-releasing substituent, then it is possible to imagine oxidation of the adduct radical by a one-electron transfer to the peroxide. The radical is thus converted into the corresponding cation and the peroxide into the radical-anion. The latter rapidly collapses into a carboxylate anion and a carboxylic radical, which undergoes extrusion of carbon dioxide to furnish an undecyl radical capable of propagating the chain. The efficiency of the now modified chain process depends largely on the efficacy of the electron transfer step.

These considerations are detailed in Scheme 25 for the case of a 5-endo ring closure from an enamide xanthate [56]. In the example shown, this step leads to an intermediate tertiary radical 40, which is too stabilised to propagate the chain but which can be easily oxidised to the cation 41. Loss of a proton provides bicyclic derivative 42 as a mixture of regioisomers. Heating of this product with *p*-toluenesulfonic acid induces an intramolecular Friedel–Crafts reaction of the Pictet–Spengler type to give tetracycle 43 possessing the erythrina skeleton in a highly concise manner [56].

The cascade depicted in Scheme 26 represents another instance of a passage from a radical to the cation, with the sequence this time starting from an amidyl radical and ending by ring-closure to a furan ring [57]. The last radical 44 is electron rich and easily oxidised to the corresponding cation 45. Quenching with methanol finally provides cyclic acetal 46, as a mixture of diastereoisomers, the characterisation of which is simplified by oxidation to lactone 47 by Jones' reagent. This uncommon transformation of the furan ring provides rapid access to complex and unusual structures.

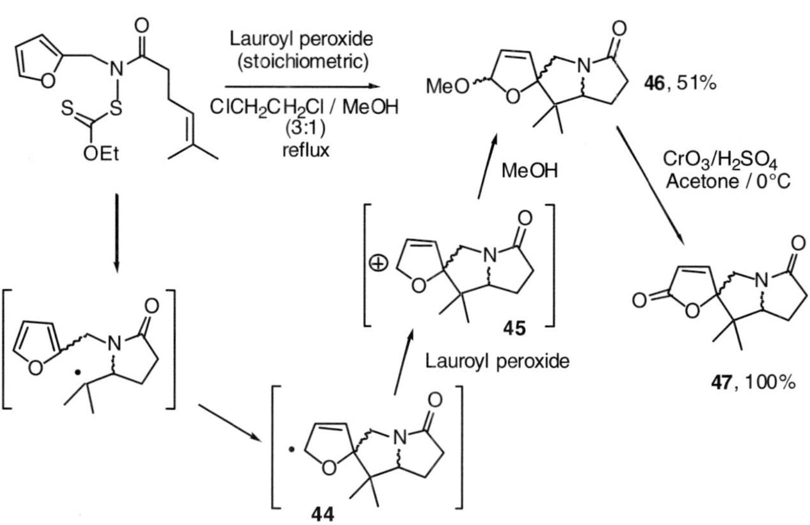

Scheme 26 The formation of lactones from furans

6
Synthesis of Aromatic Derivatives

The possibility of a radical to cation crossover adds another significant and exciting dimension to the synthetic potential of the xanthate-based technology. Thus, addition to aromatic or heteroaromatic rings also leads to stabilised radicals that are easy to oxidise by the peroxide. Loss of a proton then regenerates the aromaticity. Indeed, the long life of the intermediate radicals created through the xanthate process provides a simple, efficient and often convergent approach to a very broad variety of aromatic derivatives. This approach nicely complements traditional routes, such as the Friedel–Crafts and related reactions or more recent transition metal-mediated couplings. The following examples will serve to highlight the diversity of the accessible structures as well as the flexibility of the method.

Oxindoles [58] and indolines [59, 60] can be obtained by direct cyclisation for the former and by an addition–cyclisation for the latter, as shown in Scheme 27. Both electron-withdrawing and electron-donating groups can be present on the aromatic ring and a large number of indolines can be made, merely by modifying the xanthate component. Furthermore, the substrates are trivial to prepare, consisting on one hand of displacing a chloride or a bromide

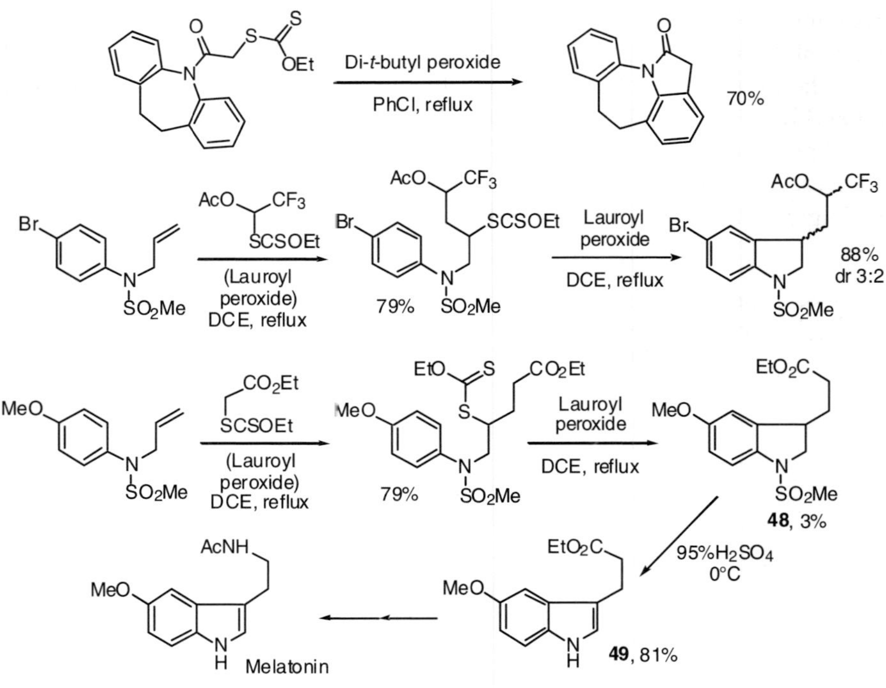

Scheme 27 Synthesis of oxindoles, indolines and indoles

from a chloro- or bromo-acetanilide and, on the other, of a simple allylation of an aniline. The last sequence summarises a synthesis of melatonin that exploits a serendipitous finding, whereby elimination of methylsulfinic acid to give an indole can be accomplished in some cases with cold sulfuric acid [60].

In the case of indolines, the reaction leads naturally to 3-substituted derivatives. The introduction of a substituent in the 2-position can be accomplished in principle by incorporating the desired substituent in the *N*-allyl aniline component. There is, however, an alternative approach, outlined in Scheme 28. It consists of using an *N*-vinylsulfonyl aniline as the olefinic trap [61]. The addition and cyclisation steps in the example shown occur in the same pot and lead to a cyclic sulfonamide **50** which, upon heating, undergoes a retro-cheletropic elimination of sulfur dioxide leading ultimately to compound **51**. This sequence represents a convenient *ortho*-functionalisation of the starting aniline. If a base, such as DBU, is added either at the beginning or at the end of the reaction, then a base catalysed migration of the double bond takes place, followed by an intramolecular Michael addition to give the 2-substituted indoline **52**.

The closure onto aromatic cycles is not limited to 5-membered rings. Remarkably, various 6-membered rings fused to aromatics can be made by a similar general strategy. An assortment of such transformations is presented in Scheme 29. The first, taken from a study by Guillaumet and collaborators [62], illustrates the synthesis of an aminochroman. The second, also involving direct cyclisation, corresponds to an unusual synthesis of homophthalimides [63]. These deceptively simple compounds and their derived homophthalic anhydrides are surprisingly difficult to prepare, and only a handful have been described in the literature, despite their synthetic utility as a basic building block in the syntheses of various biologically active substances. Interestingly, the desired product crystallised upon cooling and the yield given corresponds to the material isolated by simple filtration.

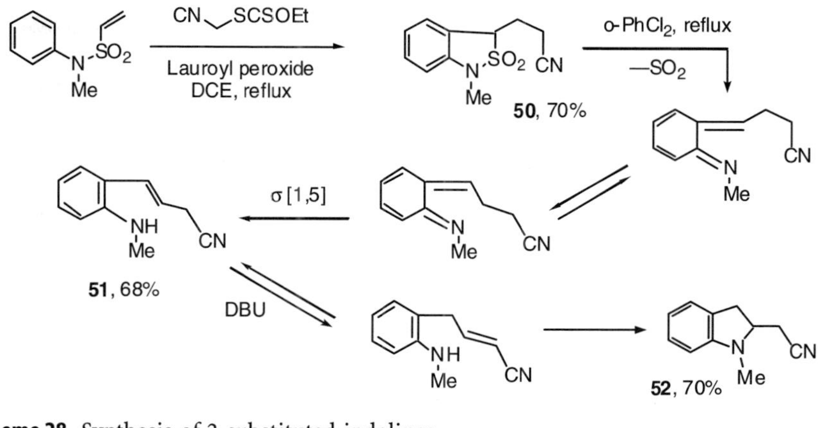

Scheme 28 Synthesis of 2-substituted indolines

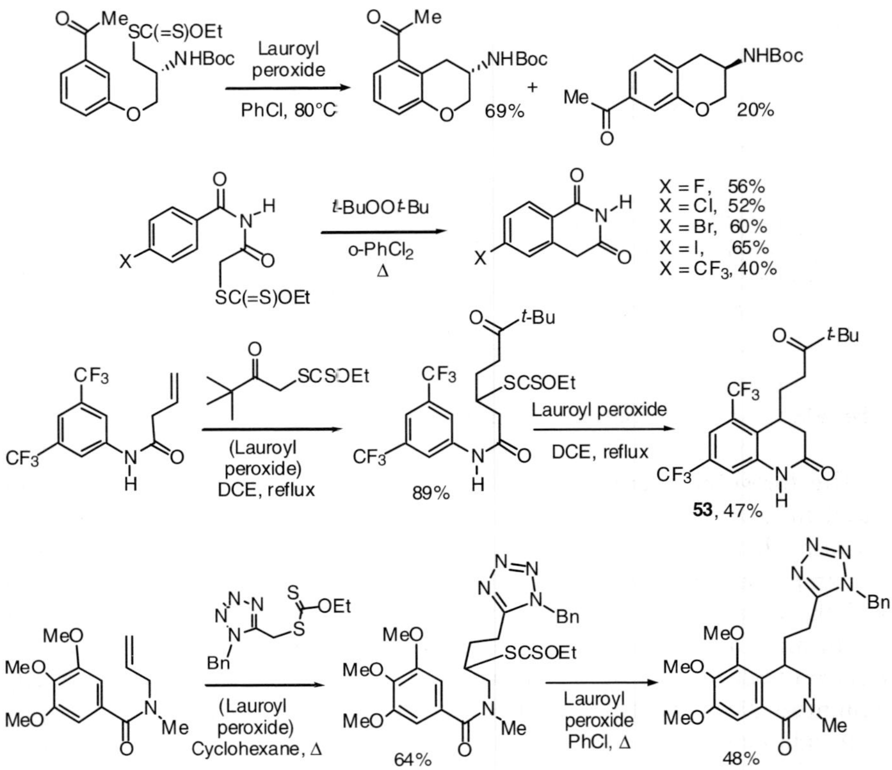

Scheme 29 6-membered ring-closure onto aromatics

A noteworthy feature of this cyclisation is that it occurs even with no substituent on the nitrogen, albeit a high temperature is required. A related, surprisingly efficient transformation that takes place without the need for a substituent on the anilide nitrogen is also displayed in Scheme 29. It involves the synthesis of a dihydroquinolone **53** by an intermolecular addition to an *N*-vinylacetanilide followed by ring-closure in the same pot [64]. The isomeric dihydroisoquinolones can be prepared by using an *N*-allyl benzamide as the olefinic substrate. This is exemplified by the last transformation in Scheme 29 where the xanthate partner bears a pharmacologically attractive tetrazole ring [65].

Unexpectedly, and very pleasingly, even the construction of 7-membered rings proved possible. Two examples are shown in Scheme 30 illustrating both a direct cyclisation and an addition–cyclisation sequence [66]. The first derivative **54**, being ultimately derived from tryptophan, can be obtained in principle as an optically pure substance. The second is a simple approach to very rare polycyclic structures such as **55**, only one example of which was documented in the Beilstein data bank.

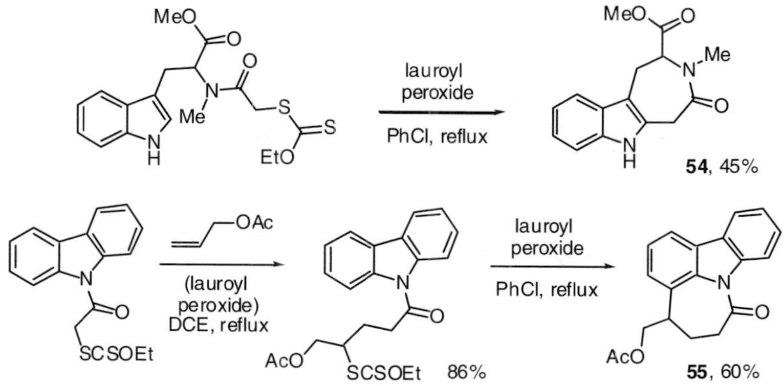

Scheme 30 Formation of benzazepinones by direct cyclisation

The foregoing examples summarise how various nitrogen heterocycles fused to aromatic rings can be assembled. Numerous derivatives of interest to medicinal and agrochemical chemists thus become readily accessible. It is important to underscore the functional group tolerance of the method and the facility with which fluorinated derivatives can be made. It is also worth mentioning the fact that the presence of one or more of the strongly electron-withdrawing trifluoromethyl groups on an aromatic ring essentially eliminates the possibility of using Friedel–Crafts type reactions.

Pyridines represent another family of aromatic derivatives that are also generally resistant to attack by electrophiles, but not by radicals. Hence, a similar strategy can be applied to the construction of nitrogen containing heterocycles adjoining pyridine rings. The synthesis of a number of such derivatives is detailed in Schemes 31 and 32. Azaoxindoles and azaindolines, as well as tetrahydroazaquinolines are readily obtained from the logical precursors [67]. The possibility of keeping an iodine atom throughout the radical sequence is noteworthy and allows the subsequent use of organometallic reactions to introduce further diversity. In the third example shown, a Sonogashira coupling has been used to attach a phenylacetylene group. Furthermore, it is not necessary to separate the addition and cyclisation steps, even though this is possible, as can be seen by comparing the second and third examples in Scheme 31.

Two approaches to building 7-membered rings around a pyridine nucleus are pictured in Scheme 32 [67]. Such structures are of great recent interest, because of their potential activity as inhibitors of tyrosine kinases, but are exceedingly tedious to obtain by traditional routes [68]. It is significant to note, in the first transformation, that the *intermolecular* addition to allyl acetate leading to **57** is faster than direct ring-closure to oxindole **56**, a possibly reversible process which nevertheless takes place efficiently in the absence of the external olefin. In all of these examples, the presence of the chlorine atom in the 6-position on the pyridine ring can be exploited to introduce

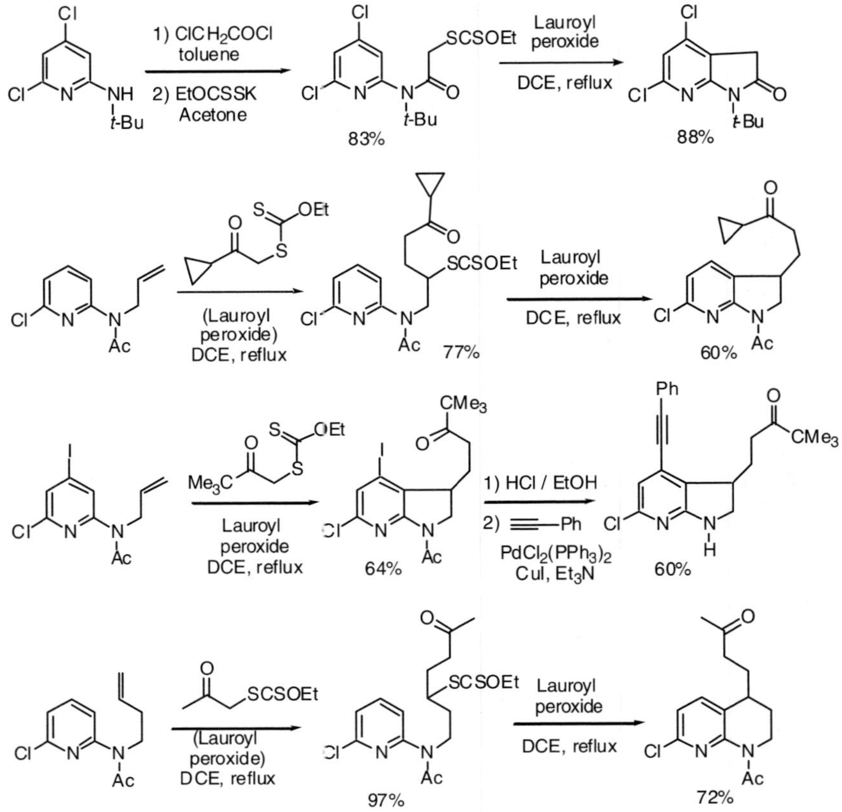

Scheme 31 Synthesis of bicyclic aza-pyridine derivatives

other functionality but also serves to block the nucleophilicity of the pyridine nitrogen and to suppress unwanted ionic side reactions.

The problem of interference by ionic side reactions is even more acute in the case of imidazole derivatives [57]. Nucleophilic attack by the imidazole on the xanthate group leads to the formation of thiols and other sulfur-containing side products, which inhibit the radical addition process. This complication can be circumvented by protonating the imidazole nitrogen with a strong acid. The xanthate group is relatively resistant to anhydrous acid and the addition of an equivalent amount of camphorsulfonic acid suppresses any undesired nucleophilic interactions with the thiocarbonyl group of the xanthate. Intermolecular additions and cyclisations onto the imidazole and other related ring systems such as triazoles can thus be accomplished efficiently, as demonstrated by the transformations in Scheme 33 [57]. The choice of camphorsulfonic acid (CSA) was dictated by the need to have an easily prepared anhydrous salt that is soluble in the organic medium. Other strong acids can no doubt be employed with equal efficiency.

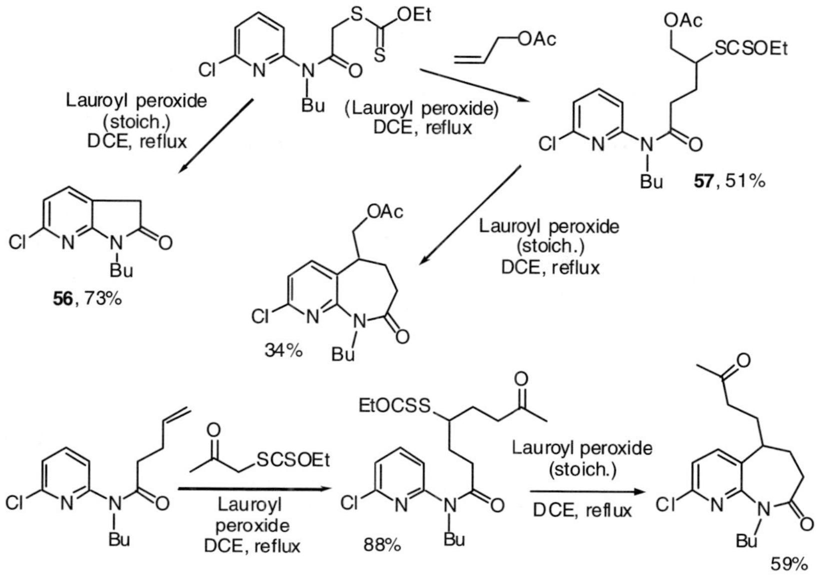

Scheme 32 Synthesis of pyridoazepinones

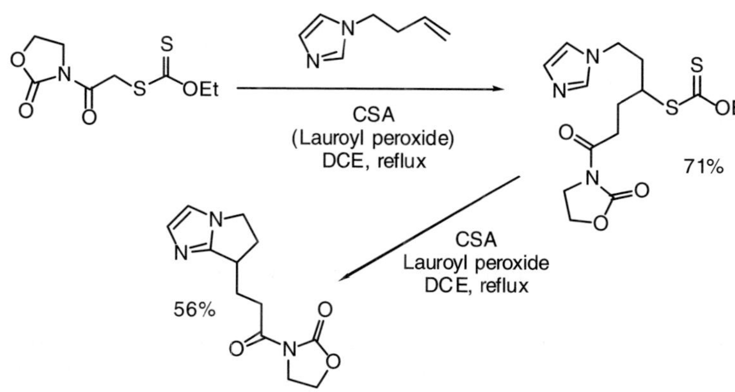

Scheme 33 Ring-closure onto imidazoles in an acidic medium

The xanthate transfer process provides a simple and uniquely powerful route to α-tetralones, another family of important aromatic derivatives [69–71]. α-Tetralones are starting materials for the synthesis of a host of medicinally important compounds. They are precursors to naphthalenes, naphthols, naphthylamines, and ring expansion through the Beckmann rearrangement provides access to benzazepine derivatives. The two examples in Scheme 34 illustrate, on one hand, the possibility of preparing a tetralone with a carbohydrate-derived appendage [69] and, on the other, the synthesis of a substituted naphthol **59** by aromatisation of tetralone **58** through acid

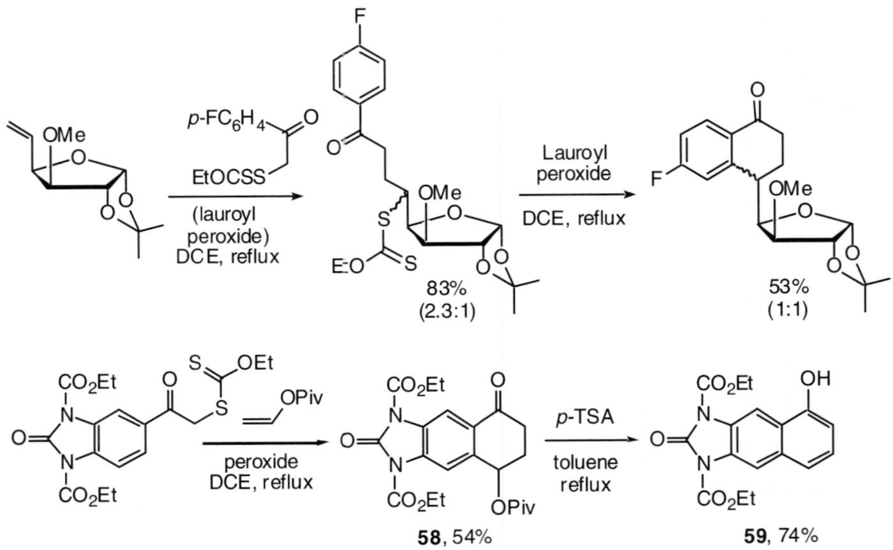

Scheme 34 Formation of α-tetralones and naphthalenes

catalysed elimination of the pivalate group [72]. The addition–cyclisation to vinyl pivalate was performed in one pot in this case. There are numerous variations of this approach and naphthalenes with a broad range of substitution patterns can be readily constructed.

This strategy was used in a concise synthesis of norparvulenone and *O*-methyl asparvenone, two natural products possessing a tetralone motif [73]. The former has been claimed to be an anti-viral and anti-influenza agent, whereas the latter is the first non-nitrogenous serotonin uptake inhibitor. The synthesis depicted in Scheme 35 started with a Friedel–Crafts reaction on methoxyphenol using chloroacetyl chloride as the acylating agent [73]. Next, the chlorine was displaced by the xanthate, and the phenolic group protected as the acetate to afford compound **60** in good overall yield. Radical addition to vinyl pivalate, ring closure to the aromatic ring, and mild aminolysis removed the acetate group to give **61**, without affecting the bulkier and less reactive pivalate. To complete the synthesis, all that remained to do was to add the missing hydroxymethyl group. This was initially attempted using formaldehyde and acid but this approach failed. Instead, the interesting exo-methylene derivative **62** was obtained in surprisingly good yield. Eventually, the use of a combination of methyl dichloromethyl ether and titanium tetrachloride delivered the desired aldehyde **63** in good yield, which could be selectively reduced to the desired alcohol with sodium cyanoborohydride. The last step to norparvulenone was a simple saponification of the pivalate ester. Replacing the reduction of the aldehyde by a Wittig reaction furnished **64** and catalytic hydrogenation converted the vinyl into the ethyl group present in

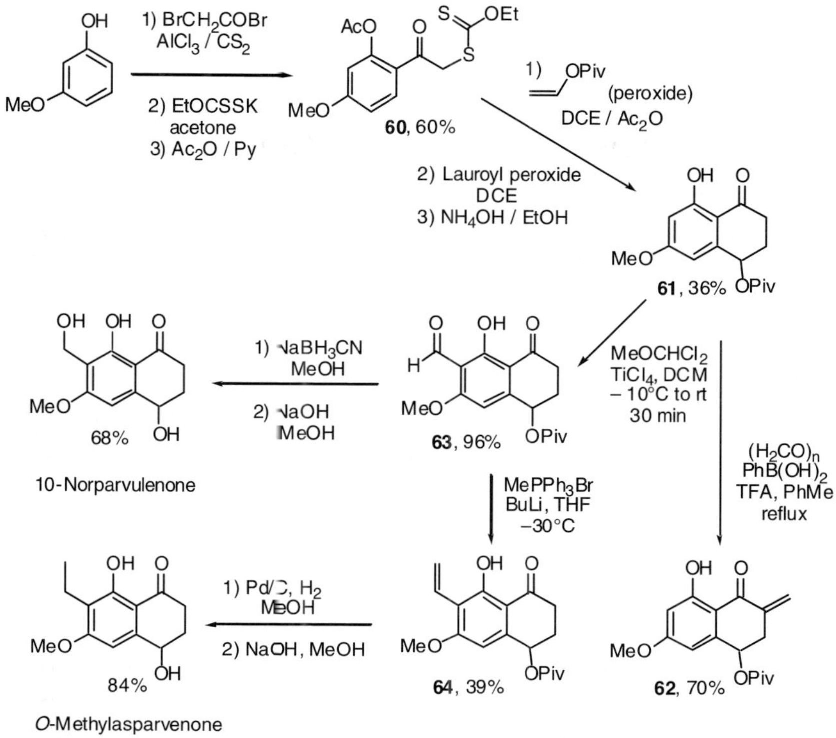

Scheme 35 Total synthesis of norparvulenone and *O*-methylasparvenone

O-methylasparvenone. The last step was the removal of the pivalate by saponification. The pivalate is a very convenient group since it could be carried intact through the various steps of the synthesis. It is perhaps useful to point out to the high flexibility of this approach if the preparation of analogues is desired: the vinyl pivalate can be replaced by other olefins; the exo-methylene in **62** is highly reactive as a Michael acceptor towards numerous nucleophiles; and the aldehyde in **63** can be subjected to a reductive amination using the plethora of commercially available primary or secondary amines.

The association of the xanthate-based tetralone synthesis with the alkylative Birch reduction and other reactions allowing the modification of the aromatic core constitutes a particularly potent combination. This is illustrated by the expedient synthesis of tricyclic system **65** pictured in the top part of Scheme 36 (Cordero-Vargas et al, personal communication). Routes to highly complex architectures can be conceived through modification of the olefinic partner, the substitution pattern of the aromatic moiety, or the nature of the alkylating agent associated with the Birch reduction.

It is appropriate to end this section by mentioning the possibility of accomplishing *intermolecular* additions to certain heteroaromatic structures [74].

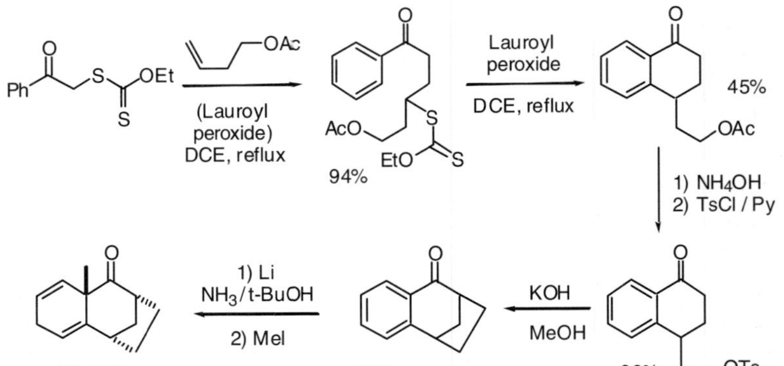

65, 80%

Scheme 36 Polycyclic structures via the alkylative Birch reduction

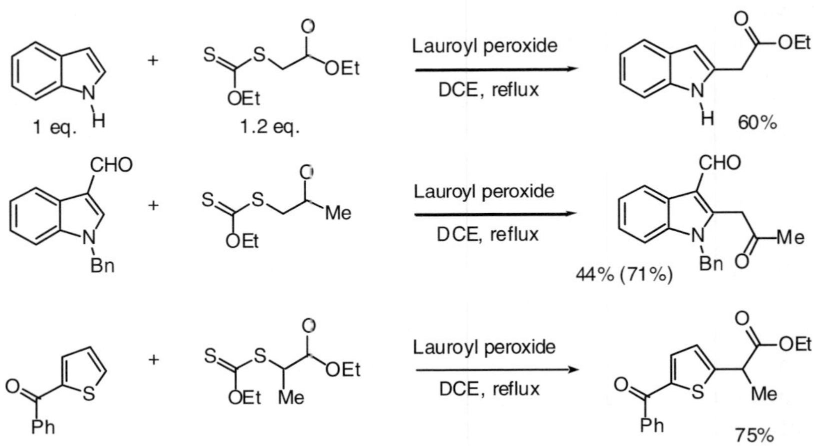

Scheme 37 Intermolecular addition to heteroaromatics

Some examples are sketched in Scheme 37 (the yield in parenthesis is based on recovered starting material). This approach may lack the generality of the intramolecular variant but nevertheless allows an expedient access to structures of great utility to medicinal chemists. The one-step introduction of the propionate side chain to 2-benzoylthiophene in the last example is especially noteworthy in this respect.

7 Concluding Remarks

The radical chemistry of xanthates and related derivatives has proved to be extraordinarily powerful for the creation of carbon–carbon bonds. In addition to being capable of mediating the common cyclisations and additions to activated olefins associated with traditional radical methods, such as those based on organotin, organosilicon, or organomercury hydrides, it is also capable of accomplishing unusual cyclisations or intermolecular additions to non-activated olefins and other reputedly difficult radical processes that are beyond the scope of traditional technology. Many different functional groups can be brought together under the mild, neutral conditions of the radical addition and then allowed to interact through ionic or other pathways by applying the appropriate reagents. The transformations described and briefly discussed in this account represent only a minute fraction of what has been accomplished and give only a tiny glimpse of the enormous potential of the xanthate transfer technology for organic synthesis.

And not only for organic synthesis: the reversible addition fragmentation to the thiocarbonylthio motif found in xanthates, dithiocarbamates, dithioesters, trithiocarbonates etc., discussed in Scheme 2 for the particular case of xanthates, is now being actively exploited for the synthesis of bloc polymers. For a recent review, see [75]; for the original patents on MADIX and RAFT, see [76, 77]. The principle of this approach is summarised in Scheme 38 for the synthesis of a diblock polymer **66**. The RAFT and MADIX processes, as they are now called, are set to revolutionise the crafting of polymers with well-defined architectures. It is an extremely effective technique that can be applied to essentially all commercial monomers and is tolerant of many functional groups. Scientific papers and patents on the subject now number in the hundreds.

Like much else in science, the perception of the feasibility of the xanthate exchange process arose initially following a seemingly insignificant experimental observation. Accidental discovery or, to use the more fancy word, serendipity, has followed us throughout this work. For beside the expected and not so expected radical transformations, many of which have not been mentioned here, we have stumbled upon some amazing non-radical chemi-

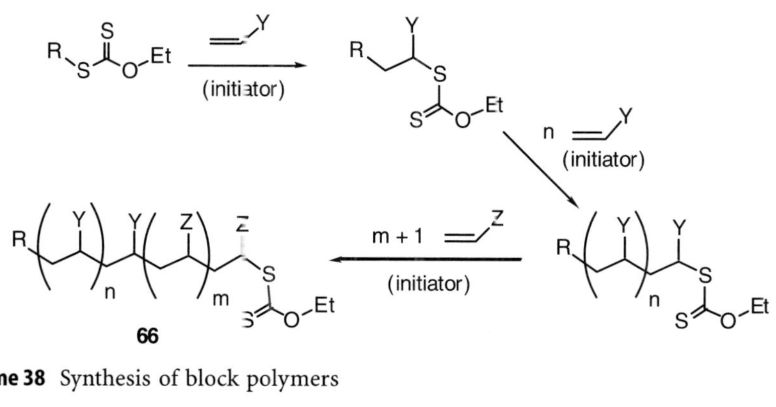

Scheme 38 Synthesis of block polymers

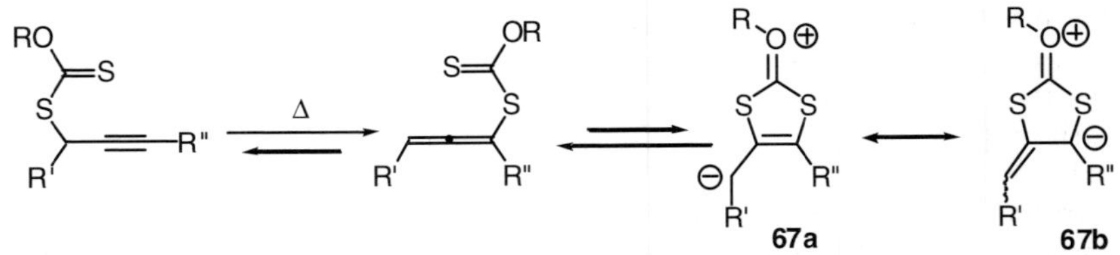

Scheme 39 Formation of an unusual betain

istry of xanthates. Perhaps the most fascinating is that related to *S*-propargyl xanthates. These derivatives undergo a [3,3] sigmatropic rearrangement upon heating to give the isomeric allenes, which exist in equilibrium with a strange betain structure **67a,b** (Scheme 39) [6]. Depending on the substituents and reagents, the betain reactivity can be channelled into formal 3 + 2 cycloadditions, Michael reactions, Mitsunobu-type ester formation (with inversion), or collapse into rigid, cisoid dienes.

Acknowledgements We are deeply grateful to our students and collaborators, whose names appear in the references, and who made this chemistry possible through their hard work and dedication. We also thank the following organisations and companies who have provided financial support over the years: Ecole Polytechnique, CNRS, DGA, MNRT, the Alfred Kastler Foundation, the Royal Commission for the Exhibition of 1851, CONACyT (Mexico), MECyD (Spain), Rhodia, and Sanofi-Aventis.

References

1. Renaud P, Sibi MP (eds) (2001) Radicals in Organic Synthesis. Wiley, Weinheim
2. Zard SZ (2003) Radical Reactions in Organic Synthesis. Oxford University Press, Oxford
3. Zard SZ (2001) Xanthates and Related Derivatives as Radical Precursors. In: Renaud P, Sibi MP (eds) Radicals in Organic Synthesis, vol 1. Wiley, Weinheim, p 90–108
4. Quiclet-Sire B, Zard SZ (1999) Phosphorus Sulfur 153–154:137
5. Quiclet-Sire B, Zard SZ (1999) J Chinese Chem Soc 46:139
6. Zard SZ (1997) Angew Chem Int Ed Engl 36:672
7. Gagosz F, Zard SZ (2003) Org Lett 5:2655–2657
8. Tournier L, Zard SZ (2005) Tetrahedron Lett 46:455
9. Denieul MP, Quiclet-Sire B, Zard SZ (1996) Chem Commun 2511
10. Bertrand F, Pevere V, Quiclet-Sire B, Zard SZ (2001) Org Lett 3:1069
11. Matsui S (1987) Bull Chem Soc Japan 60:1853
12. Gagosz F, Zard SZ (2003) Synlett 387
13. Maslak V, Cekovic Z, Saicic RN (1998) Synlett 1435
14. Gagosz F, Moutrille C, Zard SZ (2002) Org Lett 4:2707
15. Bouhadir G, Legrand N, Quiclet-Sire B, Zard SZ (1999) Tetrahedron Lett 40:277
16. Thang SH, Chong YK, Mayadunne RTA, Moad G, Rizzardo E (1999) Tetrahedron Lett 40:2435
17. Kreutzkamp N, Peschel H (1970) Pharmazie 25:322
18. Binot G, Quiclet-Sire B, Saleh T, Zard SZ (2003) Synlett 382

19. Tournier L, Zard SZ (2005) Tetrahedron Lett 46:971
20. Quiclet-Sire B, Quintero L, Sanchez-Jimenez G, Zard SZ (2003) Synlett 75
21. Quiclet-Sire B, Sanchez-Jimenez G, Zard SZ (2003) Chem Commun 1408
22. Lusinchi M, Stanbury T, Zard SZ (2002) Chem Commun 1532
23. Boivin J, Jrad R, Jugé S, Nguyen VT (2003) Org Lett 5:1645
24. Liard A, Quiclet-Sire B, Zard SZ (1996) Tetrahedron Lett 37:5877
25. Bacqué E, Pautrat F, Zard SZ (2002) Chem Commun 2312
26. Bacqué E, Pautrat F, Zard SZ (2003) Org Lett 5:325
27. Kaoudi T, Miranda L, Zard SZ (2001) Org Lett 3:3125
28. Barbier F, Pautrat F, Quiclet-Sire B, Sortais B, Zard SZ (2002) Synlett 811
29. Quiclet-Sire B, Seguin S, Zard SZ (1998) Angew Chem Int Ed Engl 37:2864
30. Bertrand F, Le Guyader F, Liguori L, Ouvry G, Quiclet-Sire B, Seguin S, Zard SZ (2001) Comptes Rendus de l'Académie des Sciences Paris II4:547
31. Bertrand F, Quiclet-Sire B, Zard SZ (1999) Angew Chem Int Ed 38:1943
32. Kalai C, Tate E, Zard SZ (2002) Chem Commun 1430
33. Georghe A, Quiclet-Sire B, Vila X, Zard SZ (2005) Org Lett 7:1653
34. Ouvry G, Zard SZ (2003) Sylnett 1627
35. Boivin J, Pothier J, Zard SZ (1999) Tetrahedron Lett 40:3701
36. Ouvry G, Zard SZ (2003) Chem Commun 778
37. Chabaud L, Landais Y, Renaud P (2005) Org Lett 7:2587
38. Chabaud L, Landais Y, Renaud P (2002) Org Lett 4:4257
39. Panchaud P, Renaud P (2004) J Org Chem 69:3205
40. Panchaud P, Ollivier C, Renaud P, Zigmantas S (2004) J Org Chem 69:2755
41. Ollivier C, Renaud P (2001) J Am Chem Soc 123:4717
42. Ollivier C, Renaud P (2000) J Am Chem Soc 122:6496
43. Kim S, Song HJ, Choi TL, Yoon JY (2001) Angew Chem Int Ed 40:2524
44. Kim S (2004) Adv Synth Catal 346:19
45. Ouvry G, Quiclet-Sire B, Zard SZ (2003) Org Lett 5:2907
46. Binot G, Zard SZ (2003) Tetrahedron Lett 44:7703
47. Cholleton N, Gillaizeau-Gauthier I, Six Y, Zard SZ (2000) Chem Commun 535
48. Udding JH, Giesselink JPM, Hiemstra H, Speckamp WN (1994) J Org Chem 59:6671
49. Blakskjaer P, Pedersen L, Skrydstrup T (2001) J Chem Soc, Perkin Trans I 910
50. Lopez-Ruiz H, Zard SZ (2001) Chem Commun 2618
51. Heinrich MR, Sharp L, Zard SZ (2005) Chem Commun 3077
52. Heinrich MR, Zard SZ (2004) Org Lett 6:4969
53. Forbes JE, Saicic RN, Zard SZ (1999) Tetrahedron 55:3791–3802
54. Grainger RS, Innocenti P (2004) Angew Chem Int Ed 43:3445
55. De Greef M, Zard SZ (2004) Tetrahedron 60:7781
56. Miranda L, Zard SZ (2002) Org Lett 4:1135
57. Gagosz F, Zard SZ (2002) Org Lett 4:4345
58. Axon J, Boiteau L, Boivin J, Forbes JE, Zard SZ (1994) Tetrahedron Lett 35:1719
59. Ly TM, Quiclet-Sire B, Sortais B, Zard SZ (1999) Tetrahedron Lett 40:2533
60. Quiclet-Sire B, Sortais B, Zard SZ (2002) Chem Commun 1692
61. Moutrille C, Zard SZ (2004) Tetrahedron Lett 45:4631
62. Pavé G, Usse-Versluys S, Viaud-Marsaud MC, Guillaumet G (2003) Org Lett 5:4253
63. Quiclet-Sire B, Zard SZ (2002) Chem Commun 2306
64. Binot G, Zard SZ (2005) Tetrahedron Lett 46:7503
65. Cholleton N, Zard SZ (1998) Tetrahedron Lett 39:7295
66. Kaoudi T, Quiclet-Sire B, Seguin S, Zard SZ (2000) Angew Chem Int Ed 39:731
67. Bacqué E, El Qacemi M, Zard SZ (2004) Org Lett 6:3671

68. Keen SP, Cowden CJ, Bishop BC, Brands KMJ, Davies AJ, Dolling UH, Lieberman DR, Stewart GW (2005) J Org Chem 70:1771
69. Cordero-Vargas A, Quiclet-Sire B, Zard SZ (2004) Tetrahedron Lett 45:7355
70. Legrand N, Quiclet-Sire B, Zard SZ (2000) Tetrahedron Lett 41:9815
71. Liard A, Quiclet-Sire B, Saicic RN, Zard SZ (1997) Tetrahedron Lett 38:1759
72. Cordero-Vargas A, Pérez-Martin I, Quiclet-Sire B, Zard SZ (2004) Org Biomol Chem 2:3018
73. Cordero-Vargas A, Quiclet-Sire B, Zard SZ (2003) Org Lett 5:3717
74. Ozornio YM, Cruz-Almanza R, Jimenez-Montano V, Miranda LD (2003) Chem Commun 2316
75. Moad G, Rizzardo E, Thang SH (2005) Aust J Chem 58:379
76. Corpart P, Charmot D, Biadatti T, Zard SZ, Michelet D (1998) WO 9858974 [(1999) Chem Abstracts 130:82018]
77. Le TP, Moad G, Rizzardo E, Thang SH. International Patent 9801478 [(1998) Chem Abstracts 128:115390]

Author Index Volumes 251–264

Author Index Vols. 26–50 see Vol. 50
Author Index Vols. 51–100 see Vol. 100
Author Index Vols. 101–150 see Vol. 150
Author Index Vols. 151–200 see Vol. 200
Author Index Vols. 201–250 see Vol. 250

The volume numbers are printed in italics

Subject Index